CASH CROP HALOPHYTES: *RECENT STUDIES*

TASKS FOR VEGETATION SCIENCE 38

SERIES EDITORS

A. Kratochwil, *University of Osnabrück, Germany*
H. Lieth, *University of Osnabrück, Germany*

The titles published in this series are listed at the end of this volume.

Cash Crop Halophytes: *Recent Studies*

10 Years after the Al Ain Meeting

Edited by

HELMUT LIETH

and

MARINA MOCHTCHENKO
University of Osnabrück, Institute for Environmental Systems Research, Osnabrück, Germany

with contributions from members of the EUCA
"Sustainable Utilisation of Halophytes
in the Mediterranean and Subtropical Region"

SPRINGER-SCIENCE+BUSINESS MEDIA, B.V.

A C.I.P. Catalogue record for this book is available from the Library of Congress.

Additional material to this book can be downloaded from http://extras.springer.com

ISBN 978-90-481-6256-7 ISBN 978-94-017-0211-9 (eBook)
DOI 10.1007/978-94-017-0211-9

Printed on acid-free paper

Table of contents

Part II. Chemical Contents

Part III. Field Scale Investigations

Part IV. Relevant Accompanying Contributions

Part V. Future Outlook for Research and Development on Cash Crop Halophytes and Foundation of ISHU, a New Scientific Society

Series editor's foreword

This volume is a landmark in the development of the halophyte ecology research and development for the following reasons:

1. The volume brings together contributions from many countries from which otherwise little information is available in the international literature.
2. The volume summarises through individual contributors the state of research possibilities in many developing countries.
3. The volume documents the latest work by the members of the European Union's Concerted Action Group "Sustainable utilisation of halophytes in the Mediterranean and subtropical dry regions".
4. The volume reports contributions from the workshop held at the beginning of the EXPO 2000, mostly from members of the abovementioned EU CA which received for its contribution the gold medal of the EXPO 2000 as shown in Part V.
5. The volume is the hard copy documentation of the scientific society initiated at the workshop held at the beginning of the EXPO 2000. Much of the achievements is still available for some time on internet via the home page http://www.usf.uni-osnabrueck.de/projects/expo2000.

The series T:VS has a reputation for reporting research in and on saline systems and halophytes. If you look at the list of T:VS publications at the end of this volume you will see that we covered these topics from the very beginning in 1981, eight volumes deal with this topic and among these the most important reports of the Al Ain/UAE conference on halophytes in 1990, where the importance and future potentials of saline irrigation systems were so successfully presented, that in the time after that meeting several new institutions for saline research were established.

Several participants of the Al Ain conference and the EU Concerted Action are now leaders in their field of halophytes research and we hope that this volume stimulates further halophyte research and the development of pilot plantations in order to demonstrate the ecological sustainability and the economic feasibility of saline irrigated production systems.

This book is intended to provide starting points for work at several levels of research from physiology through chemistry and practical applications to taxonomy and floristic analyses. The volume is structured in this way and shows the different approaches possible in different parts of the world limited by the interest of the local scientists, laboratory and monetary resources available or suggested by the type of information needed for further studies.

Towards the end of this book we report on the initiation of the new scientific society dealing with the problems to be solved to create new cash crop halophytes which is very much needed for the immediate future in many developing countries with desert or semidesert climates.

Although still small in numbers of members, the society is very active in sponsoring international meetings on halophyte research and utilisation. While GM may make great contributions to salinity tolerance of common crops in the future, the utilisation of species with natural salinity tolerance may become great resources for many purposes for which other crops may not be suitable. All members of the new society support the effort of newcomers to start halophyte research and saline production systems development.

The publications listed in this volume as well as in the other relevant T:VS volumes, the publications by the EU CA also compiled at the end of this book and the wealth of information available through the herein listed electronic media may provide a solid base for researchers worldwide. We hope that it serves this purpose. We thank the publisher, especially Mrs Noeline Gibson and Helen Buitenkamp for their efforts to present the volume in the usual Kluwer quality and hope that the volume will stimulate more colleagues to use the T:VS as an outlet for their research results relevant to saline systems.

Osnabrück, December 2002 *Helmut Lieth*

H. Lieth (ed.), Cash Crop Halophytes: Recent Studies, ix.

Introduction to the volume

This volume was compiled 10 years after the Al Ain meeting. It has several aims:

1. It should inform the scientific community about the possibilities to utilise saline water for irrigation and halophytes for sustainable agricultural production.
2. In this respect it presents the results of the research projects undertaken by the EU Concerted Action group "Sustainable Utilisation of halophytes in the Mediterranean and subtropical dry region".
3. It presents the relevant literature on these aspects and in doing so it shows:
4. Results from eco-physiological research on salt tolerance of plants.
5. Halophyte utilisation for the reclamation of soils with salinity levels just over the tolerance for common crops.
6. Halophytes with the potential to become cash crops.
7. Sustainable irrigation systems with seawater.
8. Future aspects for halophyte utilisation. The need for pilot studies demonstrating the organisational possibility, the ecological sustainability and the economic feasibility of saline irrigation systems.

The volume is structured in the following way:

1. salinity tolerance analyses,
2. chemical contents of useful halophytes and its utilisation,
3. field scale tests experiments,
4. halophytes production systems: potentials and problems,
5. future topics for research and development,
6. recent relevant publications.

The volume is for the future development of saline ecosystems especially relevant because it marks the start of the respective international association of Halophyte Utilisation ISHU which since its founding session in Osnabrueck has already sponsored or co-sponsored three international meetings, one in Huang Hua/China (2001) and the second one in Cairo/Egypt and the third one in Doha/Qatar (2002). All meetings were very successful and created new research projects as well applications in different directions.

New research and development is needed to combat salinisation and desertification. This volume will indicate new avenues.

H. Lieth (ed.), Cash Crop Halophytes: Recent Studies, xi.

List of contributors

C Abdelly Laboratoire d'Adaptation et d'Amélioration des Plantes, INRST, BP 95, 2050 Hammam-Lif, Tunisia
Nutrition Minérale, Laboratoire d'Adaptation et d'Amélioration des Plantes, I.N.R.S.T, BP. 95 2050 Hammam-Lif, Tunisia

D Bartsch Biology V (Ecology, Ecotoxicology, Ecochemistry), Aachen University of Technology – RWTH, Worringerweg 1, 52056 Aachen, Germany

S Bouzid Laboratoire de Biologie Végétale, Faculté des Sciences de Tunis, Campus Universitaire, El Manar 1060, Tunis, Tunisia

A Bustan The Institutes for Applied Research, Ben Gurion University of the Negev, P.O. Box 653, Beer Sheva 84105, Israel; Tel.: 972-7-6461996; Fax: 972-7-6472984; E-mail: abustan@bgumail.bgu.ac.il.

I Caçador Instituto de Oceanografia, Campo Grande, 1749-016 Lisboa, Portugal

W Chaibi Laboratoire de Physiologie Végétale, Faculté des Sciences de Tunis, Campus Universitaire, El Manar 1060, Tunis, Tunisia

AL Costa Instituto de Oceanografia, Campo Grande, 1749-016 Lisboa, Portugal

S Daoud Faculté des Sciences, Univ. Ibnou Zohr, Agadir, Morocco

R Davidovici The Institutes for Applied Research, Ben Gurion University of the Negev, P.O. Box 653, Beer Sheva 84105, Israel; Tel.: 972-7-6461996; Fax: 972-7-6472984.

A Debez Nutrition Minérale, Laboratoire d'Adaptation et d'Amélioration des plantes, INRST, BP 95, Hammam-Lif, 2050, Tunisia

H El Almi Laboratoire d'Adaptation et d'Amélioration des Plantes, INRST, BP 95, 2050 Hammam-Lif, Tunisia

HM El Shaer Animal and Poultry Production Division, Desert Research Center, Mataria, Cairo, Egypt

KB Hamed Nutrition Minérale, Laboratoire d'Adaptation et d'Amélioration des Plantes, I.N.R.S.T, BP. 95 2050 Hammam-Lif, Tunisia

MC Harrouni Institut Agronomique et Vétérinaire Hassan II, Agadir, Morocco

AJ Joshi Department of Life Sciences, Bhavnagar University, Bhavnagar 364 002, India

H Lieth University of Osnabrück, Institute of Environmental Systems Research, Osnabrück, Germany

MA Khan Department of Botany, University of Karachi, Karachi 75270, Pakistan

A Koocheki College of Agriculture Ferdowsi University of Mashhad, Iran

H-W Koyro Institute of Plant Ecology, Justus Liebig University Gießen, Heinrich Buff Ring 26-32, D-35392 Gießen

D Messedi Nutrition Minérale, Laboratoire d'Adaptation et d'Amélioration des Plantes, I.N.R.S.T., BP. 95; 2050 Hammam-Lif, Tunisie

DB Miled Laboratoire d'Adaptation et d'Amélioration des Plantes, INRST, BP 95, 2050 Hammam-Lif, Tunisia

M Misra Department of Life Sciences, Bhavnagar University, Bhavnagar 364 002, India

Y Mochida Yokohama National University; 79-2, Tokiwadai, Hodogaya Ward, Yokohama City, Kanagawa Prefecture, 240-8501, Japan

M Mochtchenko University of Osnabrück, Institute of Environmental Systems Research, Osnabrück, Germany

PR Moghaddam College of Agriculture Ferdowsi University of Mashhad, Iran

K Ohno Yokohama National University; 79-2, Tokiwadai, Hodogaya Ward, Yokohama City, Kanagawa Prefecture, 240-8501, Japan

J Papenbrock Universitaet Hannover, Institut fuer Botanic, Hannover, Germany

D Pasternak The Institutes for Applied Research, Ben Gurion University of the Negev, P.O. Box 653, Beer Sheva 84105, Israel; Tel.: 972-7-6461996; Fax: 972-7-6472984

V Sardo Department of Agricultural Engineering, University of Catania, Via Valdisavoia, 5 – 95123 Catania, Italy; Tel 0039-095-317303; Fax 0039-095-7312174; E-mail, sardo@mbox.fagr.unict.it

A Shmidt Institut Agronomique et Vétérinaire Hassan II, Agadir, Morocco

Y Sitrit The Institutes for Applied Research, Ben Gurion University of the Negev, P.O. Box 653, Beer Sheva 84105, Israel; Tel.: 972-7-6461996; Fax: 972-7-6472984

N Sleimi Laboratoire d'Adaptation et d'Amélioration des Plantes, INRST, BP 95, 2050 Hammam-Lif, Tunisia
Nutrition Minérale, Laboratoire d'Adaptation et d'Amélioration des Plantes, I.N.R.S.T., BP. 95; 2050 Hammam-Lif, Tunisie Faculté des Sciences de Bizerte, Jarzouna 7021, Bizerte; Tunisie

A Smaoui Laboratoire d'Adaptation et d'Amélioration des Plantes, INRST, BP 95, 2050 Hammam-Lif, Tunisia

C Vale Instituto de Investigaçiones das Pescas e do Mar, Av. Brasília, 1400 Lisboa, Portugal

L Weiqiang Shijiazhuang Institute of Agricultural modernization, CAS, Shijiazhuang, Hebei 050021, People's Republic of China

L Xiaojing Shijiazhuang Institute of Agricultural modernization, CAS, Shijiazhuang, Hebei 050021, People's Republic of China

B Yajun Shijiazhuang Institute of Agricultural modernization, CAS, Shijiazhuang, Hebei 050021, People's Republic of China

I Yokoyama Yokohama National University; 79-2, Tokiwadai, Hodogaya Ward, Yokohama City, Kanagawa Prefecture, 240-8501, Japan

NB Youssef Laboratoire d'Adaptation et d'Amélioration des Plantes, INRST, BP 95, 2050 Hammam-Lif, Tunisia

M Zaccai The Institutes for Applied Research, Ben Gurion University of the Negev, P.O. Box 653, Beer Sheva 84105, Israel; Tel.: 972-7-6461996; Fax: 972-7-6472984

M Zarrouk Laboratoire d'Adaptation et d'Amélioration des Plantes, INRST, BP 95, 2050 Hammam-Lif, Tunisia

Part I

Salinity Tolerance Analysis

Introduction to Part I

H. LIETH

The success of saline irrigation systems rests largely on the knowledge about the ecophysiology of the plants to be used in the development of halophyte production systems.

The key issue is the salinity tolerance of the plants to be used. For this reason we start this volume with chapters which deal with that topic. The second most important parameter is the chemical composition of the adult plant at the time of harvesting. Simultaneously, the level of harvestable production is also important. After that determines the amount of water needed for an economically feasible level of harvestable crop the usability of any species.

Of major importance, therefore, is the rapid salinity tolerance checking for all species with potential inclusion in cropping systems. The standardised quick check for salinity tolerance is, therefore, the first chapter in Part I. All halophytes included in the list on the enclosed CD should be screened in the way described in this section of the book.

Halophytes have developed different methods to cope with increasing levels of salinity. The salinity tolerance may differ greatly between species and within species between subspecies or ecotypes. The salt exclusion at different positions in the plant, root, stem, leaf or fruit is of great importance for its practical usability in saline production systems.

Salinity in nature can have quite different salt compositions. In this volume we deal primarily with sodium chloride salinity, this is the salinity type which can be tolerated at increasing levels by animals while other types of salinity, e.g. sodicity is poisonous to them at far lower salinity levels.

Future types of salinity analyses and tolerance tests must include the differentiation between salinity and sodicity as well as other salts contained in the plant body and the soil water. The information about the ecophysiological parameters is of different importance depending on the utilisation of the plant matter. While a green cover for aesthetic purposes may need only information about salinity and drought tolerance and the production level needed for complete ground cover, its use as forage or vegetable requires a detailed analysis of the chemical content of the biomass. This will be the focal point of Part II.

H. Lieth (ed.), Cash Crop Halophytes: Recent Studies, 3.

Study of potential cash crop halophytes by a quick check system: *Determination of the threshold of salinity tolerance and the ecophysiological demands*

HANS-WERNER KOYRO

SUSTAINABLE USE OF HALOPHYTES

The sustainable use of halophytic plants is a promising approach to valorize strongly salinised zones unsuitable for conventional agriculture and mediocre waters (Boer and Gliddon, 1998; Lieth et al., 1999). There are already many halophytic species used for economic interests (human food, fodder) or ecological reasons (soil desalinisation, dune fixation, CO_2-sequestration). However, the wide span of halophyte utilisation is not jet explored even to a small degree.

DEFINITIONS OF THE TERMS HALOPHYTE AND SALINITY TOLERANCE

Halophytes are plants, which are able to complete their life cycle in a substrate rich in NaCl (Schimper, 1891). One of the most important property of halophytes is their salinity tolerance (Lieth, 1999). This substrate offers for obligate halophytes advantages for the competition with salt sensitive plants (glycophytes). There is a wide range of tolerance among the 2,600 known halophytes (Pasternak, 1990; Lieth and Menzel, 1999). However, informations about these halophytes need partially careful checking. A precondition for a sustainable utilisation of suitable halophytes is the precise knowledge about their salinity tolerance and the various mechanisms enabling a plant to grow at (their natural) saline habitats (Marcum, 1999; Warne et al., 1999; Weber and Dántonio, 1999; Winter et al., 1999). The many available definitions especially for salinity tolerance or threshold of salinity tolerance impede a uniform description and complicate the comparison between species.

(a) Phytosociologists are using this term only for plants growing natural in saline habitats. In order to get first information on salinity tolerance, a phytosociological vegetation analysis is very helpful and salinity tolerance numbers are widely applied for qualitative approximations (Ellenberg, 1974; Landolt, 1977).

(b) Another group of scientists describes salinity tolerance with polygonal diagrams of the mineral composition in the plants.

(c) The threshold level of salinity tolerance is described in a further definition as the point (salt concentration) when the ability of plants to survive and to reproduce is no longer given (Pasternak, 1990). This definition is a modification of the definition of halophytes presented above. Survival and reproduction of a plant are not always impeded at the same salinity level (Tazuke, 1997). However, the definition of Pasternak (1990) is still important for the interpretation of the ecological dissemination and can be used as a solid basis for physiological studies concerning the survival strategies of plants.

(d) Generally, classification of the salinity tolerance (or sensitivity) of crop species is based on the threshold electrical conductivity (EC) and the percentage of yield decrease beyond threshold (Greenway and Munns, 1980; Marschner, 1995). The substrate-concentration leading to a growth depression of 50% (refer to fresh weight, in comparison to plants without salinty) is widely used by ecophysiologists as a definition for the threshold of salinity-tolerance (Kinzel, 1982). This definition is based on the same background as the Michaelis/Menten factor, because it is as difficult to fix the upper limit of salinity tolerance as it is to determine the lowest substrate concentration

H. Lieth (ed.), Cash Crop Halophytes: Recent Studies, 5–17.

for an optimal enzymatic activity. The agreement to the above-mentioned growth depression is comparatively arbitrary, but it leads to a precise specification of a comparative value for halophytic species and is especially expressive for applied aspects such as economic potentials of suitable halophytes.

(e) It is worth mentioning that there is another definition of salinity tolerance in use for glycophytic species. Especially in agriculture it is very common to speak about salinity tolerance if a variety of a glycophyte such as *Hordeum vulgare* survives at a slightly higher salinity level than another variety of the same species. However, the tolerated NaCl-substrate concentrations are in both varieties far beyond sea water salinity (Amzallag, 1994; Jeschke et al., 1995).

QUICK CHECK SYSTEM FOR THE SELECTION OF USEFUL PLANTS AND THE PHYSIOLOGICAL CHARACTERIZATION OF SALINITY TOLERANCE

It is – without doubt – necessary to develop sustainable biological production systems which can tolerate higher water salinity because freshwater resources will become limited in near future (Lieth, 1999). A precondition is the identification and/or development of salinity tolerant crops. An interesting system approach lines out that after halophytes are studied in their natural habitat and a determination of all environmental demands has been completed, the selection of potentially useful plants should be started (Lieth, 1999). The first step of this identification list contains the characterization and classification of the soil and climate, under which potentially useful halophytes grow.

The measurement of the EC in (μS cm^{-1}) offers a simple method for characterizing the salt content (Osmotic potential MPa = EC −0.036; Koyro and Lieth, 1998). A saline soil has an EC greater than 4 mmho cm^{-1} (equivalent to 40 mol m^{-3} NaCl; U.S. Salinity Staff, 1954; Koyro and Lieth, 1998) and is widely used for this purpose. Spatial variability in salt-affected fields is normally very high. Since the habitats are often complex and the concentrations varies with water content, the EC of saturation extract is only an insufficient indicator for salinity tolerance. Plant growth in saline soils can be influenced although the EC indicates no changes, because the actual salt concentration at the root surface can differ to the bulk soil. The EC characterises only the total salt content but not changes in its spatial composition. The importance of micro-heterogeneity of salinity and fertility for maintenance of the plant diversity was shown, for example, from several authors (Igartua, 1995; Abdelly et al., 1999). The study in the natural habitat represents a mean behaviour but the major constraints can vary this much that a precise definition of the salinity tolerance of a species (and a selection of useful plants) is not possible.

Only artificial conditions in sea water irrigation systems in a growth cabinet under photoperiodic conditions offers the possibility to study potentially useful halophytes under reproducible experimental growth and substrate conditions. The supply of different degrees of sea water salinity [0%, 25%, 50%, 75%, 100% (and if necessary higher) sea water salinity] to the roots in separate systems under otherwise identical conditions gives the necessary preconditions for a quick check system (QCS) for potential cash crop halophytes. Former studies have shown that hydroponic cultures (soil-free) and soil cultures were not as reliable as a gravel/hydroponic system with drip (or flow) irrigation (Figure 1; Koyro and Huchzermeyer, 1999a). Only the latter system had the potential to work under (nearly) completely artificial test conditions with high reproducibility and under close natural conditions.

Fig. 1. Gravel/hydroponic QCS with automatic drip irrigation under photoperiodic conditions in a growth cabinet (plant species: *Beta vulgaris* ssp. *maritima*). Controls are visible in the foreground, the sea water salinisation treatment in the background.

The experiments of the QCS started off with a gravel/hydroponic system imitating the climatic conditions of subtropical dry regions (Koyro and Huchzermeyer, 1999a). It is worth mentioning that the reliability of this cultivation system depended beside the climatic conditions (light intensity, relative humidity, air temperature) directly on a constant periodical irrigation (15 min per 4 h), a high drainage capacity and constant nutrient conditions (pH, nutrient composition, temperature).

The Quick Check of potential crop halophytes comprises the following eco-physiological tests:

1. A collection of general scientific data and some special physiological examinations. General scientific data contain informations about factors such as salt-induced morphological changes, gas exchange, water relations, mineral content, content of osmotically active organic substances (such as carbohydrates and amino acids) and growth.
2. Special physiological examinations are mainly on cellular level. They include the study of the relations inside single cells such as the compartmentation between cytoplasm and vacuole, the distribution of elements in different cell types or along a diffusion zone in a root apoplast and ultrastructural changes.

This QCS (i.e. research about the physiology of salt tolerance) seems to be ideal for the selection of useful plants and it suggest itself as a first step for the controlled establishment of cash crop halophytes (see also chapter on "Some physiological and biochemical aspects of salt tolerance of *Sesuvium portulacastrum*") because it provides detailed information about the three major goals:

1. the threshold of salinity tolerance,
2. to uncover the individual mechanisms for salt tolerance,
3 the potential of utilization for the pre-selected species.

THRESHOLD OF SALINITY TOLERANCE

In correspondence with the definition for the threshold of salinity tolerance according to Kinzel (1982), the growth reaction and the gas exchange are used in the QCS for halophytes as objective parameters for the description of the actual condition of a plant (Ashraf and O'Leary, 1996). There are now reliable informations available about studies with several species such as *Beta vulgaris* ssp. *maritima* (Figure 2a), *Plantago* cf. *coronopus* (Figure 2b), *Laguncularia racemosa, Aster maritima, Batis maritima* and *Spartina townsendii* (Koyro and Huchzermeyer, 1997, 1999a; Koyro et al., 1999; Koyro, 2000). The substrate-concentration leading to a growth depression of 50% (refer to freshweight, in comparison to plants without salinity) is easy to calculate with the QCS (by extrapolation of the data) and it leads to a precise specification of a comparative value for halophytic species (Figure 2a and b). The threshold of salinity tolerance amounts to 300 mol m^{-3} NaCl in *Plantago* cf. *coronopus* and 375 mol m^{-3} in *Beta vulgaris* ssp. *maritima.*

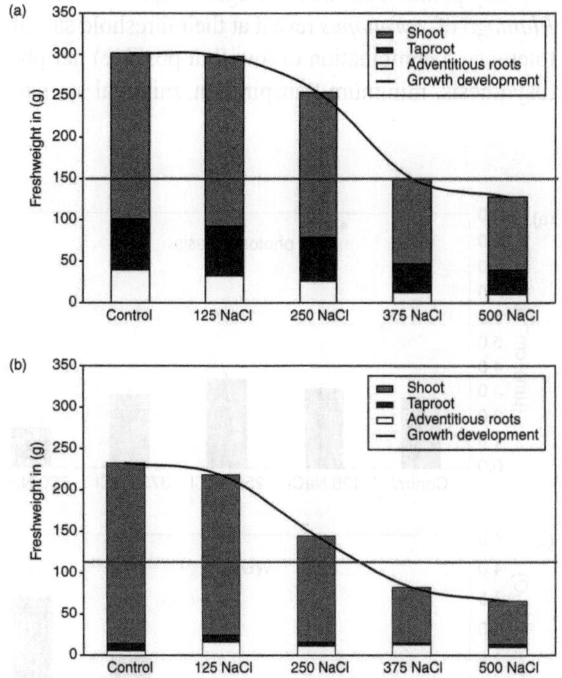

Fig. 2. Development of the plant freshweight at treatments with different percentages of sea water salinity. The crossover of the red and the black lines reflects the NaCl-salinity where the growth depression falls down to 50% of the control plant (threshold of NaCl-salinity according to Kinzel, 1982). (a) *Beta vulgaris* ssp. *maritima*: 75% sea water salinity; (b) *Plantago* cf. *coronopus*: 60% sea water salinity. Additionally it is a clear record for the influence of NaCl-salinity that the shoot growth was more depressed in both plant species than the root growth (Marschner, 1995). 0% sea water salinity = control; 25% = 125 NaCl; 50% = 250 NaCl; 75% = 375 NaCl and 100% = 500 NaCl.

BALANCE BETWEEN WATER LOSS AND CO_2-UPTAKE

Terrestric plants at saline habitats are often surrounded by low-water potentials in the soil solution and atmosphere. Plant water loss has to be minimized under these circumstances, since biomass production depends mainly on the ability to keep a high net photosynthesis by low water loss rates. In this field of tension, biomass production of a plant has to be seen always in connection to the energy consumption and gas exchange [e.g. water use efficiency (WUE)]. A critical point for the plant is reached if the CO_2-fixation falls below the CO_2-production (compensation point). Therefore, one crucial parameter of the QCS is the study of growth reduction and net photosynthesis dependency especially at the threshold of salinity tolerance (Figure 3).

Many plants such as *Beta vulgaris* ssp. *maritima* and *Plantago* cf. *coronopus* reveal at their threshold salinity tolerance a combination of low (but positive) net photosynthesis, minimum transpiration, minimal stomatal conductivity, minimum internal CO_2-concentration and increasing WUE.

MORPHOLOGICAL STRUCTURES TO REDUCE SALT CONCENTRATIONS

In many cases various mechanisms and special morphological structures are advantageous for halophytes since they help to reduce the salt concentrations especially in photosynthetic or storage tissue and seeds. Salt glands may eliminate large quantities of salt by secretion to the leave surface. This secretion appears in complex multicellular organs, for example, in *Avicennia marina* or by simple two cellular salt glands, for example, in *Spartina townsendii* (Sutherland and Eastwood, 1916; Walsh, 1974; Koyro and Stelzer, 1988; Marcum et al., 1998). Several halophytes can reduce the salt concentrations in vital organs by accumulation in bladder hairs (*Atriplex halimus, Leptochloa fusca* (L.), *Halimione portulacoides*), enhancing the leave mass to area (LMA) ratio, for

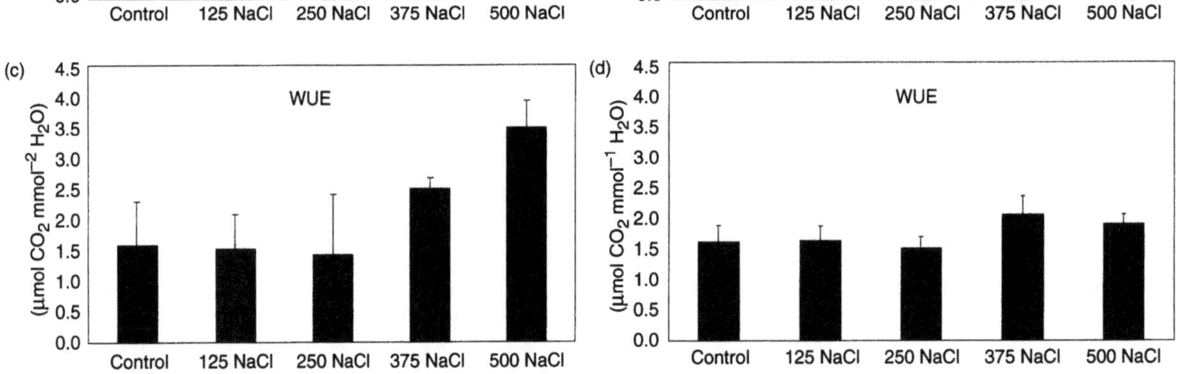

Fig. 3. Net photosynthesis rate (µmol cm^{-2} s^{-1}, Figure 3a and b), WUE of the photosynthesis (µmol CO_2 $mmol^{-1}$ H_2O, Figure 3c and d) of juvenile leaves of *Beta vulgaris* ssp. *maritima* (a and c) and *Plantago* cf. *coronopus* (b and d) at different NaCl-salinities. Many species such as *Beta vulgaris* ssp. *maritima* and *Plantago* cf. *coronopus* show at their threshold salt tolerance a combination of low (but positive) net photosynthesis and increasing water use efficiency of the photosyntheis. 0% sea water salinity = control; 25% = 125 NaCl; 50% = 250 NaCl; 75% = 375 NaCl and 100% = 500 NaCl.

example, by *Suaeda fruticosa, Salicornia europaea, Salsola kali*), translocation into other organs (z.B. *Kandelia candel* L.) or shedding of old leaves (*Beta vulgaris* ssp. *maritima*, see literature in Koyro, 2000).

SCREENING OF MECHANISMS TO AVOID SALT INJURY IN INDIVIDUAL SPECIES

Major Constraints for Plant Growth on Saline Habitats

Many halophytic species can tolerate high-sea water salinity without possessing special morphological structures (see chapter on "Salinity tolerance of *Beta vulgaris* ssp. Maritima"). The salinity tolerance of halophytic plants is in most cases multigenic and there is often a strong reliance between various mechanisms. It is the exception, that a single parameter is of major importance for the ability to survive at high NaCl-salinity. A comprehensive study with the analysis of at least a combination of several parameters is a necessity to get a survey about mechanisms constitution leading at the end to the salinity tolerance of individual species. These mechanisms are connected to the four major constraints of plant growth on saline substrates:

(a) water deficit,

(b) restriction of CO_2 uptake,

(c) ion toxicity,

(d) nutrient imbalance.

Plants growing in saline habitats face the problem of having low water potential in the soil solution and high concentrations of potentially toxic ions such as chloride and sodium. Salt exclusion minimizes ion toxicity but accelerates water deficit and diminishes indirectly the CO_2-uptake. Salt absorption facilitates osmotic adjustment but can lead to toxicity and nutritional imbalance. It should not be overlooked for the structure of the QCS that the relative contribution of the four major constraints depends also on ion relations in the substrate, duration of exposure, stage of plant development, plant organ and environmental conditions .

Major Plant Responses to High NaCl-salinity

In principle, salinity tolerance can be achieved by salt exclusion or salt inclusion. The following physiological mechanisms to avoid salt injury (and to protect the symplast) are known as major plant responses to high NaCl-salinity:

(a) Adjustment of the waterpotential, decrease of the osmotic and metric potential, enhanced synthesis of organic solutes.

(b) Regulation of the gas exchange (H_2O and CO_2), for example, high WUE (H_2O-loss per net CO_2-uptake).

(c) Ion-selectivity to maintain homeostasis especially in the cytoplasm of vital organs:
- selective uptake or exclusion (e.g. salt glands),
- selective ion-transport in the shoot, in storage organs, to the growing parts and to the flowering parts of the plants, retranslocation in the phloem,
- compartmentation of Na and Cl in the vacuole.

(d) High-storage capacity for NaCl in the entirety of all vacuoles of a plant organ, generally in old and drying parts (e.g. in leaves supposed to be dropped later) or in special structures such as hairs. The dilution of a high NaCl content can be reached in parallel by an increase in tissue water content (and a decrease of the surface area, succulence).

(e) Avoidance of ionic imbalance.

(f) Endurance of high NaCl-concentrations in the symplast.

(g) Compatibility of whole plant metabolism with high NaCl-concentrations (synthesis of NaCl-tolerant enzymes, protecting agents such as proline and glycine-betaine).

(h) Restricted diffusion of NaCl in the (root-) apoplast.

Beneficial Scientific Data

Useful parameters for a QCS of halophytes should base on the major plant responses to high NaCl-salinity (Volkmar et al., 1998). It seems to be essential that such a screening system should include salt-induced morphological changes (see chapter on "Salinity tolerance of *Beta vulgaris* ssp. Maritima", succulence, LAR: leave mass to area ratio), growth (see chapter on "Some physiological and biochemical aspects of salt tolerance"), water relations, gas exchange (see chapter on "Physiological responses and structural modifications") and composition of minerals (and compatible solutes) at different parts of the root system and in younger and older leave tissues. The measurement of such general scientific data at plant-, organ- or tissue-level reveals general trends – but since these represent a mean behaviour of several cell types, many informations on single cell adjustment are lost. They cannot give sufficient information about the compartmentation inside a cell or along a diffusion zone in a root apoplast or about ultrastructural changes. The collection of scientific data should be completed if necessary by a special physiological research at single cell level.

Collection of General Scientific Data

Information about salt-induced morphological changes, gas exchange and growth were already presented in this chapter. The list of general scientific data can be completed by information on water relations (e.g. leave water potential) and composition of organic and inorganic solutes.

Leaf water potential

Data of the leave water potentials (measured by dew point depression with a WESCOR HR-33T) demonstrated clearly that *Beta vulgaris* ssp. *maritima* (Figure 4a) and *Plantago* cf. *coronopus* (Figure 4b) have a sufficient adjustment mechanism even at high-salinity treatment, suggesting that there was no reason for growth reduction by water deficit. Thus, if the rate of supply of water to the shoot is not restricted, the depression of the shoot growth shown in Figure 1 is likely to be mineral nutrition connected. It seems to be only a matter of controversy as to whether a decrease in

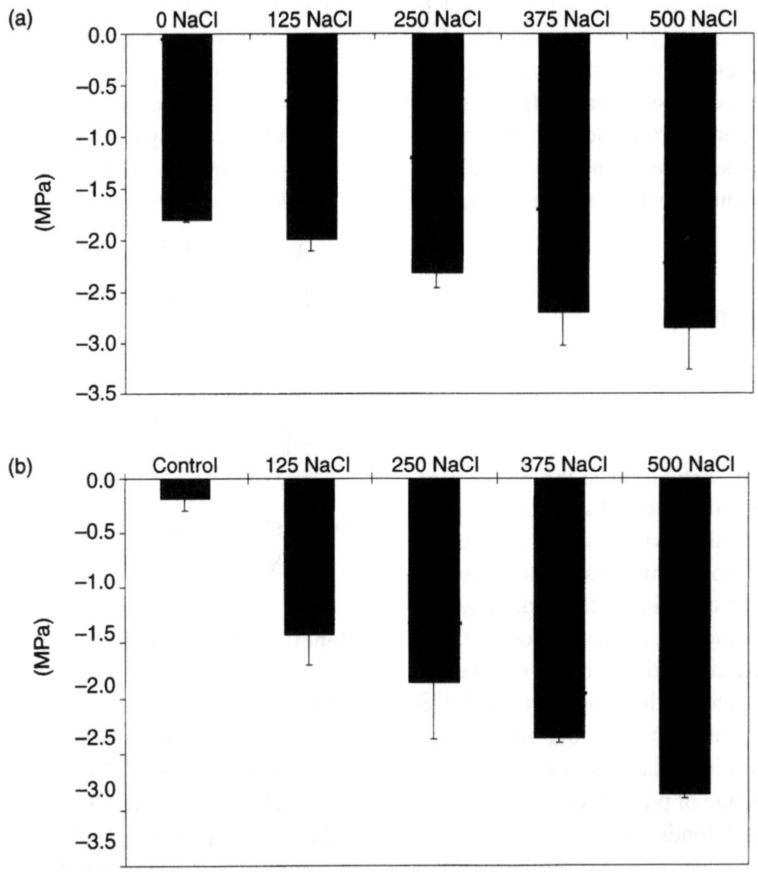

Fig. 4. Leaf water potentials (MPa) of *Beta vulgaris* ssp. *maritima* (a) and *Plantago* cf. *coronopus* (b). The red lines in the bars mark the water potentials in the nutrient solutions. Leave water potentials were always lower than in the assigned nutrient solution potential. The difference between water potentials in the leaves and in the nutrient solutions decreased with increasing NaCl-salinity. There was only one exception: *Plantago* cf. *coronopus* in the control treatment. 0% sea water salinity = control; 100% sea water salinity = sea water salinity.

the amounts of nutrients or unfavourable nutrient ratios (e.g. Na^+/K^+) are important factors for impaired leave elongation (Lynch et al., 1988; Munns et al., 1989).

Nutrients

The physiological responses of *Beta vulgaris* ssp. *maritima* and *Plantago* cf. *coronopus* were quite similar up to this point. However, data of additional scientific studies have shown that these two species exhibit very different ways of adjustment to high NaCl-salinity. For clarification further scientific data (selected minerals and sucrose) will be presented only for the sea beet (for further details see Koyro and Huchzermeyer, 1999a). The measurements were carried out from different plant tissues to evaluate their individual storage capacity and beneficial effects.

Potassium was stored in juvenile and adult leaves of salt-free controls in much higher concentrations as in the root (Figure 5a). Salinity led to a significant decrease of the potassium concentrations in all parts of the plant and to a shift of the maximum potassium concentration from adult to juvenile leaves. The potassium supply to adult leaves was reduced more than to juvenile and growing parts.

The sodium concentrations were in all plant tissues of the controls and low-salinity plants and especially in the leaves of the high salinity treatment always much higher than in the nutrient solution (Figure 5b). The comparison of the K- and Na-concentrations in juvenile and adult leaves reveals obviously that *Beta vulgaris* ssp. *maritima* uses the older parts for internal detoxification or better internal exclusion of sodium. The step inverse Na/K gradient between juvenile and adult leaves of the sea beet is a typical reaction of many halophytes to high NaCl-salinity (Wolf et al., 1991; Koyro and Huchzermeyer, 1999).

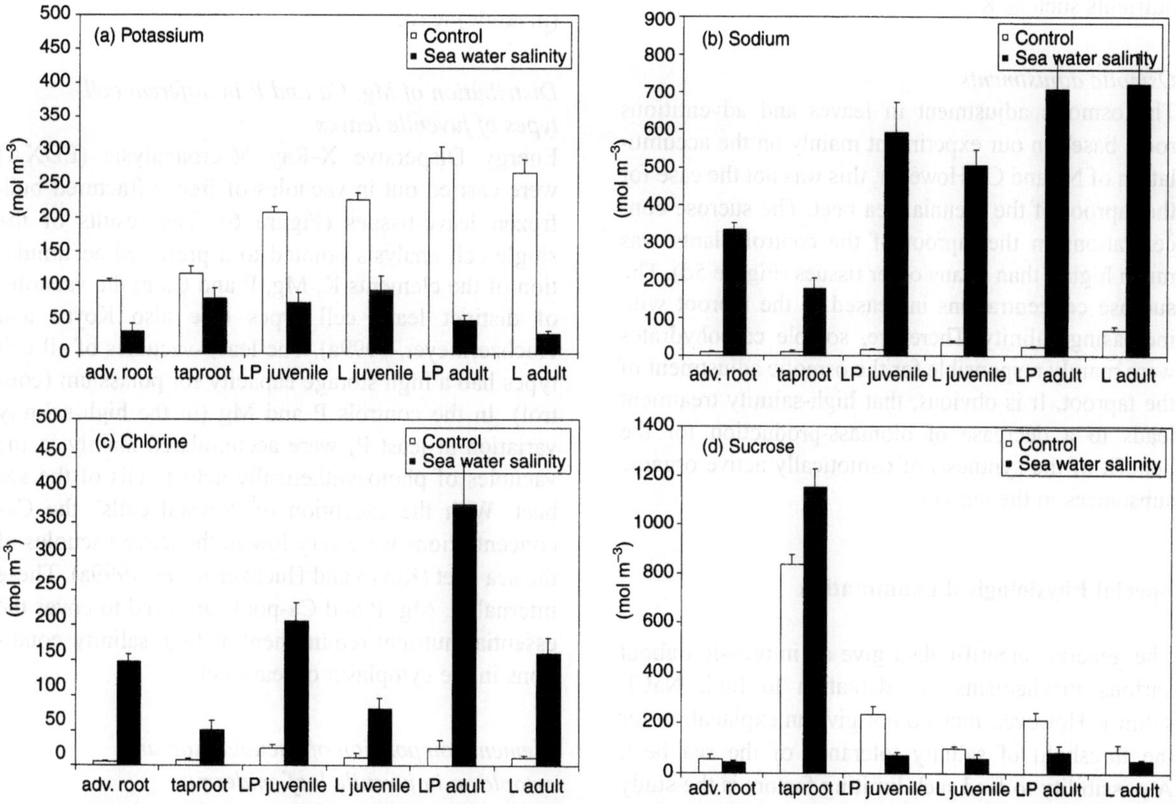

Fig. 5. Potassium- (a), sodium- (b), chlorine- (c) and sucrose-concentrations (d) in mol m^{-3} in different tissues of *Beta vulgaris* ssp. *maritima*. The osmotic adjustment in leaves and adventitious roots based mainly on the accumulation of Na and Cl. Soluble carbohydrates were mainly responsible for the osmotic adjustment of the tap root. Adv. Root = adventitious roots; LP = leave petiole; L = leave lamina.

There is scientific debate whether high potassium concentrations in young leaves and reproductive organs can be achieved by low xylem import of both potassium and sodium, and/or high phloem import from mature leaves (Wolf et al., 1991).

The chlorine-concentrations were very low in all tissues of the controls (Figure 5c). The high-salinity treatment led only to a comparatively (reference: different NaCl-concentrations in the nutrient solutions) small increase of the Cl-concentrations in all tissues. However, the chlorine were just like the sodium concentrations much higher in adult than in juvenile leaves. An effective restriction of sodium and chlorine import into young leaves compared to old leaves is also typical for *Agrostis stolonifera* (Robertson and Wainwright, 1987). Another mechanisms of sea beet to reduce internal NaCl-concentrations was shedding of old leaves (see literature in Koyro, 2000). This system successfully diminishes and controls the accumulation of Na and Cl in younger or metabolically active parts of the plant and buffers against an imbalance of nutrients such as K.

Osmotic adjustments

The osmotic adjustment in leaves and adventitious roots based in our experiment mainly on the accumulation of Na and Cl. However, this was not the case for the taproot of the biennial sea beet. The sucrose concentrations in the taproot of the control plants was much higher than in any other tissues (Figure 5d). The sucrose concentrations increased in the taproot with increasing salinity. Therefore, soluble carbohydrates were mainly responsible for the osmotic adjustment of the taproot. It is obvious, that high-salinity treatment leads to a decrease of biomass-production for the benefit of the synthesis of osmotically active organic substances in the taproot.

Special Physiological Examination

The general scientific data give an impression about various mechanisms of adaptation to high NaCl-salinity. However, they do not give an explanation for the threshold of salinity tolerance of the sea beet. A possibility to find such limiting factors is the study of the relations inside single cells such as the compartmentation between cytoplasm and vacuole, the distribution of elements in different cell types or along a diffusion zone in a root apoplast and ultrastructural changes. The analysis of the distribution of elements (compartmentation) in a distinct cell type can be used to answer several questions such as:

(a) Is there a compartmentation and a division of labour in cytoplasm and vacuole ?

(b) Is it necessary to look more in detail to toxic effects caused by high cytoplasmic sodium or chlorine concentrations?

(c) Is there a compatibility of the metabolism with high NaCl-concentrations in the symplast (cytoplasm)?

(d) Is there a high ionic imbalance in the symplast (cytoplasma)?

Supportive data were given in this chapter that sodium and chloride were stored mainly in the shoot of the sea beet and that the growth reduction of the above ground parts were much higher than of the root. Therefore, further investigations are presented based on the (juvenile) leave.

Distribution of Mg, Ca and P in different cell types of juvenile leaves

Energy Dispersive X-Ray Microanalysis (EDXA) were carried out in vacuoles of freeze-fractured bulk frozen leave-tissues (Figure 6). The results of the single-cell analysis pointed to a preferred accumulation of the elements K, Mg, P and Ca in the vacuoles of distinct leave cell types (see also Koyro and Huchzermeyer, 1999a). The leave-vacuoles of all cell types had a high storage capacity for potassium (control). In the controls P and Mg (in the high-salinity variation at least P) were accumulated mainly in the vacuoles of photosynthetically active cells of the sea beet. With the exception of "crystal cells" the Ca-concentrations were very low in the leave-vacuoles of the sea beet (Koyro and Huchzermeyer, 1999a). These internal K, Mg, P and Ca-pools are used to cover the essential nutrient-requirement at high-salinity conditions in the cytoplasm of leave cells.

Element composition of vacuolar sap and cytoplasm in juvenile leaf epidermis

The single cell data of the vacuolar and cytolasmatic composition in cells of the upper leaf-epidermis are summarized for the controls and the high-salinity treatments in Figure 7. It is obvious that NaCl-salinity led to

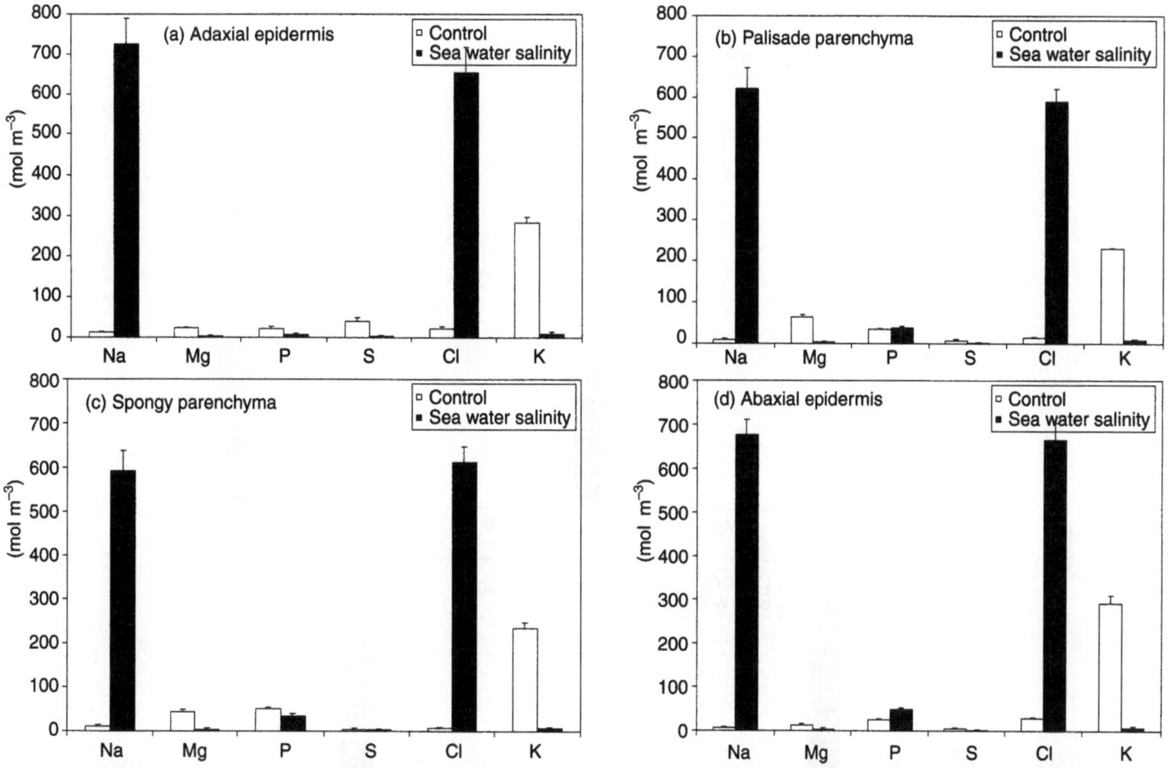

Fig. 6. Vacuolar sodium-, magnesium-, phosphor-, sulphur-, chlorine- and potassium-concentrations in mol m^{-3} (measured with EDX-analysis in bulk frozen tissues) in different leaf tissues of *Beta vulgaris* ssp. *maritima*. The vacuolar potassium, magnesium and phosphor pools cover the essential nutrient-requirement at high-salinity conditions in the cytoplasm of leaf cells. Sodium and chloride are used for the osmotic adjustment in the vacuoles of all leaf tisssues. White bars = controls at 0% sea water salinty, black bars = 100% sea water salinity. (a) adaxial epidermis, (b) palisade parenchyma, (c) spongy parenchyma, and (d) adaxial epidermis.

a decrease of P, Cl, Ca, Mg and K (measured only with EDXA) in this compartment. The K-concentrations were in the epidermal cytoplasm of control plants in an ideal range for enzymatic reactions (Wyn Jones et al., 1979; Wyn Jones and Polard, 1983; Koyro and Stelzer, 1988). However, NaCl led to a significant decrease of the K and P concentrations. This result points at a deficiency for both elements in the cytoplasm and supports the hypothesis that the major reason for the threshold of the salinity tolerance in the sea beet is not ion-toxicity but ion-deficiency! This result demonstrates the necessity of single-cell measurements for the study of characteristic salinity tolerance mechanisms and threshold levels of individual species.

Summary of the single cell data

In all leaf cell types, the vacuolar K-concentration in control treatments was significantly higher in comparison to the cytoplasm. The leaf-vacuoles in its entirety can be described as a voluminous potassium-pool with high-storage capacity for sodium and chloride. This pool is needed in case of high NaCl-salinity for the maintenance of the K-homeostasis in the cytoplasm. High NaCl-salinity leads to a breakdown of the K-homeostasis because of K-deficiency.

Evaluation of the Screening Procedure

The results presented in this chapter contain a lot of informations about the essential eco-physiological needs of the sea beets at high salinity. One possible consequence is the supply of sufficient fertilizers (especially K and P) at high NaCl-salinity to reduce the symptoms of K- and P-deficiency. This QCS is the basis for a very variable screening of

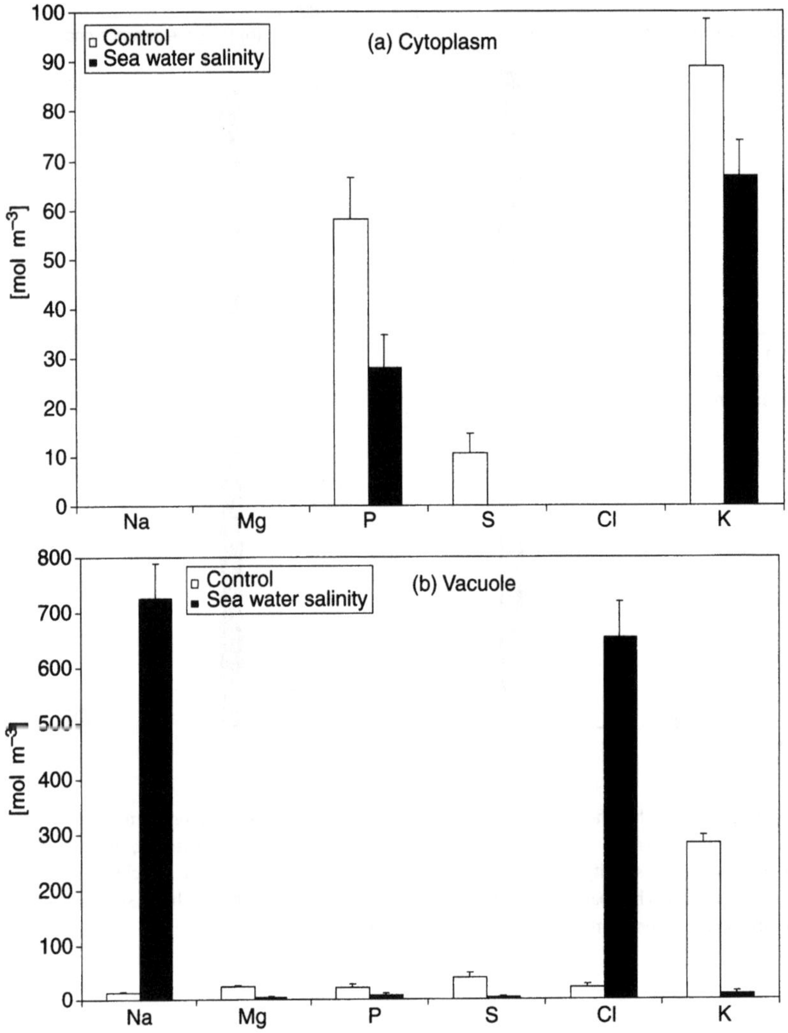

Fig. 7. Sodium-, magnesium-, phosphor-, sulphur-, chlorine- and potassium-concentrations in mol m^{-3} (measured with EDX-analysis in bulk frozen tissues) in the vacuoles and in the cytoplasm of adaxial epidermis cells of *Beta vulgaris* ssp. *maritima*. High NaCl-salinty leads to a decrease of P, Cl, Ca, Mg and K in the vacuole and to a significant decrease of the K and P concentrations in the cytoplasm. White bars = controls at 0% sea water salinty, black bars = 100% sea water salinity.

individual species. It enables to study the characteristic combination of mechanisms against salt injury and the threshold of salinity tolerance. The QCS can also be used with distinct modifications like different irrigation systems. It can be modified to the special characteristics and needs of other species and is therefore useful to study a wide range of suitable halophytes. This OCS is a practical first step on the selection of economically important cash crop halophytes.

SELECTION OF SALINITY TOLERANT SPECIES WITH PROMISING TOLERANCE AND YIELD CHARACTERISTICS

For future studies on utilisation potentials of halophytes precise data about the ecological demands of halophtic species are required. Comparative physiological studies about salinity tolerance are essential. A precondition for this demand is a precise specification of a comparative value for halophytic species as shown in this chapter. Four steps are prerequisites for the selection of appropriate salinity tolerant plants with promising tolerance (and yield) characteristics.

(a) The literature has to be screened prior to the selection of priority species (potentially useful species) in order to get first-order information about their natural occurrence in dry or saline habitats, existing utilisation (because of their structure, chemical content or other useful properties), natural climatic and substrate conditions, water requirement and salinity tolerance.

(b) Soon after the selection of a priority species, the threshold of salinity should be determined according to Kinzel (1982). The characteristic major plant responses has to be evaluated for precise informations of ecophysiological demands. The data can build up a well-founded basis for the improvement of the utilisation potential.

(c) Additionally, research about the genetic composition of chromosomes mastering saline environment is also needed and bases on quantitative precise determination.

BREEDING PROGRAMS AND OTHER POSSIBILITIES FOR THE SELECTION OF ADEQUATE YIELD AND TOLERANCE CHARACTERISTICS

Selection and breeding programs designed to improve the adaptation of crop plants to saline soils have to consider the various mechanisms responsible for salinity tolerance or sensitivity. Efforts to develop new crop cultivars with improved salinity tolerance have been intensified over the past 20–25 years. However, there is only a limited number of cultivars that have been developed with improved salinity tolerance and for all of them selection has been based on agronomic characters such as yield or survival. Plants regenerated from selected cells of cell or tissue culture (for improving salinity tolerance) have not shown an unequivocal increase in salinity tolerance (Winicov and Bastola, 1999). It is a matter of question whether cell cultures nor breeding of crop plants are appropriate for selection because salinity tolerance is in most cases multigenic and it depends on the structural and physiological integrity of plants.

On the background that genetic variability exists within crop species it seems promising to introduce important traits of salinity tolerance into crop species from their wild relatives through interspecific hybridization [e.g. *Beta vulgaris* ssp. *vulgaris* (sugar beet) and *Beta vulgaris* ssp. *maritima* (sea beet)]. These are valuable tools available for screening and breeding for higher salinity tolerance.

During recent years gene modification technology has been applied to obtain genetically improved crop plants (Yeo, 1998; Winicov and Bastola, 1999). However, genetical as well as molecular biological methods should be tested according to their possible applications. The development of halophytes to useful plants ("cash crop halophytes") should go ahead with the development of glycophytes to salinity tolerant crops (Figure 8). The selection and breeding for salinity tolerance are important issues for traditional agricultural production systems especially in semiarid and arid regions. It is conceivable, that an improvement of the utilisation potential can be reached in near future by gene transfer of profitable features to a

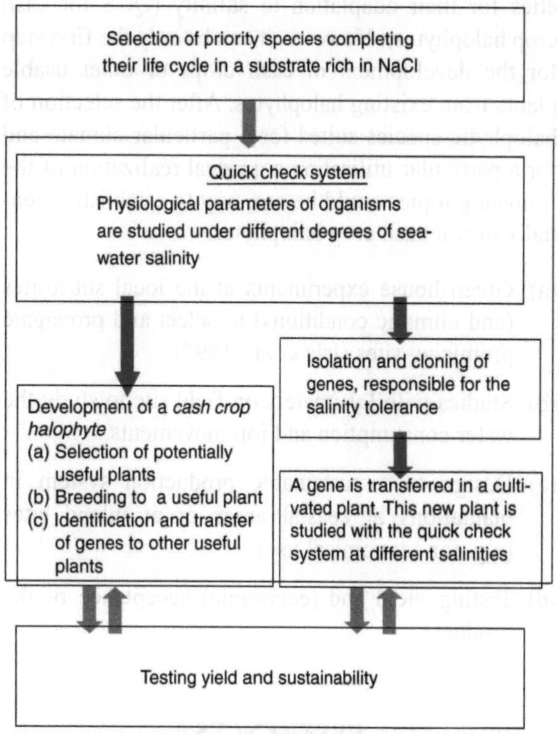

Fig. 8. Scheme about the development of halophytic crops as a systems approach. Only halophytes were selected with high utilisation potential on saline substrates.

halophyte or of tolerance-enhancing features to a glycophyte (Winicov, 1998). Recent studies have shown, that a relatively small number of quantitative trait loci (QTL) may govern complex physiological characters. After isolation and cloning of genes from a (glycophytic) cultivated plant, responsible, for example, for yield-enhancing characteristics, the gene could be transferred into a halophyte. Alternatively, the transfer of genes, responsible for salinity tolerance could occur from a halophyte to a glycophytic crop. It is highly probable, that at least one of these ways is leading to the development of sea water tolerant crops.

DEVELOPMENT OF CASH CROP HALOPHYTES

The physiological studies with the sea water irrigation system have the potential to provide highly valuable means of detecting individual mechanisms of species against NaCl stress, and may also provide opportunities for the comparison and screening of different varieties for their adaptation to salinity (QCS for cash crop halophytes). However, it can be only the first step for the development of cash crops or other usable plants from existing halophytes. After the selection of halophytic species suited for a particular climate and for a particular utilisation a gradual realization of the following topics could be one way to establish potentially useful cash crop halophytes:

(a) Green house experiments at the local substrates (and climatic conditions) to select and propagate promising sites (Isla et al., 1997).

(b) Studies with Lysimeters on field site to study the water consumption and ion movements.

(c) Design of a sustainable production system in plantations at coastal areas or at inland sites (e.g. for economical use).

(d) Testing yield and (economic) acceptance of the product.

REFERENCES

Abdelly, C., Lachaal, M. and Grignon, C. Importance of microheterogeneity of salinity and fertility for maintenance of the plant diversity. In Lieth, H., Moschenko, M., Lohmann, M., Koyro, H.-W. and Hamdy, A., editors. Progress in biometeorology Vol. 13. Leiden, Backhuys Publishers, 1999; 65–76.

Amzallag, G.N. Influence of parental NaCl treatment on salinity tolerance of offspring in *Sorghum bicolor* (L) *Moench*. New Phytologist 1994; 128:715–23.

Ashraf, M. and O'Leary, J.W. Effect of drought stress on growth, water relations, and gas exchange of two lines of sunflower differing in degree of salt tolerance. Int J Plant Sci 1996; 157:729–32.

Boer, B. and Gliddon, D. Mapping of coastal ecosystems and halophytes (case study of Abu Dhabi, United Arab Emirates). Marine Freshwater Res 1998; 49:297–301.

Greenway, H. and Munns, R. Mechanisms of salt tolerance in nonhalophytes. Annu Rev Plant Physiol 1980; 31:149–90.

Ellenberg, H. Zeigerwerte der Gefäßpflanzen Mitteleuropas. Scripta Geobotanica 1974; 9:97.

Igartua, E. Choice of selection environment for improving crop yields in saline areas. Theoret Appl Sci 1995; 91:1016–21.

Jeschke, W.D., Klagges, S., Hilpert, A., Bhatti, A.S. and Sarwar, G. Partitioning and flows of ions and nutrients in salt-treated plants of *Leptochloa fusca* L Kunth .1. Cations and chloride. New Phytologist 1995; 130:23–35.

Isla, R., Royo, A. and Aragues, R. Field screening of barley cultivars to soil salinity using a sprinkler and a drip irrigation. Plant Soil 1997; 197:105–17.

Kinzel, H. Pflanzenökologie und Mineralstoffwechsel. Eugen Ulmer Publ., Stuttgart, 1982.

Koyro, H.-W. and Stelzer, R. Ion concentrations in the cytoplasm and vacuoles of rhizodermal cells from NaCl treated *Sorghum, Spartina* and *Puccinellia* plants. J Plant Physiol 1988; 133:441–6.

Koyro, H.-W. and Lieth, H. Salinity conversion table. 2nd enlarged edn. © H. Lieth ISSN 09336-3114, Osnabrück, 1998.

Koyro, H.-W. and Huchzermeyer, B. The physiological response of *Beta vulgaris* ssp. *maritima* to sea water irrigation. In: Lieth, H., Hamdy, A. and Koyro, H.-W., editors. Water management, salinity and pollution control towards sustainable irrigation in the mediterranean region. Salinity problems and halophyte use. Tecnomack, Bari, Italy, 1997; 29–50.

Koyro, H.-W. and Huchzermeyer, B. Influence of high NaCl-salinity on growth, water and osmotic relations of the halophyte *Beta vulgaris* ssp. *maritima*. Development of a quick check In Lieth, H., Moschenko, M., Lohmann, M., Koyro, H.-W. and Hamdy, A., editors. Progress in Biometeorology, Vol. 13. Backhuys Publishers, Leiden, NL, 1999a; 87–101.

Koyro, H.-W. and Huchzermeyer, B. Salt and drought stress effects on metabolic regulation in maize. In Pessarakli, M., editor. Handbook of plant and crop stress 2nd edn. Marcel Dekker Inc., New York, USA, 1999b; 843–78.

Koyro, H.-W., Wegmann, L., Lehmann, H. and Lieth, H. Adaptation of the mangrove *Laguncularia racemosa* to high NaCl salinity. In Lieth, H., Moschenko, M., Lohmann, M., Koyro, H.W. and Hamdy, A., editors. Progress in Biometeorology, Vol. 13. Backhuys Publishers, Leiden, 1999; 41–62.

Koyro, H.-W. Untersuchungen zur Anpassung der Wildrübe (*Beta vulgaris* ssp. *maritima*) an Trockenstreß oder NaCl-Salinität. Habilitation, Justus-Liebig-University Giessen, Germany, 2000.

Landolt, E. Ökologische Zeigerwerte zur Schweizer Flora. Veröffentlichungen des geobotanischen Instituts der Eidgenössischen Technischen Hochschule in Zürich 64, Stiftung Ruebel, 1977; 208.

Lieth, H. Development of crops and other useful plants from halophytes. In Lieth, H., Moschenko, M., Lohmann, M., Koyro, H.-W. and Hamdy, A., editors. Halophytes uses in different climates, ecological and ecophysiological studies. Backhuys Publishers, Leiden, 1999; 1–18.

Lieth, U. and Menzel, U. Halophyte database Vers. 2. In Lieth, H., Moschenko M., Lohmann, M., Koyro, H.-W. and Hamdy, A., editors. Halophytes uses in different climates, ecological and ecophysiological studies. Backhuys Publishers, Leiden, 1999; 159–258.

Lieth, H., Moschenko, M., Lohmann, M., Koyro, H.-W. and Hamdy, A. Halophyte uses in different climates I. Ecological and ecophysiological studies. In Progress in biometeorology 13. Backhuys Publishers, Leiden, 1999; 258.

Lynch, J., Thiel, G. and Läuchli, A. Effects of salinity on the extensibility and Ca availability in the expanding region of growing barley leaves. Botanica Acta 1988; 101:355–61.

Marcum, K.B., Anderson, S.J. and Engelke, M.C. Salt gland ion secretion: A salinity tolerance mechanism among five zoysiagrass species. Crop Sci 1998; 38:806–10.

Marcum, K.B. Salinity tolerance mechanisms of grasses in the subfamily *Chloridoideae*. Crop Sci 1999; 39:1153–60.

Marschner, H. Mineral nutrition of higher plants. Academic Press, London, 1995; 889

Munns, R., Gardner, P.A. Tonnet, M.L. and Rawson, H.M. Growth and development in NaCl-treated plants. II Do Na^+ or Cl^- concentrations in dividing or expanding tissues determine growth in barley. Aust J Plant Physiol 1989; 15:529–40.

Pasternak, D. Fodder production with saline water. The institute for applied research, Ben Gurion University of the Negev. Project report BGUN-ARI-35-90. Beer-Sheva/Israel, 1990; 173

Robertson, K.P. and Wainwright, J.J. Photosynthetic responses to salinity in two clones of *Agrostis stolonifera*. Plant Cell and Environ 1987; 10:45–52.

Schimper, A.F.W. Pflanzengeographie auf physiologischer Grundlage. Fischer Publ., Jena, 1891.

Sutherland, G.K. and Eastwood, A. The physiological anatomy of *Spartina townsendii*. Ann Bot 1916; 30:333–51.

Tazuke, A. Growth of cucumber fruit as affected by the addition of NaCl to nutrient solution. J Jpn Soc Horticult Sci 1997; 66:519–26.

U.S. Salinity Laboratory Staff. Diagnosis and improvement of saline and alkali soils. In: L.A. Richards, editor. Agricultural handbook of the U.S. Vol. 60. Department of Agriculture, 1954.

Volkmar, K.M., Hu, Y. and Steppuhn H. Physiological responses of plants to salinity: A review. Can J Plant Sci 1998; 78:19–27.

Walsh, G.E. Mangroves. A review. In Reimold, R.J. and Queen, W.H., editors. Ecology of halophytes. Academic Press, New York, 1974; 51–174.

Warne, T.R., Hickok, L.G., Sams, C.E. and Vogelien, D.L. Sodium/potassium selectivity and pleiotropy in stl2, a highly salt-tolerant mutation of *Ceratopteris richardii*. Plant Cell Environ 1999; 22:1027–34.

Weber, E. and D'Antonio, C.M. Germination and growth responses of hybridizing *Carpobrotus* species (*Aizoaceae*) from coastal California to soil salinity. Am J Bot 1999; 86:1257–63.

Winicov, I. and Bastola, D.R. Salt tolerance in crop plants: New approaches through tissue culture and gene regulation. Acta Physiologiae Plantarum 1997; 19:435–49.

Winicov, I. New molecular approaches to improving salt tolerance in crop plants. Ann Bot 1998; 82:703–10.

Winicov, I. and Bastola, D.R. Transgenic overexpression of the transcription factor Alfin1 enhances expression of the endogenous MsPRP2 gene in alfalfa and improves salinity tolerance of the plants. Plant Physiol 1999; 120:473–80.

Winter, U., Kirst, G.O., Grabowski, V., Heinemann, U., Plettner, I. and Wiese, S. Salinity tolerance in *Nitellopsis obtusa*. Aust J Bot 1999; 47:337–46.

Wolf, O., Munns, R., Tonnet, M.L. and Jeschke, W.D. The role of the stem in the partitioning of Na+ and K+ in salt treated barley. J Exp Bot 1991; 42:697–704.

Wyn Jones, R.G., Brady, C.J. and Speirs, J. Ionic and osmotic relations in plant cells. In Laidman, D.L. and Wyn Jones, R.G., editors. Recent Advances in the biochemistry of cereals. Academic Press, New York, 1979.

Wyn Jones, R.G. and Pollard, A. Proteins, enzymes and inorganic ions. In Läuchli, A. and Bieleski, R.L., editors. Inorganic plant nutrition. Encyclopedia of plant physiology, Vol. 15b. Springer Verlag, 1983; 528–55.

Yeo, A. Molecular biology of salt tolerance in the context of whole-plant physiology. J Exp Bot 1998; 49:915–29.

Physiological responses and structural modifications in *Atriplex halimus* L. plants exposed to salinity

A. DEBEZ, W. CHAIBI AND S. BOUZID

INTRODUCTION

The deleterious effects owing to the presence of salt in the plants' environment could be resumed in three points: (i) a water stress, resulting from the decrease of the water availability, following to the lowering of water potential in the medium with regard to the plant's tissues, (ii) a toxic action which disrupts the metabolic activity of the cell and (iii) a nutritional stress generated by high-salt concentrations (Bajji et al., 1998). For example, Na^+ competes for the absorption sites with K^+ and Ca^{++}, and Cl^- with nitrates and phosphate. On the other hand, salinity affects several and important metabolic processes in the plant, as the absorption of water and nutrients, the osmotic adjustment, photosynthesis, but also the protein synthesis and enzyme activity (Levigneron et al., 1995).

Halophytes, known as the plants which accomplish their vital cycle in an environment rich in salts, have developed during their evolution several mechanisms which preserve them from the depressive effects of salinity, enabling them to grow well, sometimes even better in the presence of salt (Flowers et al., 1977; Osmond et al., 1980; Breckle, 1995). Salt-tolerance is a complex phenomenon requiring the intervention of several mechanisms at the whole plant level (Yeo, 1983). On the other hand, there are no clear limits to classify halophytes in function of their level of salt-tolerance, the latter varying with families, genera, species, and with varieties as well (Levigneron et al., 1995). However, one estimates that optimal growth of halophytes is generally reached within the concentrations ranging from 20 to 500 mM NaCl (Flowers et al., 1977). The positive evolution of halophytes growth in presence of salt has already been reported (Yeo and Flowers, 1986; Marcum and Murdoch, 1992; Freitas and Breckle, 1993b; Köhl, 1997).

Arid and semi-arid zones, located in the Mediterranean basin, are among the most severely salt-affected lands. The climatic and edaphic conditions prevailing there, both are contributing to this situation. These regions could be rehabilitated by introducing tolerant species of economical interest, capable of developing on such hostile surroundings, which is the case with the *Atriplex*. The important polymorphism characterizing this genus in its response to the saline stress in particular (Osmond et al., 1980), makes it of higher interest, because it would permit the selection of high salt-tolerant individuals.

Our objective, using this biologic diversity, is to perform a comparative study of the effect of NaCl on two provenances of *Atriplex halimus* L. The work consists of a physiological and structural study. Moreover, the evolution of total soluble peroxidases and foliar phosphoenolpyruvate carboxylase (PEPC) activities in function of salinity is discussed.

MATERIAL AND METHODS

Plant Material and Culture Conditions

The plant material we used, represents the sub-species *Atriplex halimus* cv. halimus. Seeds of Amilcar provenance, a littoral locality near Tunis, were harvested from the same stock plant growing at the seashore. Those from El Alam, a region close to Kairouan (in the centre of Tunisia), were collected from the same stock plant, too. Seeds were sown in sand, at 1.5 cm of depth and 1 cm apart. They were watered with distilled water during the first week, and with a KNOP solution for another 3 weeks. After one month, seedlings were grown hydroponically in a controlled environment chamber: 25 ± 5°C, with a photoperiod

H. Lieth (ed.), Cash Crop Halophytes: Recent Studies, 19–30.

of 16 h, 55–85% of relative humidity and 35 W/m^2 of light intensity. KNOP solutions (pH 4.9) were continuously aerated and changed every week. Saline stress (NaCl at 5, 10, 20 and 30 g/L) was progressively applied after 7 days until reaching the final concentration. Prior to salinization, initial dry weights of 10 randomly harvested seedlings were measured.

Once harvested, 2 month-plants were divided in leaves, stems and roots. Plant growth parameters were determined by measuring fresh and dry weights (FW and DW, respectively), foliar area (cm^2), relative growth rate (RGR) and shoots DW/roots DW ratio. Water status relations assessed were water content and succulence rating (leaf FW/leaf DW).

$$RGR = \frac{\text{final DW} - \text{initial DW}}{(\text{final DW} + \text{initial DW})\, t/2}$$

where t corresponds to the period of treatment (days).

After an extraction in 0.1 N HNO_3, cations (Na^+, K^+ and Ca^{++}) contents were determined by an emission flame spectrophotometer (Eppendorf), while Chloride was measured by choulometry using a chloridometer (Buchler-Cotlove).

Enzyme Activity

Total soluble peroxidases
For the extraction, 1 g of leaf fresh matter was ground in 2 ml of 100 mM phosphate buffer (pH 7) at 4°C. The homogenates were centrifuged at 13,000 g for 20 min. Supernatants were then used for peroxidases activity measurements which were performed in a reaction mixture containing a 50 mM phosphate buffer (pH 7), 10 mM guaiacol and 5 mM H_2O_2. Guaiacol oxidation was monitored at 470 nm using a spectrophotometer (Beckman Of 640).

Phosphoenolpyruvate carboxylase (PEPC)
(a) Enzyme preparation. 1 g of leaf fresh matter was ground with 5 ml of extraction medium containing 100 mM Tris-HCl buffer (pH 8), 5% PVP insoluble, 10% Glycerol, 5 mM $MgCl_2$, 1 mM EDTA, 1 mM DTT, 1 mM PMSF, 1 mM Leupeptine. The homogenates were centrifuged at 10,000 g for 5 min. Supernatants were used for activity measurements. The above operations were carried out at 4°C.

(b) Enzyme assays. In C_4 plants, the foliar PEPC catalyses the phosphoenolpyruvate carboxylation to oxaloacetic acid, which is then converted into malate, a reaction catalysed by the NAD malate dehydrogenase. The reaction was run at 30°C and the PEPC activity was measured by following the NADH oxidation at 340 nm by a spectrophotometer (Beckman Of 640). The reaction mixture contained (in μM) Tris-HCl (pH 8), 100; $MgCl_2$, 10; EDTA, 1; DTT, 1; $NaHCO_3$, 40; NADH, 0.2; ATP, 4; PEP, 2 and 100 μl of the crude extract.

Chlorophyll Contents

The chlorophyll a and b contents were measured using a spectrophotometer (Beckman Of 640), at 645 and 663 nm (Arnon, 1949).

Cyto-histological Investigations

In order to determine the *Atriplex halimus* L. caryotype, seeds were germinated for 24 h at 25°C in the dark. Then, root apexes were colored with orceine and observed with a light microscope.

The structural survey consisted in achieving cross sections of newly formed or aged leaves. After their inclusion with paraffin and coloration with hematoxylin, safranin and anilin blue (triple coloration), samples were observed with a light microscope. For the scanning electronic microscope (SEM) study, leaves were fixed in glutaraldehyde and washed with a cacodylate buffer. This operation was followed by a dehydration in ethanol increasing concentrations and by a passage in pure acetone.

RESULTS

Plants Growth

El Alam Control plants were more productive with regard to Amilcar, while NaCl addition, up to 10 g/l, significantly improved the growth of both provenances (Figure 1). At optimal concentrations (5 and 10 g/l), NaCl promoting effect on growth was less pronounced in Amilcar provenance than El Alam. At 20 g/l, NaCl stimulating effect disappeared, and a significative decrease of the growth was observed at 30 g/l NaCl, but with differences between the two provenances. This depressive action was more marked in

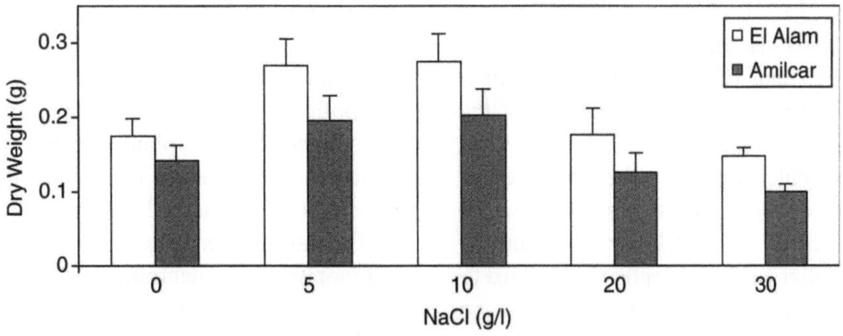

Fig. 1. Effect of NaCl concentration on biomass production of 2-month old plants of *Atriplex halimus* L. El Alam and Amilcar provenances. Plants were treated for 3 weeks (means of 15 measurements, 5% confidence interval).

Table 1. Evolution of leaf area, shoot DW/root DW ratio, RGR, plants content and succulence rating as a function of NaCl in *A. halimus* L. El Alam and Amilcar provenances (means of 15 measurements, 5% confidence intervals)

	Leaf area (cm^2)		Shoot DW/root DW		Shoot RGR		Water content (ml/g DW)		Foliar succulence rating (FW/DW)	
NaCl (g/l)	El Alam	Amilcar	El Alam	Amilcar	El Alam	Amilcar	El Alam	Amilcar	El Alam	Amilcar
0	1.95 ± 0.15	1.51 ± 0.11	6.94	6.87	0.05	0.049	7.81	6.35	9.19	7.38
5	2.25 ± 0.19	1.76 ± 0.12	9.675	8.382	0.056	0.054	10.69	8.41	12.26	10.28
10	2.72 ± 0.2	2.22 ± 0.16	9.2	8.265	0.056	0.054	10.47	8.35	12.33	10.31
20	1.86 ± 0.11	1.17 ± 0.1	8.816	7.27	0.051	0.048	8.90	7.40	10.43	8.42
30	1.38 ± 0.1	1.06 ± 0.08	8.84	7.33	0.048	0.044	7.67	5.94	9.44	6.9

Amilcar, which appeared less tolerant to high NaCl concentrations.

The same trend was noted for the other growth parameters, as indicated in Table 1. Leaf area rised up to 10 g/l NaCl, before sloping down at higher concentrations, notably at 30 g/l. In the same way, the shoot DW to root DW ratio evolved positively up to 10 g/l NaCl, ranging from 6.94 in control to 9.2 in plants grown with 10 g/l NaCl for El Alam and from 6.87 to 8.26 in Amilcar.

Comparison of growth performance between the both provenances were assessed by RGR. The growth activity was slightly higher in El Alam relative to Amilcar controls. The presence of NaCl up to 10 g/l, enhanced root (data not shown) and shoot growth activity for the both provenances. Reductions observed at above concentrations were more marked in Amilcar, notably at 30 g/l. These preleminary results confirm the halophilic character of *Atriplex halimus* L. as well as the variability of the plant response to salt, El Alam provenance distinguishing by a better behaviour with regard to Amilcar, at optimal (5–10 g/l NaCl), but also at extreme concentrations (30 g/l).

Water Relations

NaCl improved the water status of the plants (Table 1). Shoot or root water content was maximal at 10 g/l before declining at higher concentrations, suggesting the presence of a close relationship between plants growth and their water status. Nevertheless, El Alam provenance was able to maintain, at the extreme concentrations (20 and 30 g/l NaCl), a better level of hydration in its tissues in comparison with Amilcar, which would partly explain the sensivity of the latter. The amelioration of the leaves hydration by NaCl, is reflected through their succulence ratio enhancement (Table 1). In El Alam provenance, it increased substantially in presence of 5 and 10 g/l NaCl. Although decreasing at higher salt levels, this parameter remained above control level. This trend was also noted in Amilcar provenance, but with a more pronounced fall at 20 and 30 g/l NaCl. Light microscopy transverse sections of leaves show that the thickness of leaves was 2-fold higher in El Alam (Figure 2A) than Amilcar (Figure 2B), confirming the higher succulence of the first provenance.

The increased succulence permits the storage of an important quantity of water. This was observed by SEM investigations: 10 g/l NaCl increased turgescence of El Alam foliar epidermic cells with regard to control. Similarly, dimensions of stomatic cells were enhanced (Figure 2C and D), leading probably to a modification of their aperture degree.

Total Soluble Peroxidases Activity

In control plants, total soluble Peroxidases activity of the both provenances, expressed on a dry matter basis, was higher than in plants treated with 10 g/l NaCl (optimal concentration for the growth) (Figure 3). At this level, the measured activity was 50.7% of control in El Alam provenance and 53.2% in Amilcar. NaCl supraoptimal concentrations (in particular 30 g/l) induced an increase of the activity by 73% in relation to 10 g/l NaCl, in El Alam provenance, and by 71.5% in Amilcar. These results display an inverse

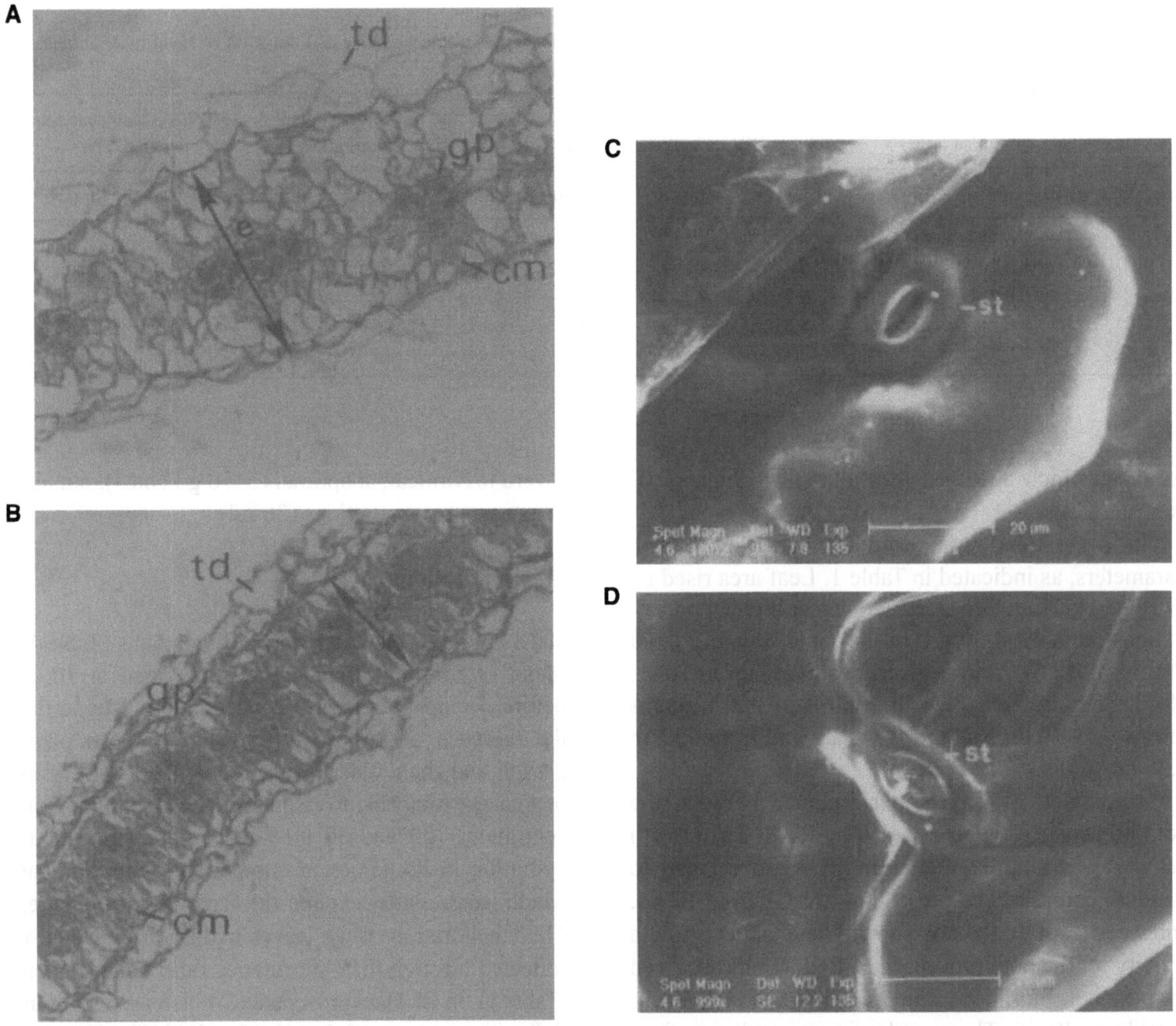

Fig. 2. Cross sections of leaves of *Atriplex halimus* L. treated with 30 g/l NaCl. The leaf thickness (e) of El Alam provenance (A) is twice more important than Amilcar (B). (td) trichomes debris; (cm) mesophyll cells; (gp) bundle-sheath cells. Scanning Electron Microscopy cross sections of leaves of *Atriplex halimus* L. treated with 10 g/l NaCl. Stomatic cells (st) of treated plants (C) are more turgescent than control ones (D).

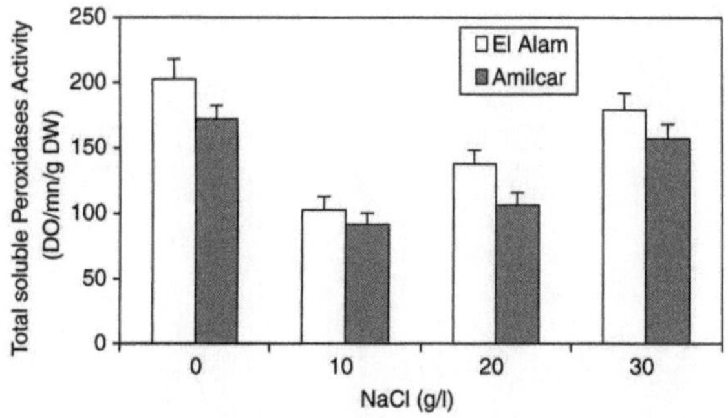

Fig. 3. Evolution of total soluble Peroxidases activity in leaves of 2-month old plants of *Atriplex halimus* L. El Alam and Amilcar provenances at different NaCl concentrations. Plants were treated for 3 weeks (means of 4 assays, 5% confidence interval).

correlation between growth and the soluble peroxidases activity.

It appears on the other hand, that at high concentrations (20 and 30 g/l NaCl), the total soluble peroxidases activity is more elevated in El Alam provenance, in spite of its better behaviour than Amilcar. This apparent contradiction could be assigned to the fact that we measured the activity of soluble peroxidases, which contain many isozymes, certain of them limiting the growth by contributing to cell walls lignification, while other having other functions in the cell.

Mineral Nutrition

Ionic accumulation

As indicated by Figure 4A and B, potassium and calcium represented the major cations in the control leaves. In El Alam provenance, NaCl enhanced Na^+ contents, concomitant with a decline of K^+ and Ca^{++} (Figure 4A). This trend continued at higher concentrations, but at a slower rate. In the other organs, one noted similar trend (data not shown). The same tendency characterized Amilcar, but with less contents for the three cations (Figure 4B). Foliar Cl^- contents also rised at 5 g/l NaCl, and then remained relatively constant (Figure 4C). Such evolution occured also in the stems and the roots, but with Cl^- contents lower than those measured in leaves (data not shown).

The structural survey we performed in leaves aimed to give informations about Na^+ and Cl^-, which were easily transported to the leaves. It was described in the literature that *Atriplex halimus* L. leaves are provided with anatomic structures, called trichomes or vesiculated hairs which represent the site of salt excretion.

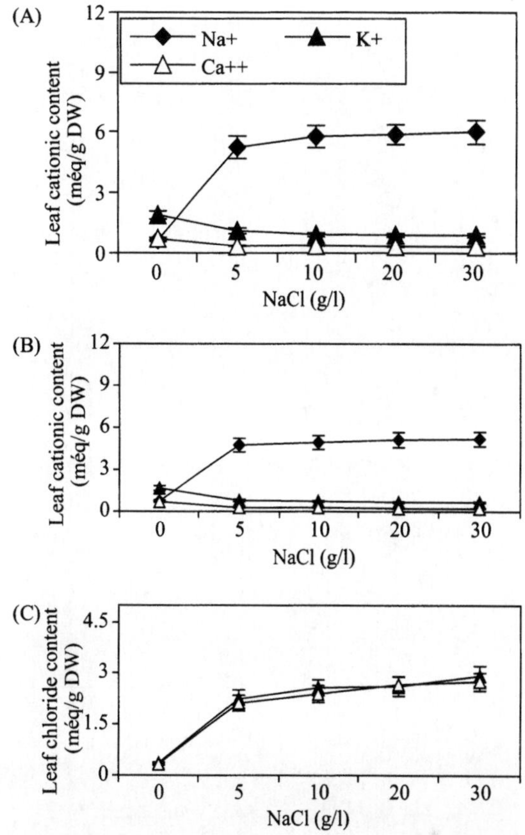

Fig. 4. Effect of NaCl on the ionic status of 2-month old plants of *Atriplex halimus* L. leaves. Cationic contents of El Alam (A) and Amilcar (B) provenances. Plants were treated for 3 weeks (means of 15 measurements, 5% confidence interval). (C) Chloride contents. Plants were treated for 3 weeks (means of 15 measurements, 5% confidence interval).

When full of salt, the trichomes burst and collapse over the leaf surface (Smaoui, 1971; Osmond et al., 1980). This means prevents the excessive accumulation of sodium in the apoplast and the cell dehydration it is responsible for. Several authors indicated that although trichomes ontogenesis continues during the leaves' life span, it is, however, more important in the young leaves. This would explain their low density or absence in the adult leaves (Osmond et al., 1980, Schrimer and Breckle, 1982; Storey et al., 1983; Freitas and Breckle, 1993a and b).

In the control juvenile leaves, the trichomes though frequent (Figure 5A), looked somewhat flattened (Figure 5B). In the plants treated by 10 or 30 g/l NaCl (Figure 6A), they appeared clearly more numerous as swollen bladders. These vesiculated cells (bladder cells) are in contact by the stalk cell with the leaf epidermis

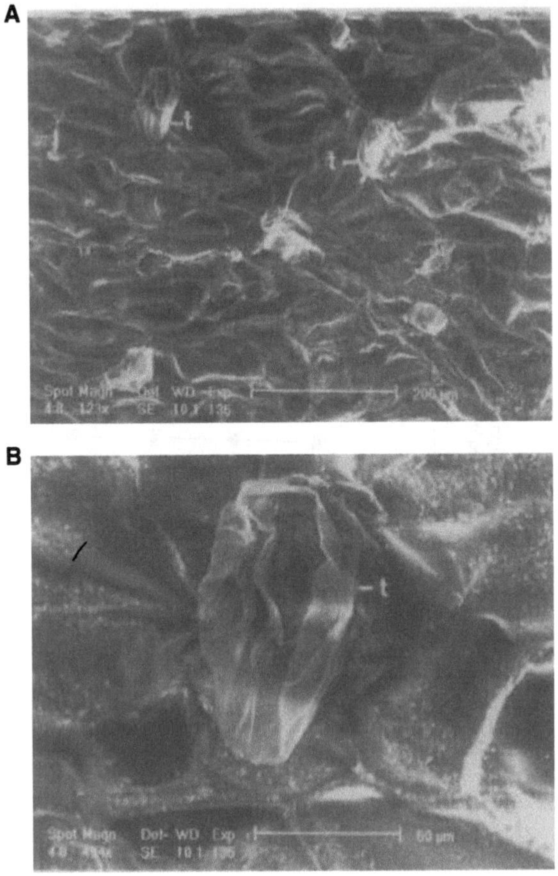

Fig. 5. Scanning Electron Microscopy of young leaves of *Atriplex halimus* L. control plants (A), showing the presence of trichomes (t) at the leaf surface. Trichomes (t) look somewhat flattened (B).

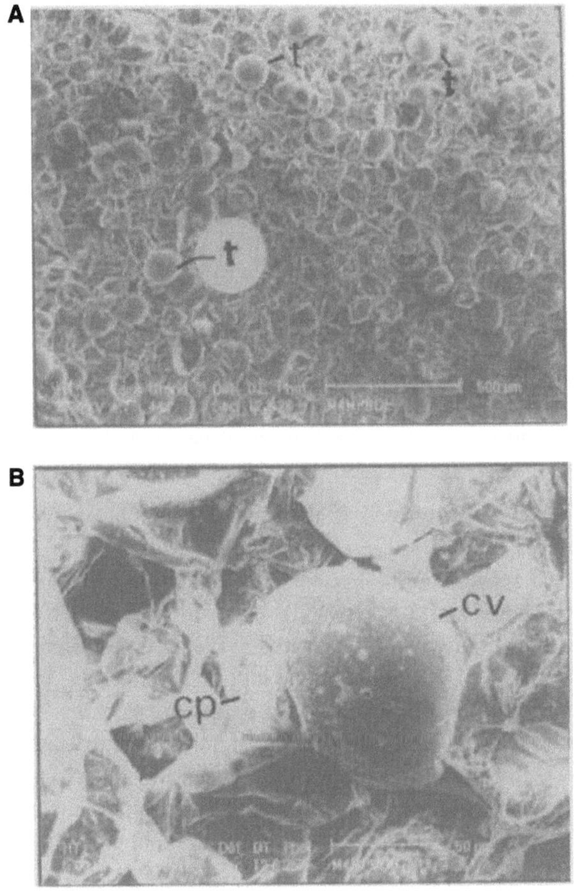

Fig. 6. Scanning Electron Microscopy of young leaves of *Atriplex halimus* L. treated with 30 g/l. Numerous trichomes (t) at the leaf surface appear like swollen bladders (A), being in contact with the leaf epidermis (B) by the stalk cell (cp) which is surmounted by the vesiculated cell (cv).

(Figure 6B). In older leaves, characterized by a lesser density of trichomes, plants of El Alam provenance treated by high NaCl concentrations (30 g/l) presented pronounced intercellular spaces, delimited by a membrane (Figure 7A and B). These vesicles located in the leaves (between the mesophyll cells), and not at their surface like trichomes, would have an accumulating function of excessive salt. In Amilcar, diffuse spaces occurred between the mesophyll cells (Figure 8A and B), suggesting that the cellular structure would be more affected in this less salt-tolerant provenance.

Potassium-sodium selectivity

It is known from halophytes that they succeed in conserving a pronounced selectivity for potassium over sodium, despite increased concentrations of this latter in the medium (Osmond et al., 1980; Köhl, 1997). In our case, we compared the medium's $K^+/(K^+ + Na^+)$ ratios with those of the plant organs (leaves, stems and roots).

Table 2 indicates that independent of NaCl doses, these ratios were higher in the plant than in the medium, which denotes the ability of this plant to maintain a marked selectivity for potassium, even at the extreme concentration (30 g/l NaCl). Selectivity in favour of potassium in the roots was more pronounced than in the shoots. This is in agreement with the hypothesis according to which, sodium is privileged during the long distance transport (Storey et al., 1983). Moreover, El Alam provenance was characterized by a higher selectivity than Amilcar, especially at the extreme concentrations (20 and 30 g/l NaCl). This important trait explains the better salt-tolerance of the first provenance.

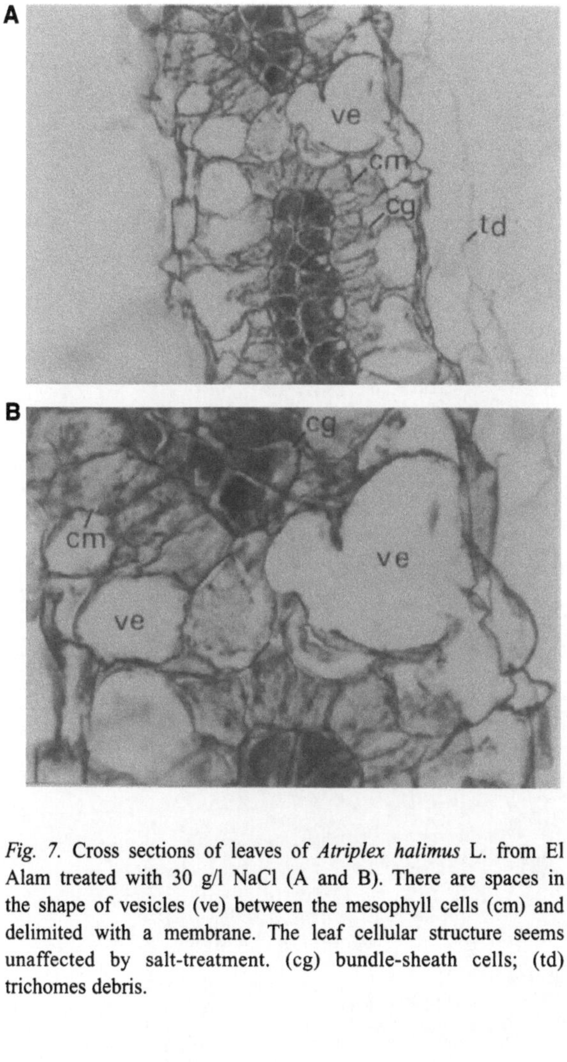

Fig. 7. Cross sections of leaves of *Atriplex halimus* L. from El Alam treated with 30 g/l NaCl (A and B). There are spaces in the shape of vesicles (ve) between the mesophyll cells (cm) and delimited with a membrane. The leaf cellular structure seems unaffected by salt-treatment. (cg) bundle-sheath cells; (td) trichomes debris.

Photosynthesis

The *Atriplex* species are characterized by a variability in their photosynthesis pathway. According to Osmond et al. (1980), 60% are C_4 plants, such *Atriplex nummularia* and 40% are C_3 like *A. hortensis* and *A. patula.* Our observations indicated that *A. halimus* L. leaf anatomy was characterized by an arrangement typical of C_4 plants (Figure 7) showing: (i) an inner layer of concentric cells called bundle-sheath cells, where C_3 reactions occur, surrounding the vascular tissues, and (ii) an adjacent outer cylinder of mesophyll cells, the cytoplasm of which contains the PEPC, the responsible enzyme for the primary fixation of atmospheric CO_2.

Foliar PEPC activity rised in the both provenances up to 10 g/l NaCl (Table 3). The higher concentrations inhibited enzyme activity with a considerable fall at 30 g/l. One also noted that PEPC activity was higher in El Alam leaves. Similarly, chlorophyll contents increased up to 10 g/l NaCl, but declined at the extreme concentrations (Table 3). This was true for the two provenances.

Caryotype Study

The variability that we have noted in the response of *A. halimus* L. to salinity as a function of its origin area could be related to a diversity in ploïdy. According to Osmond et al. (1980), though the majority of the *Atriplex* are diploïd ($2n = 18$ chromosomes), other levels of ploïdy can take place. The observation of metaphases in El Alam provenance indicated that the majority of cells were tetrapoid ($4n = 36$) (Figure 9A). In Amilcar provenance, all the chromosome counts revealed $2n = 18$ chromosomes (Figure 9B). We think that further observations should be done in order to determine precisely the caryotype of the two provenances. These results suggest that the variability in the salt-tolerance between the two provenances could be linked to the difference found in their caryotype, indicating an ecophysiological adaptation to the conditions of their original area.

DISCUSSION AND CONCLUSION

Our results confirm the halophilic status of *A. halimus* L. through the positive impact of NaCl-treatment on the plants growth, with an optimum at 5–10 g/l. This species remained alive at 30 g/l NaCl, which is close to the sea water concentration. The comparative study between the two provenances revealed that El Alam was more salt-tolerant. The NaCl stimulating effect on the growth of *Atriplex halimus* L. has often been mentioned. Ben Ahmed et al. (1995) noted that the maximal growth occured in the concentrations ranging from 50 to 200 mM NaCl. They also displayed an important polymorphism for various characters, among which, the biomass production. Bajji et al. (1998) reported too that low doses of NaCl (150 mM) promoted shoot growth of *A. halimus* L.

The same behaviour characterizes other halophytes (Storey and Wyn Jones, 1979; Yeo and Flowers, 1986; Marcum and Murdoch, 1992; Freitas and Breckle, 1993b; Köhl, 1997). On the other hand, we noted the increase of roots and shoots RGR up to 5–10 g/l NaCl, in accordance with Bajji et al. (1998), before declining

A

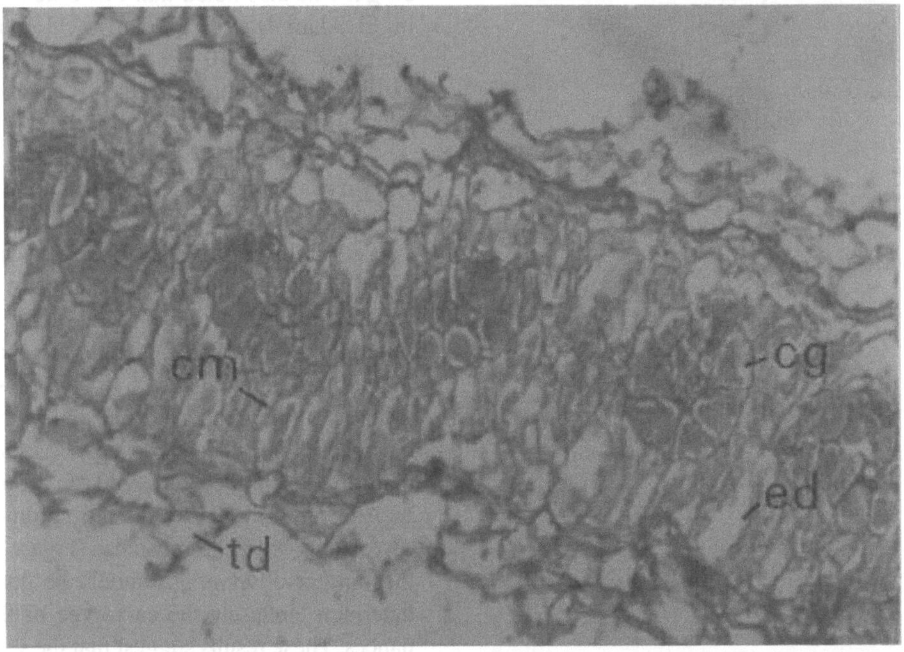

B

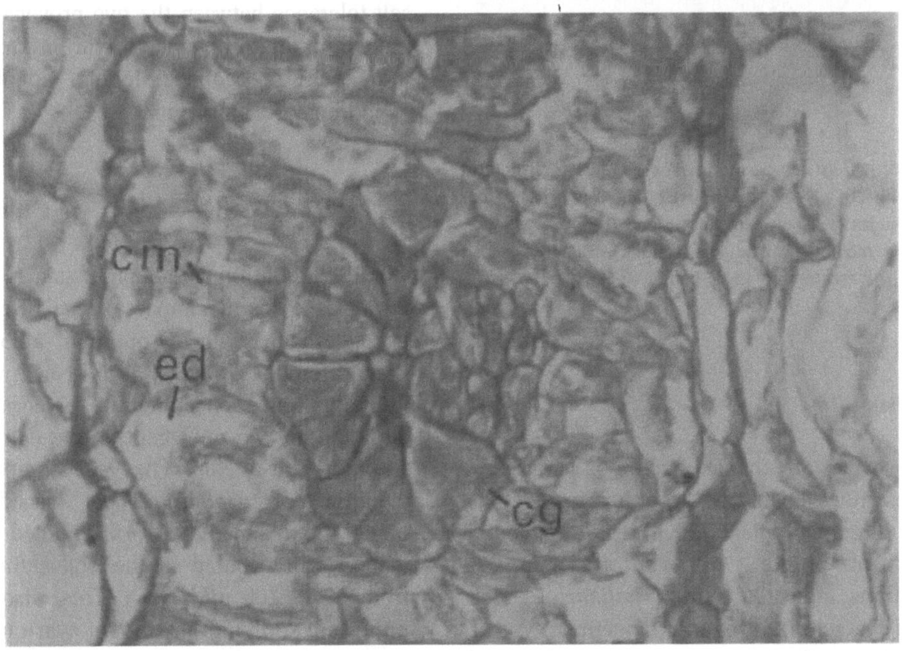

Fig. 8. Cross sections of leaves of *Atriplex halimus* L. from Amilcar treated with 30 g/l NaCl (A and B), showing the presence of diffuse intercellular spaces (ed) between the mesophyll cells (cm). In this case, the cellular structure is altered by salt-treatment. (cg) bundle-sheath cells; (td) trichomes debris.

at the higher levels, notably at 30 g/l. The shoots growth activity was however more elevated in El Alam provenance, when compared with Amilcar.

Our results also show that shoots growth was most stimulated by treatment with 5–10 g/l NaCl. According to Osmond et al. (1980), NaCl induces the preferential allocation of assimilates rather to the shoots, as reflected by a higher shoot DW/root DW ratio. In *Armeria maritima*, Köhl (1997) found that this ratio was 2- to 3-fold higher at 100 mM NaCl than in the

Table 2. $K^+/K^+ + Na^+$ selectivity ratios in the culture medium and the plant tissues of *Atriplex halimus* L. provenances El Alam and Amilcar as a function of NaCl level (means of 15 measurements)

		Leaves		Stems		Roots	
NaCl (g/l)	Medium	El Alam	Amilcar	El Alam	Amilcar	El Alam	Amilcar
5	0.023	0.188	0.143	0.177	0.128	0.247	0.233
10	0.012	0.138	0.120	0.143	0.106	0.203	0.173
20	0.0058	0.126	0.105	0.130	0.108	0.187	0.131
30	0.0039	0.120	0.099	0.126	0.097	0.171	0.123

Table 3. PEPC activity and chlorophyll contents in leaves of *Atriplex halimus* L. provenances El Alam and Amilcar as a function of NaCl level (means of 4 assays)

	PEPC (μM/mn/g DW)		Chlorophyll a (mg/g DW)		Chlorophyll b (mg/g DW)		Total chlorophyll (mg/g DW)	
NaCl (g/l)	El Alam	Amilcar	El Alam	Amilcar	El Alam	Amilcar	El Alam	Amilcar
0	21.53	17.74	8.27	4.96	6.61	4.09	14.9	9.07
10	38.14	28.51	11.55	8.99	7.89	5.13	19.46	14.05
20	18.26	21.85	6.44	6.28	6.55	1.87	13.00	8.13
30	17.21	14.74	5.66	3.08	5.67	1.41	11.29	4.48

A

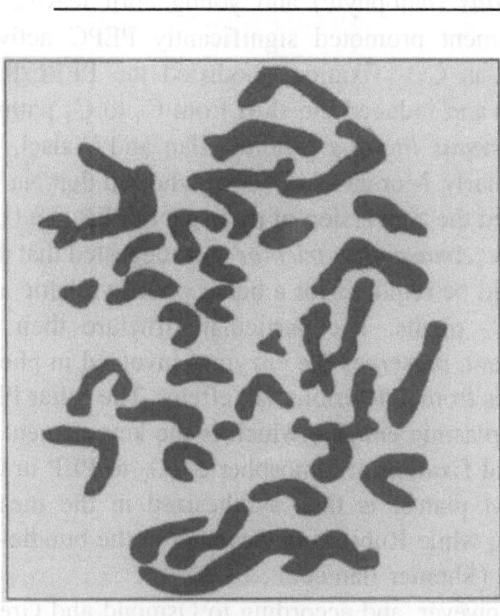

B

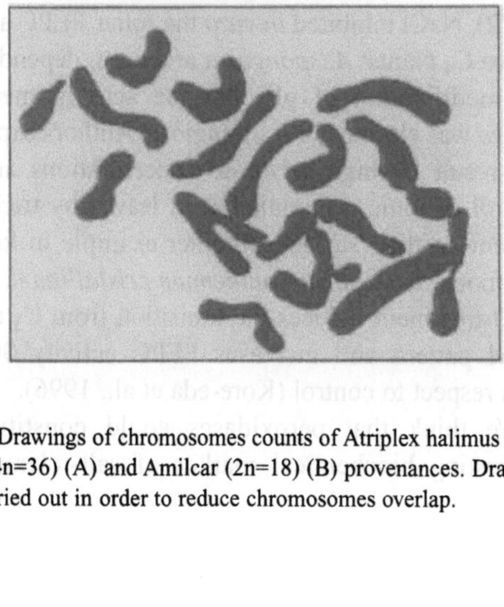

Fig. 9. Drawings of chromosomes counts of Atriplex halimus L. El Alam (4n=36) (A) and Amilcar (2n=18) (B) provenances. Drawing was carried out in order to reduce chromosomes overlap.

control. Besides, 5–10 g/l NaCl improved the leaves water status, which had consequences on the osmotic and metabolic (photosynthetic activity) levels. This confirms previous studies on other species (Handley and Jennings, 1977; Storey and Wyn Jones, 1979; Benzioni et al., 1992; Ayala and O'leary, 1995). Authors consider that leaf succulence conjugated to the ability to conserve hypertony, are among the most important features of halophytes (Osmond et al., 1980; Heller et al., 1993). Sodium is involved in this phenomenon, being highly accumulated in the leaf vacuoles (Ayadi et al., 1980; Mc Hue and Hanson, 1990).

The inversely proportional evolution of the total soluble peroxidases activity with the growth suggests that the absence of NaCl in the medium constitutes a stress for *A. halimus* L., thus confirming its halophilic nature. In the cell, peroxidases occur in different isoforms with multiple roles, among which, lignification and the tightening of cell walls, leading to the restriction of the cellular growth (Sato et al., 1993). Peroxidases are found in the cytoplasm, but also are associated to cell walls, mitochondries or ribosomes (Lee, 1973), and their activity is sensitive to many stimuli (temperature, drought, infection). Thiyagarajah et al. (1996) showed that parietal peroxidases were tolerant up to 2 M NaCl *in vitro*. *In vivo*, Hagège et al. (1988) noted the higher activity of the total soluble peroxidases in untreated plants of *Suaeda maritima* with respect to those grown with 170 mM NaCl. This suggests the presence of a negative relationship between the growth and the shoot lignification degree, and corroborates our results by the same. Similar patterns were reported by Wang et al. (1997) in *Atriplex prostrata*. Flowers (1972), showed for his part, that NaCl did not modify the peroxidases activity in *Suaeda maritima*, unlike to the sensitive plant *Pisum sativum*.

The mineral nutrition survey displayed the preferential transport of Na^+ and Cl^- to the leaves. This does not constitute a disadvantage for *A. halimus* L., unlike to what happens in salt-sensitive plants. This typical behaviour of a halophyte has been already described (Marcum and Murdoch, 1992; Glenn et al., 1996; Köhl, 1997). According to Flowers et al. (1977), about 90% of the sodium found in the halophytes tissues, is localized in the shoots. It is mainly squestred in vacuoles (in the adult leaves), or is removed via trichomes. This last mechanism takes extent in the young leaves because of a higher density of trichomes, like we have shown. It has been reported that despite the trichomes' formation occurs along the whole life of the leaf, it decreases with their age (Smaoui, 1971; Osmond et al., 1980; Freitas and Breckle, 1993a). The presence of such structures permits the xylemic sap desalination by accumulating important amounts of salt, reaching 50% (Osmond et al., 1980), to 80% (Freitas and Breckle, 1993a) of the quantities transported to the shoots. In aged leaves, owing to a low density of trichomes, the salt is accumulated in some intercellular spaces located in the mesophyll, participating by this way in the osmotic adjustment and protecting the metabolic activity in the cytoplasm. In the most tolerant provenance (El Alam), these spaces were delimited by a membrane unlike to Amilcar in which, they were diffuse.

Increased cellular sodium amounts induced the reduction of calcium and potassium ones. The NaCl depressive effect on the the plant nutrition in calcium would be due to a Na^+/Ca^{++} competition during absorption and transport (Le Saos, 1976; Stassart et al., 1981), leading to the displacement of Ca^{++} apoplasmic fraction to the profit of Na^+ (Zid and Grignon, 1985). Botella et al. (1997) reported that NaCl negatively affected the supply of potassium in corn. This effect was more marked with low K^+ concentrations. Recently, Banuls et al. (1997) recorded, however, that Ca^{++} addition to the medium attenuated the Na^+ and Cl^- depressive effects by limiting their contents in *Citrus* leaves and preserved selectivity for potassium, owing to the calcium role in maintaining the cellular membrane integrity. We also showed that *A. halimus* L. was characterized by a pronounced selectivity in favour of potassium, despite the inhibiting effect of Na^+ on its accumulation. In fact, though the plant supply in Ca^{++} and K^+ was disturbed, the contents of this last were maintained to a minimal level, even at 30 g/l NaCl.

Our results indicated a NaCl stimulating effect at optimal concentrations on the foliar PEPC activity, in agreement with Shomer-Ilan and Waisel (1973), Passera and Albuzio (1977) and Kore-eda et al. (1996). On the contrary, other authors have mentioned the inhibitory effect of this salt (Osmond and Greenway, 1972). Clipson (1987) and Wang et al. (1997) noted that salt impaired the photosynthesis in *Suaeda maritima* and *A. prostrata*, likely in relation with increased mesophyll and stomatic resistance (Gale and Poljakoff-Mayber, 1970; Kaplan and Gale, 1972; Longstreth and Nobel, 1979; Osmond et al., 1980; Ayala and O'Leary, 1995). However, C_4 plants are generally more salt-tolerant, owing to better water use and CO_2 assimilation (Binet, 1989). In *Aeluropus litoralis* (halophyte) and young corn leaves, NaCl-treatment promoted significantly PEPC activity as well as CO_2 fixation, modified the PEPC/Rubisco ratio and induced the shift from C_3 to C_4 pathway in *Aeluropus litoralis* (Shomer-Ilan and Waisel, 1973). Similarly, Murata et al. (1992) showed that Na^+ stimulated the conversion of pyruvate to PEP in a C_4 halophyte: *Amaranthus tricolor* and suggested that this ion would be required for a better photosynthetic activity in C_4 plants. The particular structure their leaves present, preserves the enzymes involved in photosynthesis from deleterious salt effetcs. The foliar PEPC, a cytoplasmic enzyme which is the key element in the initial fixation of atmospheric CO_2 to PEP in C_4 and CAM plants, is thus synthesized in the mesophyll cells, while Rubisco is localized in the bundle-sheath cells (Shomer-Ilan et al., 1979).

However, and according to Osmond and Greenway (1972), NaCl inhibited *in vitro* the foliar PEPC activity in two C_4 plants: *A. spongiosa* and corn, dependent on the modification of pH, but the activity measured *in vivo* was elevated in *A. spongiosa*. Authors explained this result by high substrate concentrations and the control of ionic accumulation in leaves by trichomes present on their surface. Another example in favor of this model is *Mesembryanthemum cristallinum*, where NaCl-treatment induces the transition from C_3 toward CAM pattern and increases PEPC activity 36-fold, with respect to control (Kore-eda et al., 1996).

We think that peroxidases could constitute an interesting biochemical marker of salt-tolerance in halophytes. Further researches would aim to the extraction and the characterization of isoforms involved in the plant answer to salinity.It would be interesting also to investigate the salt-induced ultrastructural modifications in shoots and roots. In the leaves, this study would help to explain the vesiculization process in the mesophyll cells.

ACKNOWLEDGEMENTS

The present work was supported by the European Economic Community STD program (Convention No TS3-CT94–0264). Authors would like to thank Dr. C. Abdelly for his helpful suggestions, as well as Pd. Drs. B. Huchzermeyer and H.W. Koyro for their critical reading.

REFERENCES

Arnon, D.I. Copper enzymes in isolated chloroplasts. Polyphenoloxidase in *Beta vulgaris*. Plant Physiol 1949; 241–15.

Ayadi, A., Monnier, A., Demarty, M. and Thellier, M. Echanges ioniques cellulaires: cas des plantes en milieu salé. Rôle particulier des parois cellulaires. Physiol Vég 1980; 18:89–104.

Ayala, F. and O'Leary, J.W. Growth and physiology of *Salicornia bigelovii* Torr. at sub-optimal salinity. Int J Plant Sci 1995; 156:197–205.

Bajji, M., Kinet, J.M. and Lutts, S. Salt stress effects on roots and leaves of *Atriplex halimus* L. and their corresponding callus cultures. Plant Sci 1998; 137:131–42.

Banuls, J., Serna, M.D., Legaz, F., Talon, M. and Primo-Millo, E. Growth and gas exchange parameters of *Citrus* plants stressed with different salts. J Plant Physiol 1997; 150:194–9.

Ben Ahmed, H., Zid, E., El Gazzah, M. and Grignon, C. Croissance et accumulation ionique chez *Atriplex halimus* L. Cahiers Agricult 1995; 5:367–72.

Benzioni, A., Nerd, A., Rosengärtner, Y. and Mills, D. Effect of NaCl salinity on growth and development of Jojoba clones: I. Young plants. J. Plant Physiol 1992; 139:731–6.

Binet, P. Métabolisme et adaptations des végétaux supérieurs aux contraintes hydriques, thermiques et salines. Bull Ecol 1989; 20:41–9.

Botella, M.A., Martinez, V., Pardines, J. and Cerda, A. Salinity induced potassium deficiency in Maize plants. J Plant Physiol 1997; 150:200–5.

Breckle, S.W. How do halophytes overcome salinity? In Khan, M.A. and Ungar, I.A., editors. *Biology of salt tolerant plants*. Dept. of Botany, University of Karachi, Pakistan, 1995; 199–213.

Clipson, N.J.W. Salt tolerance in the halophyte *Suaeda maritima* L. Dum. Growth, ion and water relations and gas exchange in response to altered salinity. J Exp Bot 1987; 38:1996–2004.

Flowers, T.J. Salt tolerance in *Suaeda maritima* (L.) Dum. The effect of sodium chloride on growth, respiration, and soluble enzymes in a comparative study with *Pisum sativum* L. J Exp Bot 1972; 23:310–21.

Flowers, T.J., Troke, P.F. and Yeo, A.R. The mechanism of salt tolerance in halophytes. Ann Rev Plant Physiol 1977; 28:89–121.

Freitas, H. and Breckle, S.W. Importance of bladder hairs for seedlings of some *Atriplex* species. Mésogée 1993a; 53:47–54.

Freitas, H. and Breckle, S.W. Accumulation of nitrate in bladder hairs of *Atriplex* species. Plant Physiol Biochem 1993b; 31:887–92.

Gale, J. and Poljakoff-Mayber, A. Interrelations between growth and photosynthesis of salt bush (*Atriplex halimus* L.) grown in saline media. Aust J Sci 1970; 23:947–52.

Glenn, E., Pfister, R., Brown, J., Lewis Thompson, T. and O'Leary, J. Na and K accumulation and salt tolerance of *Atriplex canescens* (Chenopodiaceae) genotypes. Am J Bot 1996; 83:997–1005.

Hagège, D., Kevers, C., Boucaud, J. and Gaspar, T. Activités peroxydasiques, production d'éthylène, lignification et limitation de croissance chez *Suaeda maritima* cultivé en l'absence de NaCl. Plant Physiol Biochem 1988; 26:609–14.

Handley, J.F. and Jennings, D.H. The effect of ions on growth and leaf succulence of *Atriplex hortensis* var. *cupreata*. Ann Bot 1977; 41:1109–12.

Heller, R., Esnault, R. and Lance, C. Physiologie Végétale. 1. Nutrition, Masson Ed, Paris, 1993.

Kaplan, A. and Gale, J. Effect of sodium chloride salinity on the water balance of *Atriplex halimus*. Aust J Biol Sci 1972; 25:895–903.

Köhl, K.I. The effect of NaCl on growth, dry matter allocation and ion uptake in salt marsh and inland populations of *Armeria maritima*. New Phytol 1997; 135:213–25.

Kore-Eda, S., Yamashita, T. and Kanai, R. Induction of light dependent pyruvate transport into chloroplasts of *Mesembryanthemum crystallinum* by salt stress. Plant Cell Physiol 1996; 37:257–62.

Lee, T.T. On extraction and quantification of plant Peroxidase isoenzymes. Physiol Plant 1973; 29:198–203.

Le Saos, J. Migration du calcium vers les organes aériens chez une halophyte *Cochlearia anglica*. Effets du NaCl. Physiol. Vég 1976; 14:381–90.

Levigneron, A., Lopez, F., Vansuyt, G., Berthomieu, P., Fourcroy, P. and Casse-Delbart, F. Les plantes face au stress salin. Cahiers Agricult 1995; 4:263–73.

Longstreth, D.J. and Nobel, P.S. Salinity effects on leaf anatomy. Consequences for Photosynthesis. Plant Physiol 1979; 63:700–3.

Marcum, K.B. and Murdoch, C.L. Salt tolerance of the coastal marsh grass *Sporobulus virginicus* (L.) Kunth. New Phytol 1992: 120:281–8.

Mc Cue, K.F. and Hanson, A.D. Drought and salt tolerance: towards understanding and application. Tibtech 1990;8:359–62.

Murata, S., Kobayashi, M., Matoh, T. and Sekiya, J. Sodium stimulates regeneration of Phosphoenolpyruvate in mesophyll chloroplasts of *Amaranthus tricolor*. Plant Cell Physiol 1992: 33:1247–50.

Osmond, C.B., Bjorkmann, O. and Anderson, D.J. *Physiological processes in plant ecology: towards a synthesis with Atriplex*, Springler-Verlag, Berlin, 1980.

Osmond, C.B. and Greenway, H. Salt responses of carboxylation enzymes from species differing in salt tolerance. Plant Physiol 1972; 49:260–3.

Passera, C. and Albuzio, A. Effect of salinity on photosynthesis and photorespiration of two wheat species (*Triticum durum* cv. PEPE 2122 and *Triticum aestivum* cv. *Marzotto*). Can J Bot 1977; 56:121–26.

Sato, Y., Sugiyama, M., Gorecki, R.J., Kukuda, H. and Komamine, A. Interrelationship between lignin deposition and the activities of Peroxidase isoenzymes in differentiating tracheary elements of *Zinnia*. Planta 1993; 189:584–89.

Schrimer, U. and Breckle, S.W. The role of bladders for salt removal in some Chenopodiaceae (mainly *Atriplex* species). In Sen, D.N. and Rajpurohit, K.S., editors. *Contributions to the ecology of halophytes*. Tasks for vegetation science, Vol. 2, Dr. W.J. Junk Publishers, The Hague. 1982; 215–31

Shomer-Ilan, A., Neumann-Ganmore, R. and Waisel, Y. Biochemical specialisation of photosynthetic cell layers and carbon flow paths in *Suaeda monoica*. Plant Physiol 1979; 64:963–5.

Shomer-Ilan, A. and Waisel, Y. The effect of sodium chloride on the balance between the C_3 and C_4 carbon fixation pathways. Physiol Plant 1973; 29:190–3.

Smaoui, A. Différenciation des trichomes chez *Atriplex halimus* L. CR Acad Sci Paris 1971; 273:1268–71.

Stassart, J.M., Neirinckx, L. and Dejaegere, R. The interactions between monovalent cations and calcium during their adsorption on barley roots. Ann. Bot 1981; 47:647–52.

Storey, R., Pitman, M.G., Steltzer, R. and Carter, C. X-Ray microanalysis of cells and cell compartments of *Atriplex spongiosa*. I. Leaves. J Exp Bot 1983; 34:778–94.

Storey, R. and Wyn Jones, W.G. Responses of *Atriplex spongiosa* and *Suaeda monoïca* to salinity. Plant Physiol 1979; 63:156–62.

Thiyagarajah, M., Fry, S.C. and Yeo, A.R. *In vitro* salt tolerance of cell wall enzymes from halophytes and glycophytes. J Exp Bot 1996; 47:1717–24.

Wang, L.W., Showalter, A.M. and Ungar, I.A. Effect of salinity on growth, ion content, and cell wall chemistry in *Atriplex prostrata* (Chenopodiaceae). Am J Bot 1997; 84:1247–55.

Yeo, A.R. Salinity resistance: Physiologies and prices. Physiol Plant 58:214–22.

Yeo, A.R. and Flowers, T.J. Ion transport in *Suaeda maritima*: its relation to growth and implications for the pathway of radial transport of ions across the root. J Exp Bot 1986; 37:143–59.

Zid, E. and Grignon, C. Sodium-Calcium interactions in leaves of *Citrus aurantium* grown in the presence of NaCl. Physiol Vég (1985); 23:895–903.

Some physiological and biochemical aspects of salt tolerance in two oleaginous halophytes: *Cakile maritima* and *Crithmum maritimum*

A. DEBEZ, K. BEN HAMED AND C. ABDELLY

INTRODUCTION

Increasing salinity levels in soils and/or irrigation water constitute a significant environmental problem that can lead to a loss of agricultural production in arid and semi-arid lands in the world (Kennedy and De Filippis, 1999). At the Mediterranean basin scale, salt-affected lands amount to about 15×10^6 ha and are mainly located in Northern Africa and the Near and Middle East (Le Houérou, 1986). Salinity can affect plant survival, biomass, plant height and plant form. Such changes in morphology affect the capacity of a plant to collect light, water and nutrients (Locy et al., 1996). In fact, salt interaction with physiological and metabolic processes in the plant is complex, depending on salt type and dose, plant genotype and developmental stage (Meneguzzo et al., 1999).

Halophytes, plants that have been naturally selected to grow in saline environments (Flowers et al., 1986), can withstand the deleterious effects by ion exclusion, regulation of ion transport and accumulation of compatible cytoplasmic osmotica (Greenway and Munns, 1980). These plants show ecological (landscaping, rehabilitatation of damaged areas, ...) and economic (production of oil for industrial application ...) interests as well. Halophytes provide excellent models to understand extreme salt-tolerance mechanisms (Jefferies and Rudmik, 1984). They contain also valuable gene pools for use in crop amelioration (Shannon and Noble, 1990). Much effort has been devoted toward elucidating the mechanisms of plant salt-tolerance aiming to improve the performance of crop plants in saline soils (Binzel and Reuveni, 1994). However, what implies salt tolerance to plants remains unresolved.

There are some indications that salt excess can induce conditions of oxidative stress. Yet, oxygen radicals (Reactive Oxygen Species) generated during plant metabolism, need to be scavenged by antioxidant system, so as to alleviate salt-induced oxidative injury (Elstner, 1982; Hernandez et al., 1999). The primary components of this antioxidant system include carotenoids, ascorbate, glutathione and tocopherols as well as antioxidant enzymes such as superoxide dismutase, catalase and peroxidases (Hernandez et al., 1999).

This chapter deals with salt tolerance in *Crithmum maritimum* (Umbellifereae) and *Cakile maritima* (Crucifereae), species of great commercial interest. Indeed, these two indigenous and oleaginous halophytic species appeared to be useful with regard to the biochemical properties of oil extracted from their seeds.

In our study, the physiological responses (growth, water relations and photosynthesis activity) of both species to NaCl were investigated. We also paid attention to some antioxidant enzymes changes in relation to increasing salinity owing to their involvement in the defence mechanisms of salt-stressed plants and thus in their salt tolerance.

MATERIALS AND METHODS

Plant Material and Experimental Conditions

C. maritimum and *C. maritima* seeds collected from their native salty ecosystem, close to the Mediterranean sea: Korbous and Raoued, respectively. Seeds were sown in pots (two seeds per pot) filled with inert sand and irrigated with distilled water until germination. At the early development stages, seedlings were watered every day with Long Ashton nutrient solution. Two months old *C. maritimum* seedlings were divided into

H. Lieth (ed.), Cash Crop Halophytes: Recent Studies, 31–39.

6 lots of 10 plants each, and submitted to increasing NaCl concentrations: 0, 50, 100, 150, 200, and 300 mM. NaCl doses (0, 50, 100, 200, 300, 400 and 500 mM) were progressively applied on 6-week-old *C. maritima* seedlings divided into 7 lots of 18 plants each. Salt treatment was adjusted with increasing NaCl concentrations by 50 mM per day.

Experiments were performed in a glass greenhouse under controlled conditions: 25 ± 5°C temperature and 60 ± 10% relative humidity. Two harvests were made: the first one at the beginning of treatment (initial harvest) and the second (final harvest), after a period of salt application ranging from 5 to 10 weeks, respectively for *C. maritima* and *C. maritimum*. Once harvested, plants were separated into leaves, stems and roots. A part of the leaf fresh material was used for enzyme analysis.

Growth and Water Relations Parameters

Growth parameters estimated were fresh and dry mass (respectively FW and DW), water content (ml H_2O/g DW), leaf area (LA), leaf succulence ratio (FW/DW), relative growth rate (RGR) and net assimilation ratio (NAR).

$$RGR = \frac{\ln(DW_2 - DW_1)}{\Delta t}$$

DW_1 and DW_2 correspond to initial (at the initial harvesting) and final dry mass respectively. Δt corresponds to treatment duration.

$$NAR = \frac{DW_2 - DW_1}{\text{mean LA}\,\Delta t}$$

Enzyme Extractions and Assays

All operations were carried out at 0–4°C. Samples were prepared for catalase and total protein analysis by homogenizing 0.5 g of crude leaf material in an ice-cold solution containing: 0.1 M Tris-HCl, 10 mM KCl, 1 mM $MgCl_2$, 0.5 M Sucrose, 3 mM cysteine 5% (v/v), TritonX-100. After filtration through four gauze layers, the extract was centrifuged at 4°C for 20 min at 16,000 × *g*. The supernatant was immediately used for total protein and enzyme analysis.

All spectrophotometric analysis were conducted at 25°C on a BECKMAN UV/visible light recording spectrophotometer. The activity of catalase was determined by monitoring the disappearance of H_2O_2, by measuring the decrease in absorbance at 240 nm of a reaction mixture (2 ml final volume) containing M/15 phosphate buffer (pH 7), 12.5 mM H_2O_2, and crude extract (Luck, 1965). For catalase, one unit of enzyme was defined as the amount necessary to decompose 1 μmole of H_2O_2/min at 25°C.

Total protein content was determined by the Bradford reagent method using bovine serum albumine (BSA) as a standard (Bradford, 1976).

RESULTS

Growth

Growth of *C. maritimum* was estimated throughout dry mass production of the whole plant and its different tissues (roots and shoots). Biomass production was not significally affected by 50–100 mM NaCl, but was considerably reduced (−56%) at 150 mM, when compared with plant controls (Figure 1). Above doses accentuated this tendency. For instance, dry mass production represented just 9% of the control values at 300 mM. Growth of roots and shoots were similarly affected by salt concentrations from 150 mM NaCl.

The response to salinity displayed by *C. maritima* was different to some extent from *C. maritimum*. Indeed, when expressed on a fresh (Figure 2a) or a dry mass (Figure 2b) basis, growth of the whole plant was lower in the medium lacking salt. Exposure to NaCl at low doses (50–100 mM) enhanced fresh and dry mass production, therefore confirming the halophilic status of *C. maritima*. At 100 mM NaCl, dry mass reached 124% of the control value (Figure 3). The growth stimulation disappeared at 200 mM, without affecting

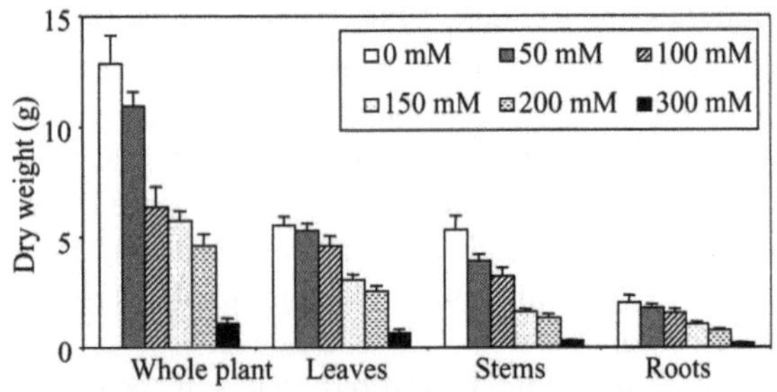

Fig. 1. Effect of NaCl on growth (DW) of *C. maritimum* (means of 10 plants and confidence limits for $p = 0.05$).

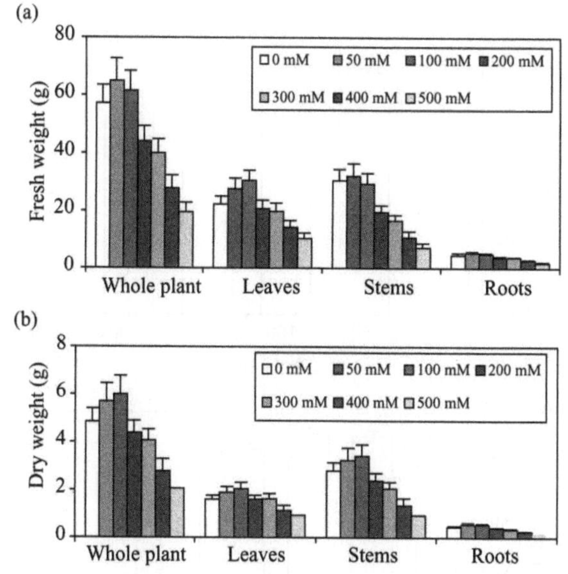

Fig. 2. Effect of NaCl on (a) growth (FW) and (b) growth (DW) of *C. maritima* (means of 18 plants and confidence limits for $p = 0.05$).

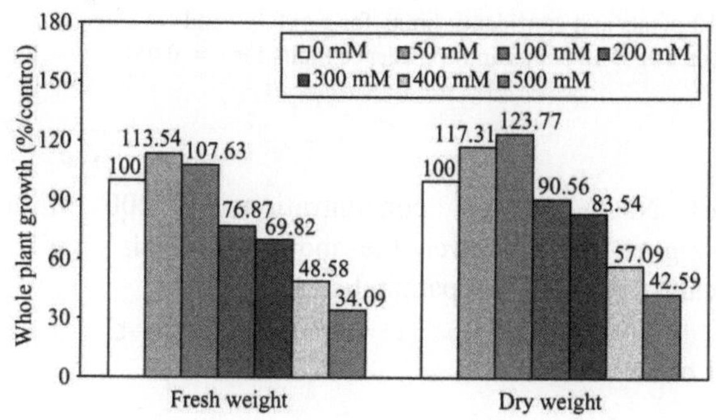

Fig. 3. Effect of NaCl on whole plant growth (%/control) in *C. maritima* (means of 18 plants).

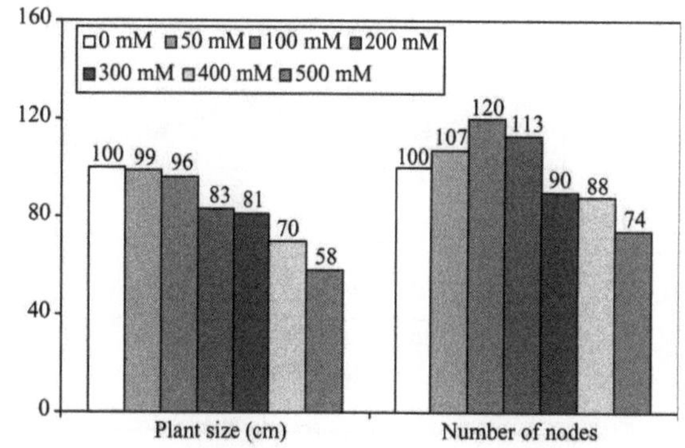

Fig. 4. Effect of NaCl on plant size and number of nodes (%/control) in *C. maritima* (means of 18 plants).

the biomass production level with regard to controls, but it was significantly decreased at supra-optimal concentrations (from 300 mM NaCl), being 57% lower than control value at the highest NaCl dose (500 mM). Trends in biomass production of stems were similar to those observed for whole plant, whereas leaves and roots dry mass did not differ from control values up to 400 mM NaCl.

To assess whether dry mass modification resulted from changes in production of nodes or size of internodes, plant size and number of nodes in *C. maritima* were measured. Control plants were taller than salt-treated ones (Figure 4). This was particularly true for plants grown at NaCl concentrations exceeding 100 mM. Salt-treatment lead to an increase of the number of nodes up to 200 mM NaCl. The fall recorded at higher salt levels was less pronounced than for plant size, indicating that the decline in plant size was mainly due to a reduction of the cells elongation rather than cells division.

To investigate whether roots or shoots were the mostly affected by salinity, root/shoot DW ratio was calculated for both species. Figure 5a shows that this parameter increased up to the 150 mM NaCl treatment for *C. maritimum* plants, indicating that shoot growth declined in favour of roots as a result of salt stress. This was corroborated by the increasing of roots dry mass until 150 mM NaCl treatment, when expressed as percent of the whole plant dry mass (Figure 5b).

Concerning *C. maritima*, salinity did not modify the resource allocation pattern within the plant tissues. Root/shoot DW ratio did not significantly differ among all treatments, as indicated by Figure 6a. The fact that root growth, when expressed as percent of whole plant dry mass production, was unaffected by salinity (Figure 6b), confirmed the above-mentioned results.

Plant RGR was calculated to give more information about the growth activity in relation with increasing salinity. Previous data relative to *C. maritimum* behaviour were supported by the progressive negative evolution of the plant RGR from 100 mM NaCl (Figure 7a), unlike *C. maritima*, which exhibited a slight stimulation in growth activity in the limit

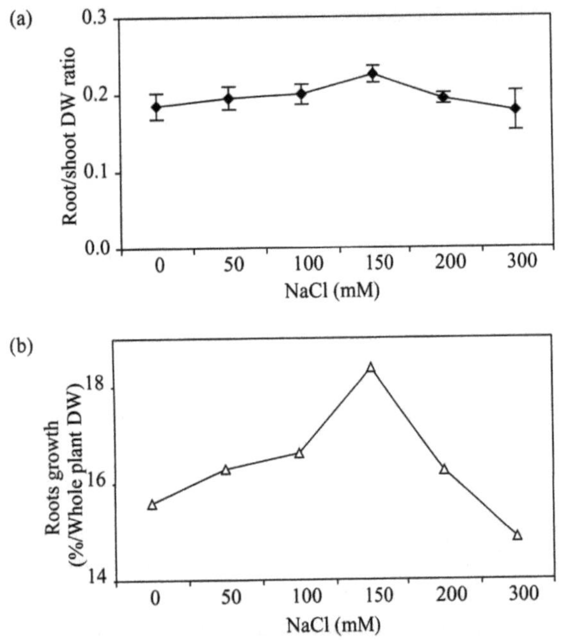

Fig. 5. Effect of NaCl on (a) root/shoot DW ratio in *C. maritimum* (means of 10 plants and confidence limits for $p = 0.05$); (b) roots growth (%/whole plant DW) in *C. maritimum* (means of 10 plants).

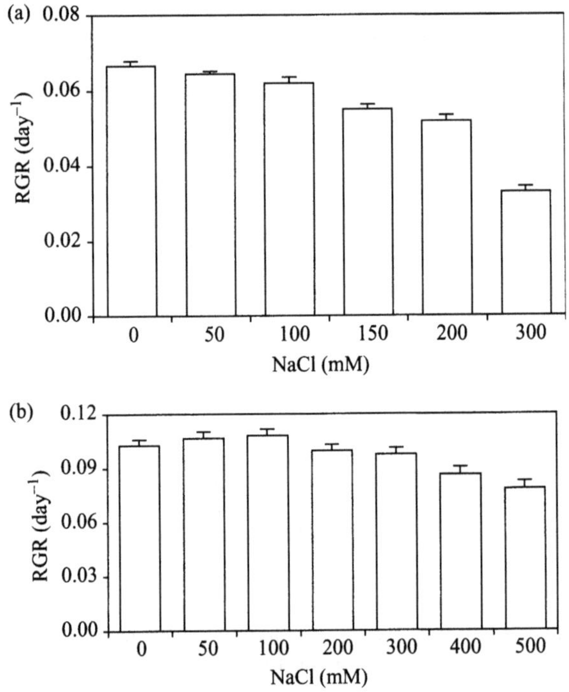

Fig. 7. Effect of NaCl on RGR in (a) *C. maritimum* (means of 10 plants and confidence limits for $p = 0.05$) and (b) *C. maritima* (means of 18 plants and confidence limits for $p = 0.05$).

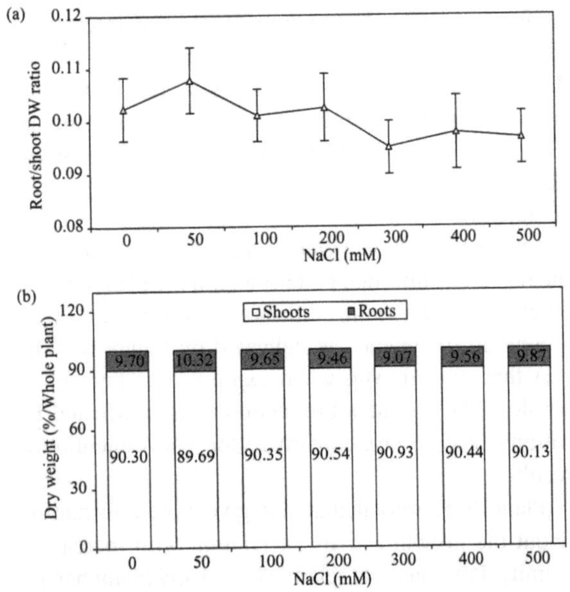

Fig. 6. Effect of NaCl on (a) root/shoot DW ratio and (b) root growth (%/whole plant DW) in *C. maritima* (means of 18 plants and confidence limits for $p = 0.05$).

of NaCl optimal concentrations (0–200 mM) (Figure 7b). However, the above salt levels significantly reduced this parameter.

Leaf Parameters

The number of leaves of *C. maritimum* plants submitted to salt stress significantly decreased (Figure 8a). Leaf area was also affected from 150 mM NaCl dose (until 60% of reduction at 300 mM). As shown by Figure 8b, salinity (50–100 mM NaCl) had a positive effect on these two parameters in *C. maritima* (9% and 30% greater than control values, respectively for leaf area and number of leaves). However, they were severely affected at higher salt doses (−61% and −57% respectively for leaf area and number of leaves).

On the other hand, a close relationship was found between the whole plant growth and leaf area for both species, *C. maritimum* (Figure 9a) and *C. maritima* plants (Figure 9b). The leaf area could, therefore, highly influence the evolution of plant biomass production.

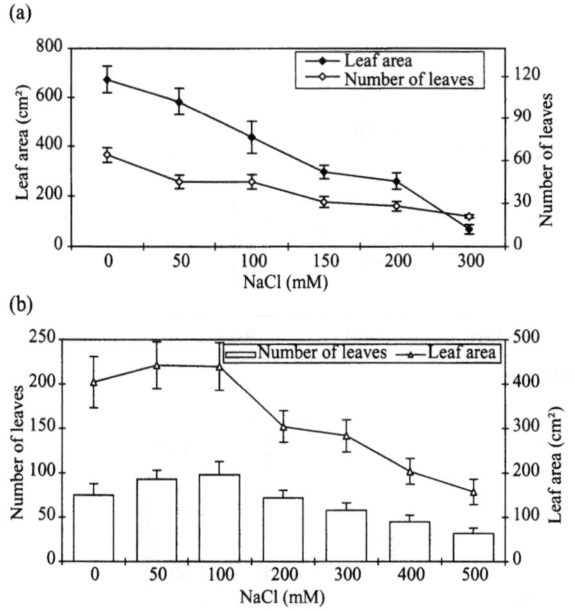

Fig. 8. Effect of NaCl on number of leaves and leaf area in (a) *C. maritimum* (means of 10 plants) and (b) *C. maritima* (means of 18 plants) and confidence limits for $p = 0.05$.

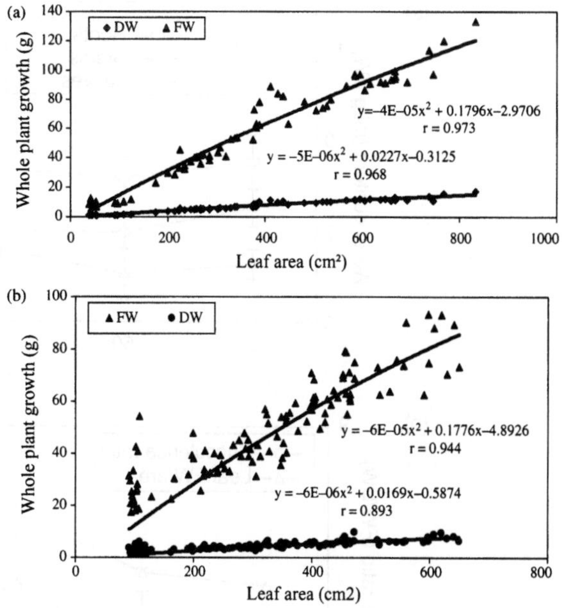

Fig. 9. Relationship between whole plant growth and leaf area in (a) *C. maritimum* plants and (b) *C. maritima* plants exposed to increasing NaCl concentrations.

Water Relations

Plant water relations were studied in order to relate them with growth parameters pattern under salt stress. In *C. maritimum*, shoot water content was maintained at high levels (close to control values) despite increasing salt concentrations, while it decreased from 200 mM NaCl in roots (Figure 10a). Leaf succulence is an important feature of halophytes. The accumulation of water in this compartment permits to the plant to grow well in the presence of salt. Neither succulence rating, nor leaf FW/area ratio were affected by salt stress (Figure 10b).

For *C. maritima*, whole plant water status was not significantly impaired by salinity up to 400 mM NaCl. Leaves and roots were less affected than stems by increasing salt levels, and were characterized by water contents close to the control values, even at the extreme salt dose (500 mM) (Figure 11). Leaf succulence parameters also appeared to be unaffected by enhanced salt levels in the medium (data not shown).

Our results show that the reduction in growth under saline conditions was not clearly concomitant with an impairment of the water status, suggesting the possible involvement of other mechanisms in the response of both species to salt stress.

Photosynthesis is one of the major processes altered by salt. Net assimilation rate (NAR), which measures the dry mass accumulation per unit of leaf area and per day, is an indicator of the photosynthetic activity. NAR was stimulated up to 100 and 400 mM NaCl, respectively for *C. maritimum* (Figure 12a) and *C. maritima* (Figure 12b), before declining at higher salt concentrations.

Moreover, this parameter was not strongly correlated to whole plant growth in both species (Figure 13). This leads to think that the both species behaviour would be mainly determined by leaf area rather than its photosynthetic activity, which was unaffected, even at extreme NaCl levels.

Total Protein Content

Data relative to growth study indicated that 150 mM NaCl resulted in a severe growth reduction of *C. maritimum*. This did not appear to be a result of the decrease in total protein content, which was maintained at higher levels than in control (Figure 14). Our results indicate that dry matter production was more affected than nitrogen assimilation and protein synthesis. They also suggest that NaCl inhibited proteolytic activity.

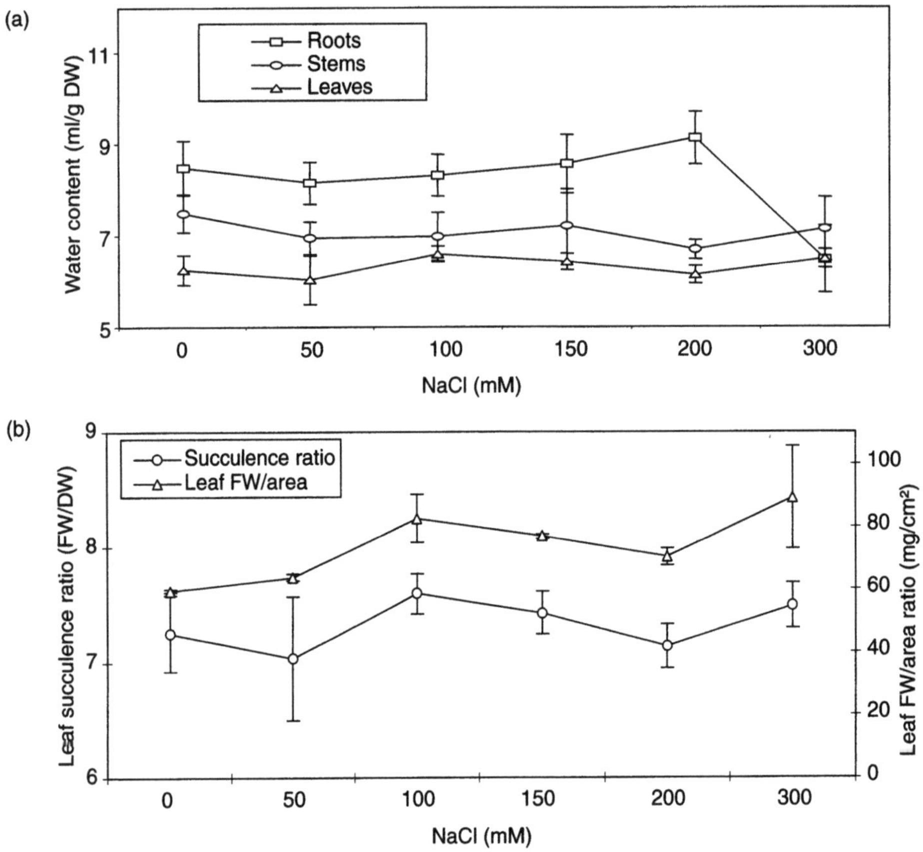

Fig. 10. Effect of NaCl on (a) shoots, stems and roots water content and (b) leaf succulence ratios in *C. maritimum* (means of 10 plants and confidence limits for $p = 0.05$).

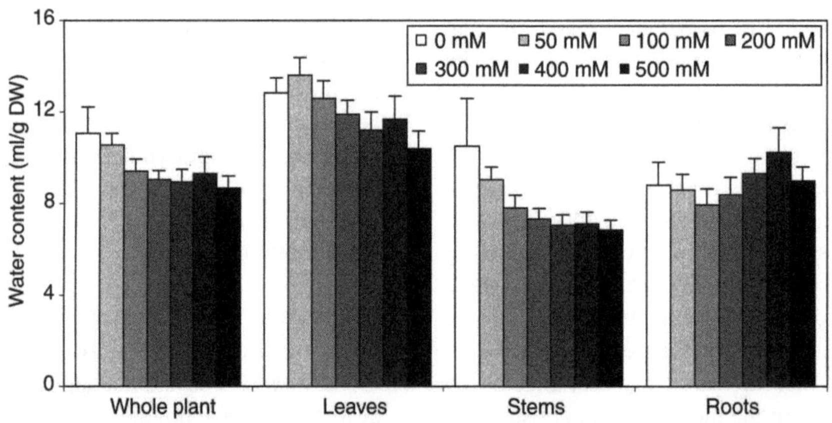

Fig. 11. Effect of NaCl on whole plant, shoots, stems and roots water content in *C. maritima* (means of 18 plants and confidence limits for $p = 0.05$).

Antioxidant Enzyme Activities

Catalase catalyses the dismutation of H_2O_2 into H_2O and O_2, preventing the damaging effects of H_2O_2 accumulation and protecting cells from oxidative stress. Catalase activity, when expressed on a total protein content basis, was significantly stimulated in leaves of *C. maritimum* plants exposed to 50–100 mM NaCl (Figure 15). Higher salt doses lead to a decline in catalase activity. This trend was more pronounced when catalase activity was referred to leaf fresh mass.

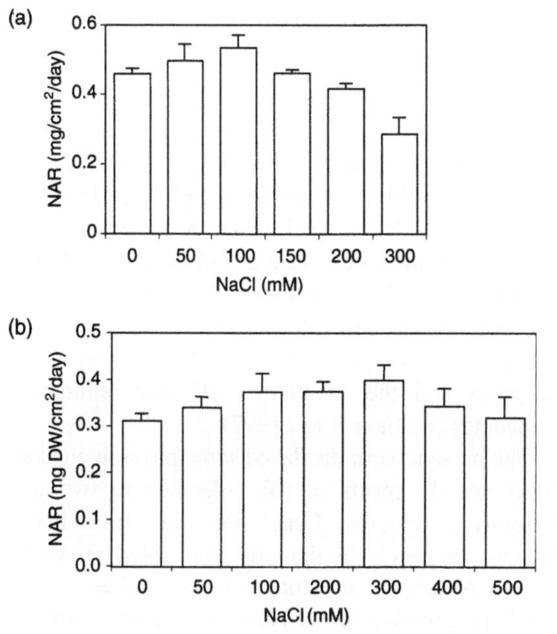

Fig. 12. Effect of NaCl on NAR in (a) *C. maritimum* and (b) *C. maritima* (means of 10 plants and confidence limits for $p = 0.05$).

DISCUSSION

The both species were able to produce biomass under all NaCl treatments, though responding somewhat differently to salinity. *C. maritimum* behaved like a facultative halophyte, with a growth activity progressively decreasing as NaCl levels raised in the medium, unlike to what happens in their native littoral area (salt concentrations are about 500 mM there), where they grew well. This apparent contradiction would be related with the sandy texture of the soil in this region, which permits the leaching of salts, thus avoiding their accumulation at high amounts. On the contrary, *C. maritima*'s biomass production was stimulated up to 100 mM NaCl and was unaffected at 200 mM. NaCl-induced depressive effect was visible from 300 mM, owing essentially to the reduction in cells size rather than cells division. However, the plant was still alive at 500 mM NaCl. It is generally assumed that shoot growth is more affected than root by salinity (Marcelis and Van Hooijdonk, 1999). This was the case in *C. maritimum* up to 150 mM.

Deleterious effects of salt stress are thought to result partly from water stress (Wang et al., 1997). Our results show that salinity did not affect the water status in *C. maritimum* plants. Plant water content was conserved at levels significantly close to control values, even at the highest NaCl doses. Leaf succulence was not particularly modified by salt treatment. Similar trends were recorded in *C. maritima*. In fact, there was

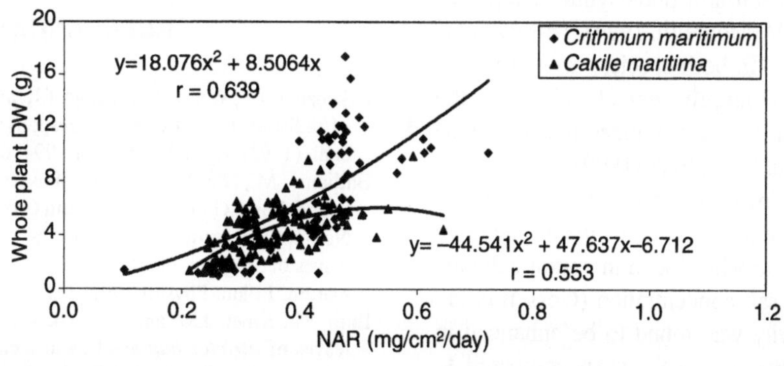

Fig. 13. Relationship between whole plant growth (DW) and net assimilation rate (NAR) in *C. maritima* and *C. maritimum* plants, exposed to increasing NaCl concentrations as shown in Figure 12.

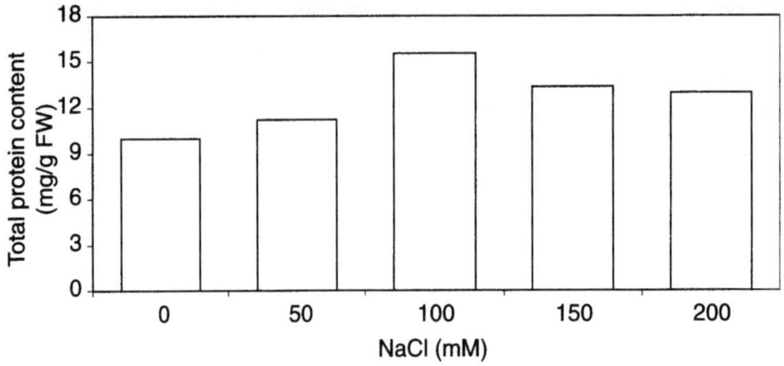

Fig. 14. Effect of NaCl on leaves total protein content in *C. maritimum* plants (means of 6 replications).

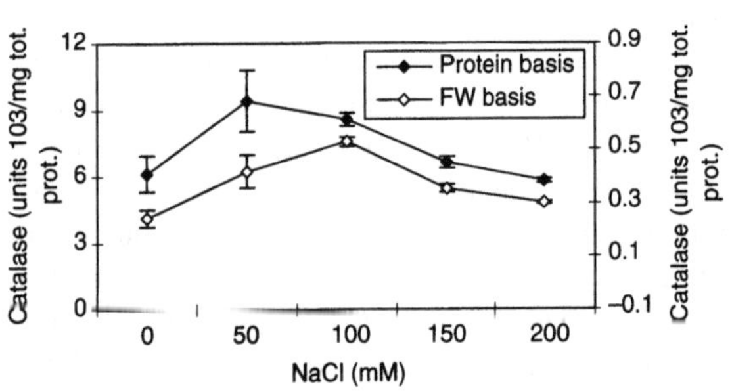

Fig. 15. Catalase activity changes in leaves of *C. maritimum* plants exposed to increasing NaCl concentrations (means of 6 replications).

not a strong relationship between growth activity and plants water content, though several studies have indicated that biomass production was highly influenced by the plant water status in halophytes (Naidoo and Rugunanan, 1990; Glenn and Brown, 1998; Bajji et al., 1998). It has also been indicated that growth is strongly related to leaf expansion and photosynthesizing area (Arkebauer et al., 1994; Monteith, 1994). This was corroborated by our data. In both species, growth reduction/stimulation was largely correlated with the reduced/enhanced leaf area and number, in accordance with Marcelis and Van Hooijdonk (1999).

From another side, increasing NaCl levels stimulated leaves total protein content in *C. maritimum* plants up to 150 mM, similarly to what found in cotton cultivars exposed to the same salt concentration (Gossett et al., 1994). Catalase activity was found to be enhanced in salt treated *C. maritimum* leaves, when compared to control. This suggests that catalase could be one key enzyme in the salt resistance of this species. In reported literature, catalase was considered as the most effective antioxidant enzyme in averting cellular damage (Scandalios, 1993). For instance, higher levels of catalase in cotton leaves lead to the development of salt tolerance in this species (Gossett et al., 1994; 1996).

However, it should be taken into consideration that the mechanisms of protection against salt stress are quite more complicated, depending on numerous enzymes and the production of many antioxidant molecules (Badiani et al., 1997).

The present work displayed some physiological and biochemical aspects of salt tolerance in two local halophytic species. Leaf area and antioxidative enzymes appear to be the most implicated parameters in this phenomenon. More detailed studies are currently progressing in our laboratory to investigate the salt-induced effects on mineral nutrition and to better understand the roles of the antioxidative systems in the response to salinity.

REFERENCES

Arkebauer, T.J., Weiss, A., Sinclair, T.R. and Blum, A. In defence of radiation use efficiency: a response to Demetriades-Shah et al. (1992). Agric For Meteor 1994; 68:221, 227.

Badiani, M., Paolacci, A.R., Fusari, A., D'Ovidio, R., Scandalios, J.G., Porceddu, E. and Giovannozzi-Sermanni, G. Non-optimal growth temperatures and antioxidant in the leaves of *Sorghum bicolor* (L.) Moench. II. Short-term acclimation. J Plant Physiol 1997; 151:409–421.

Bajji, M., Kinet, J.M. and Lutts, S. stress effects on roots and leaves of *Atriplex halimus* L. and their corresponding callus cultures. Plant Sci 1998; 137:131–42.

Binzel, M.L. and Reuveni, M. Cellular mechanisms of salt tolerance in plant cells, Hort Rev 1994; 16:33–69.

Bradford M. A rapid and sensitive method for the quantification of microgram quantities of protein utilizing the principle of protein-dye binding. Anal Biochem 1976; 72:248–54.

Elstner, E.F. Oxygen activation and oxygen toxicity. Ann Rev Plant Physiol 1982; 33:73–96.

Flowers, T.J., Hajibagheri, M.A. and Clipson, N.J.W. Halophytes, Quart Rev Biol 1986; 61:313–37.

Glenn, P.G. and Brown, J.J. Effects of soil salt levels on the growth and water use efficiency of *Atriplex canescens* (CHENOPODIACEAE) varieties in drying soil. Am J Bot 1998; 85:10–16.

Gossett, D.R., Banks, S.W., Millhollon, E.P. and Lucas, C. Antioxidant response to NaCl stress in a control and NaCl-tolerant cotton cell line grown in the presence of paraquat, buthionin sulfoximine, and exogenous glutathione. Plant Physiol 1996; 112:803–9.

Gossett, D.R., Millhollon, E.P. and Lucas, C. Antioxidant response to NaCl stress in salt-tolerant and salt-sensitive cultivars of cotton. Crop Sci 1994; 34:706–14.

Greenway, H. and Munns, R. Mechanisms of salt tolerance in non-halophytes. Ann Rev Plant Physiol 1980; 31:149–90.

Hernandez, J.A., Campillo, A., Jimenéz, A., Alarcon, J.J. and Sevilla, F. Response of antioxidant systems and leaf water relations to NaCl stress in pea plants. New Phytol 1999; 141:241–51.

Jefferies, R.L. and Rudmik, T. The responses of halophytes to salinity: an ecological perspective. In Staples, R.C. and Toenniessen G.H., editors. Salinity tolerance in plants: strategies for crop improvement. J. Wiley & Sons, New York, 1984; 213–27.

Kennedy, B.F. and De Filippis, L.F. Physiological and Oxidative Response to NaCl of the salt tolerant *Grevillea ilicifolia* and the salt sensitive *Grevillea arenaria*. J Plant Physiol 1999; 155:746–54.

Le Houérou, H.N. Salt tolerant plants of economic value in Mediterranean basin. Reclamat Reveget Res 1986; 5: 319–41.

Locy, R.D., Chang, C.C., Nielsen, B.L. and Singh, N.K. Photosynthesis in salt-adapted heterotrophic tobacco cells and regenerated plants. Plant Physiol 1996; 110:321–8.

Luck, H. Catalase. In Bergmeyereditor. Methods of enzymatic analysis. Academic Press, New York, 1965; 885–8.

Marcelis, L.F.M. and Van Hooijdonk, J. Effect of salinity on growth, water use and nutrient use in radish (*Raphanus sativus* L.). Plant Soil 1999; 215:57–64.

Meneguzzo, S., Navari-Izzo, F. and Izzo, R. Antioxidative responses of shoots and roots to increasing NaCl concentrations. J Plant Physiol 1999; 155:274–80.

Monteith, J.L. Validity of the correlation between intercepted radiation and biomass. Agric For Meteor 1994; 68: 213–20.

Naidoo, G. and Rughunanan, R. (1990) Salt tolerance in the succulent, coastal halophyte, *Sarcocornia natalensis*. J Exp Bot 1990; 41:497–502.

Scandalios, J.G. Oxygen stress and superoxide dismutase. Plant Physiol 1993; 101:7–12.

Shannon, M.C. and Noble, C.L. Genetic approaches for developing economic salt-tolerant crops. In Tanji, K.K., editor. Agricultural salinity assessment and management. American Society of Civil Engineers, New York, 1990; 161–85.

Wang, L.W., Showalter, A.M. and Ungar, I.A. Effect of salinity on growth, ion content, and cell wall chemistry in *Atriplex prostrata* (CHENOPODIACEAE). Am J Bot 1997; 84: 1247–55.

Salinity tolerance of *Beta vulgaris* ssp. *maritima*. Part I. Biomass production and osmotic adjustment

S. DAOUD, H.-W. KOYRO AND M.C. HARROUNI

INTRODUCTION

In arid and semi-arid regions saline soils are of more frequent occurrence both under irrigated and non-irrigated conditions (Greenway, 1962). In the Mediterranean region alone some 15 million ha have become seriously salt affected (Le Houerou, 1986). This phenomenon has substantial adverse social and economic effects in Morocco deteriorating agricultural productivity in many irrigated perimeters (Choukr-Allah et al., 1994). In order to overcome this problem, it is important to domesticate promising salt-tolerant plants that already have the requisite level of salt tolerance and are highly productive at high external salinity levels, and to use appropriate new methods of agro-management to have a high productivity (Aronson, 1986; O'Leary, 1986; Lieth and Al Massoum, 1993; Koyro, 1997). *Beta vulgaris* ssp. *maritima* is a halophyte with a real potential in Morocco since it exists all over the country (Fennane et al., 1999) and could be used as a cash crop. The aim of this study is to use a gravel/hydroponic quick check system with automated irrigation and drainage to determine the limit of salt tolerance of the sea beet and the optimal growth conditions which reflect the efficiency of the plant and lead to maximum yield productivity under saline irrigation.

MATERIALS AND METHODS

Plant Culture

Seeds of sea beet were stored in a refrigerator for 48 h before they were transferred into a Petri dish with wet filter paper (0.2 mol m^{-3} $CaSO_4$) for germination in a dark growth cabinet at 25°C. Seven to 14 days after germination, plants were potted into soil (type LD 80, Fa Archut, Lawkbach, Germany). After a further 2 weeks the young seedlings were transplanted into a soilless (gravel/hydroponics) culture quick check system (Koyro, 1999). The plants were irrigated with a basic nutrient solution as modified by Epstein (1972).

Salinity Trials

The stepwise addition of NaCl to the basic nutrient solution began after a period of another 2 weeks by raising salinity of the solution in steps of 50 mol m^{-3} NaCl each day. There were altogether five treatments: Control (without NaCl), 125, 250, 375 and 500 mol m^{-3} NaCl. The highest salinity treatment was reached after 9 days. The trials were conducted in an environment controlled greenhouse. Temperatures were 25 ± 2°C during the day and 15 ± 2°C during the night. Relative humidity ranged from 45% to 70%. Light intensity was in the range of 5000 lux at plant level. The quick check system was programmed by a timer to water the plants every 4 h for 15 min starting at midnight, 4 am, 8 am, 12 noon, 16 pm and 20 pm daily and allow the saline solutions to drain freely from the pots. Solutions were recycled and changed every 2 weeks to avoid nutrient depletion.

Harvest and Sample Preparation

The plants were harvested 3 weeks after the highest salinity level was established. Three plants per treatment were weighed individually and separated into laminas, petioles, tap root and adventitious roots. Leaf number and the leaf mass to area (LMA) ratio defined as weight per surface area of leaves were determined.

The dry weight was determined by oven drying at 70°C to constant weight and expressed in percent of fresh weight. The water content was determined by the

H. Lieth (ed.), Cash Crop Halophytes: Recent Studies, 41–49.

difference between fresh and dry weights. Samples of dried plants were ashed in a muffle furnace over night at 500°C. The organic matter was calculated as the difference between dry weight and ash weight.

Chemical and Physiological Parameters

Na and K were determined with the flame photometer 410 (FLM3b, Radiometer Copenhagen) after dissolving the ash in 2N HCl. The contents of Ca and Mg were determined in the same solution with atomic absorption spectrometer (3110, Perkin Elmer). Concentrations of these minerals in plant tissues are expressed on a dry weight basis.

Sugars were analysed by HPLC (Ion chromatograph, Dionex) after freeze drying of the samples (Hoffmann-Thoma et al., 2000).

Tissue osmotic potential was determined in the leaf sap with an osmometer (Osmomat 030, Gonotec, for details see also Gonzalez et al., 1996) based on the theory of freezing point depression. The water potential was determined by measuring dew point depression with a Wescor apparatus without harming the plants (Koyro, part 1 page 5 in this book).

Proline was extracted in 3% sulfosalicylacid in an ultrasonic bath. After 10 min centrifugation at 3000 rpm, 1 ml of the solution was mixed with 1 ml glacial acetic acid and 1 ml ninhydrin reagent and then heated 1 h at 60°C in a water bath. The reaction was stopped in an ice bath. 2 ml toluol were added and mixed after the samples reached lab temperature. The upper phase was pipetted after phase separation into a cuvette and measured at 546 nm in a spectrophotometer.

RESULTS

Growth in all parts of the plants was stimulated by the low salinity treatment (125 mol m^{-3} NaCl). Further increase in NaCl concentration reduced growth up to 60% in high salinity treatment (500 mol m^{-3} NaCl) (Table 1). The limit of salt tolerance (50% growth reduction, definition see Koyro, chapter # in this book) was reached between 375 and 500 mol m^{-3} NaCl.

The root to shoot ratio increased with the increase of salinity. It increased more than 50% suggesting that the shoot was more affected by salinity than the roots. The tap root to adventitious roots ratio was reduced by more than 30% in the saline treatments compared to the control (Table 1).

Water content in all parts of the plant was reduced from the control to high salinity treatment. However the reduction was not high, especially in the leaves. It was only 21.24% and 14.73% respectively in adult and juvenile leaves (Table 2). This relatively low reduction in leaf water content can explain the increase in the LMA ratio with salinity (Table 1).

The number of leaves produced was similar in the control and low salinity treatment but was reduced under high salinity where it was 43% less than the control (Table 1).

The dry weight expressed as a percentage of fresh weight (DW in Table 2) increased with salinity in the adventitious roots and in the tap root. In the other parts of the plant there was an increase of the dry weight with salinity and the highest value was obtained in 375 mol m^{-3} NaCl treatment (Table 2). The increase of the dry weight in the adventitious roots and tap root was

Table 1. Growth parameters of *Beta vulgaris* ssp. *maritima* as affected by salinity, $n = 3$, mean ± SD

Treatment	Fresh weight							LMA ratio	
	Plant (g)	Shoot (g)	Root (g)	Tap root (g)	Adv. root (g)	Root/shoot	Tap root/ Adv.root	Adult leaf (mg cm^{-2})	Leaf number
Control	26.65 ± 7.53	25.48 ± 7.57	1.31 ± 0.55	0.89 ± 0.29	0.43 ± 0.14	0.046 ± 0.007	1.99 ± 0.31	28.65 ± 5.91	14.20 ± 1.62
125 mol m^{-3} NaCl	29.99 ± 13.52	28.14 ± 12.66	1.85 ± 0.43	1.02 ± 0.42	0.83 ± 0.20	0.065 ± 0.012	1.34 ± 0.51	34.22 ± 3.82	14.20 ± 2.53
250 mol m^{-3} NaCl	18.80 ± 3.18	16.28 ± 2.16	2.52 ± 1.06	1.27 ± 0.51	1.11 ± 0.61	0.150 ± 0.04	1.34 ± 0.75	40.59 ± 1.63	17.33 ± 1.50
375 mol m^{-3} NaCl	08.37 ± 1.55	07.07 ± 1.65	1.30 ± 0.28	0.72 ± 0.15	0.53 ± 0.10	0.190 ± 0.07	1.39 ± 0.40	42.16 ± 4.31	09.40 ± 1.50
500 mol m^{-3} NaCl	10.68 ± 2.92	09.43 ± 2.49	1.25 ± 0.43	0.66 ± 0.09	0.59 ± 0.37	0.130 ± 0.01	1.39 ± 0.72	39.76 ± 3.51	08.10 ± 1.80

Table 2. Effect of salinity on the dry weight (DW expressed as % of fresh weight) and water content (WC mg mg^{-1} dry weight) of tap root, adventitious (adv.) roots, petiole (P) and lamina (L) of adult and juvenile leaves of control and salt treated plants of *Beta vulgaris* ssp. *maritima*, $n = 3$, mean ± SD

	Adv. root		Tap root		P. adult leaf	
Treatment	DW	WC	DW	WC	DW	WC
Control	14.65 ± 1.70	05.80 ± 0.75	15.33 ± 0.61	05.49 ± 0.25	07.72 ± 0.31	12.01 ± 0.52
125 mol m^{-3} NaCl	15.29 ± 2.58	05.45 ± 1.25	20.10 ± 2.47	03.82 ± 0.59	08.44 ± 1.31	10.94 ± 1.73
250 mol m^{-3} NaCl	16.60 ± 1.60	05.10 ± 0.60	27.40 ± 2.80	02.70 ± 0.40	12.80 ± 0.60	06.80 ± 0.40
375 mol m^{-3} NaCl	20.00 ± 0.90	04.00 ± 0.20	28.50 ± 0.70	02.50 ± 0.10	16.90 ± 0.60	04.90 ± 0.20
500 mol m^{-3} NaCl	22.88 ± 7.65	03.67 ± 1.81	30.12 ± 0.97	02.32 ± 0.11	14.50 ± 1.35	06.10 ± 0.63

	L. adult leaf		P. juvenile leaf		L. juvenile leaf	
Treatment	DW	WC	DW	WC	DW	WC
Control	11.44 ± 1.88	07.91 ± 1.47	8.55 ± 0.58	10.91 ± 0.81	13.56 ± 1.49	06.38 ± 0.79
125 mol m^{-3} NaCl	11.58 ± 1.01	07.60 ± 0.72	11.16 ± 1.06	08.09 ± 0.89	14.57 ± 1.04	05.84 ± 0.47
250 mol m^{-3} NaCl	15.00 ± 1.30	05.70 ± 0.60	14.10 ± 1.40	06.20 ± 0.80	17.80 ± 2.40	04.70 ± 0.80
375 mol m^{-3} NaCl	16.70 ± 0.60	05.00 ± 0.20	19.50 ± 0.10	04.10 ± 0.00	18.10 ± 1.20	04.50 ± 0.40
500 mol m^{-3} NaCl	13.91 ± 1.56	06.23 ± 0.83	17.44 ± 2.40	04.82 ± 0.86	15.43 ± 1.44	05.44 ± 0.61

due to the increase of organic matter (OM in Table 3) which accumulated in the root system; but in the shoots it can be explained by the increase of the ash content (Ash in Table 3) with salinity. Petiole and lamina of adult leaves concentrated a high amount of ash compared to juvenile leaves. The organic matter (% fresh weight) was much higher in tap root than in adventitious roots. In the shoot, organic matter in juvenile leaves was higher than in adult leaves (Table 3).

The effect of increasing NaCl concentration in the culture medium led to decreasing values of water potential in leaf petioles of the sea beet. It ranged from −2.24 MPa in the control plants to −11.48 MPa in high salinity plants (Figure 1). In the control adventitious roots had a low osmolality value compared to other parts of the plant which had similar values. In low-salinity treatment (125 mol m^{-3} NaCl) an increase of osmolality was registered in all parts of the plant compared with the control. Tap root, petioles of adult and juvenile leaves had similar osmolality values. Lamina of adult and juvenile leaves had also similar values but to some extent lower than petioles. In the high-salinity treatment (500 mol m^{-3} NaCl), the highest increase of osmolality was registered in the tap root followed respectively by petioles and lamina of juvenile and adult leaves. Adventitious roots had the lowest value (Figure 2). The increase of osmolality in all parts of the plant with the increase of salinity matches the salt-induced increase of inorganic and organic matter.

In all parts of sea beet, Na was the most abundant ion and its concentration increased simultaneously with the increase of salinity in the growing medium while K content decreased (Figures 3–5). Sodium accumulation in the shoot was much higher than in the root. In high-salinity treatment (500 mol m^{-3} NaCl), K content of tap root and adventitious roots was reduced respectively by 59% and 56%, while the reduction was more pronounced in adult leaves (79%) and juvenile leaves (69%). The Ca concentrations were generally much lower than the Na or K concentrations. The Ca content was not affected by salinity except in the tap root, where it increased with the increase of the saline treatment. However, adventitious roots contained more Ca than the other parts of the plant (Figure 6).

Table 3. Effect of salinity on the ash (expressed as % of dry weight) and organic matter (OM % fresh weight) of tap root, adventitious (adv.) roots, petiole (P) and lamina (L) of adult and juvenile leaves of control and salt treated plants of *Beta vulgaris* ssp. *maritima*, $n = 3$, mean ± SD

	Adv. root		Tap root		P. adult leaf	
Treatment	Ash	OM	Ash	OM	Ash	OM
Control	5.80 ± 0.07	13.80 ± 1.61	8.60 ± 3.07	14.00 ± 0.14	21.97 ± 0.87	6.03 ± 0.30
125 mol m^{-3} NaCl	–	15.51 ± 1.15	12.13 ± 7.58	17.78 ± 3.66	35.34 ± 6.57	5.51 ± 1.40
250 mol m^{-3} NaCl	12.40 ± 2.80	14.60 ± 1.80	9.60 ± 1.30	24.80 ± 2.70	32.70 ± 1.40	8.70 ± 0.6
375 mol m^{-3} NaCl	15.10 ± 1.00	17.00 ± 0.90	8.40 ± 1.10	26.10 ± 0.90	27.30 ± 0.60	12.30 ± 0.40
500 mol m^{-3} NaCl	16.37 ± 4.25	19.03 ± 6.19	11.75 ± 3.29	26.59 ± 1.57	37.44 ± 3.11	9.10 ± 1.30

	L. adult leaf		P. juvenile leaf		L. juvenile leaf	
Treatment	Ash	OM	Ash	OM	Ash	OM
Control	20.41 ± 3.50	9.11 ± 1.52	13.18 ± 4.87	7.41 ± 0.35	6.65 ± 2.73	12.66 ± 1.54
125 mol m^{-3} NaCl	21.86 ± 1.70	9.06 ± 1.00	25.93 ± 1.96	8.27 ± 0.88	11.27 ± 2.79	12.94 ± 1.263
250 mol m^{-3} NaCl	28.10 ± 1.90	10.80 ± 1.20	29.00 ± 2.50	10.10 ± 1.30	24.30 ± 5.20	13.60 ± 2.70
375 mol m^{-3} NaCl	27.00 ± 0.80	12.20 ± 0.50	22.30 ± 1.20	15.20 ± 0.30	24.70 ± 2.90	13.70 ± 1.40
500 mol m^{-3} NaCl	31.00 ± 2.15	9.60 ± 1.09	25.96 ± 1.73	12.94 ± 2.04	27.09 ± 0.30	11.24 ± 1.01

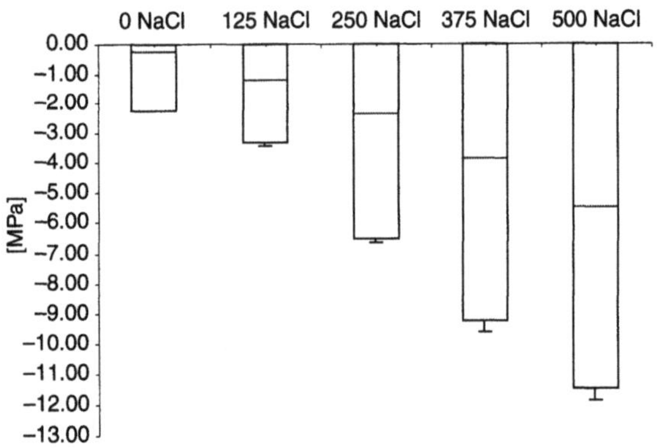

Fig. 1. Water potential (MPa) in adult leaves of controls and 4 salinity treatments (125, 250, 375 and 500 mol m^{-3} NaCl).The line in the histograms indicates the water potential of the watering solution in the five treatments.

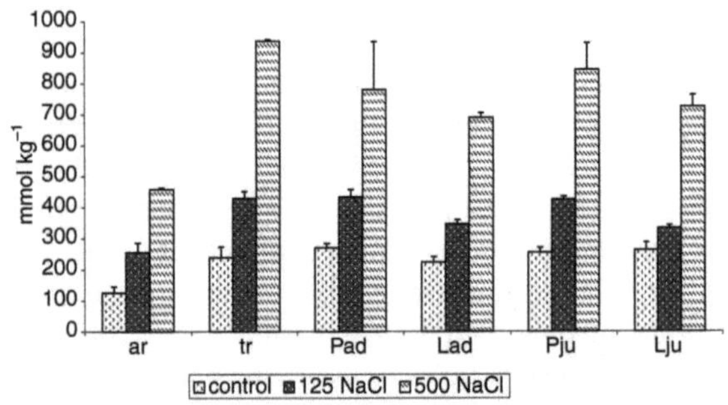

Fig. 2. Osmolality (mmol kg^{-1}) in adventitious roots (ar), tap root (tr), petiole (P) and lamina (L) of juvenile (ju) and (ad) leaves of controls, low-salinity (125 mol m^{-3} NaCl) and high-salinity plants (500 mol m^{-3} NaCl).

The Mg concentrations were generally lower than the Ca concentrations but showed a similar distribution in the tissues. Mg content was reduced by high salinity treatment in all parts of the plant except in the tap root. The reduction was more pronounced in adventitious roots, adult and juvenile leaf petioles than in lamina of adult and juvenile leaves (Figure 7).

In all parts of the plants, the total carbohydrate content increased with salinity (Figure 8). Sucrose was the most abundant organic substance followed respectively by

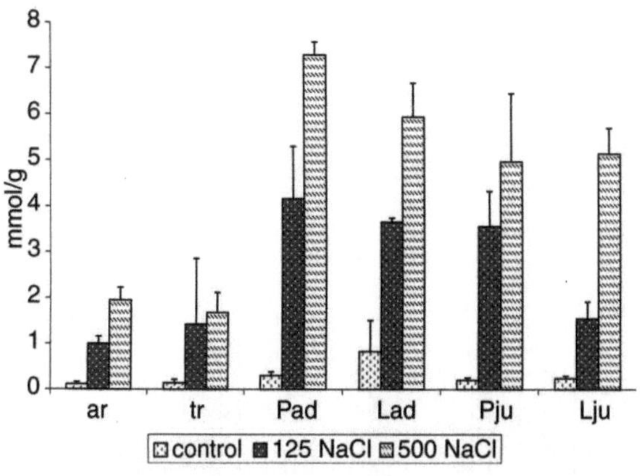

Fig. 3. Na concentrations (mmol g^{-1}) in adventitious roots (ar), tap root (tr), petiole (P) and lamina (L) of juvenile (ju) and adult (ad) leaves of control, low-salinity (125 mol m^{-3} NaCl) and high-salinity plants (500 mol m^{-3} NaCl).

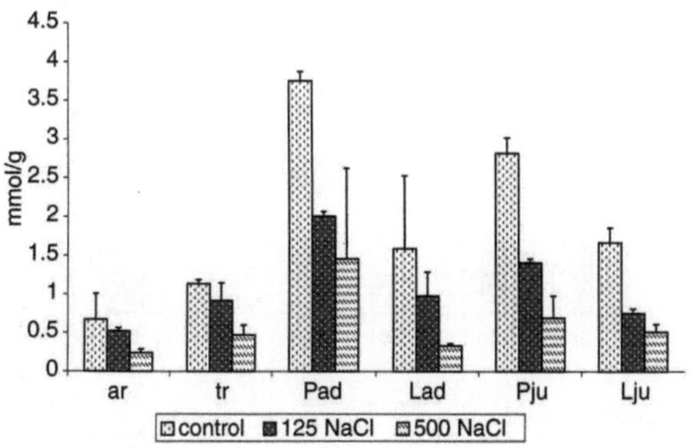

Fig. 4. K concentration (mmol g^{-1}) in adventitious roots (ar), tap root (tr), petiole (P) and lamina (L) of juvenile (ju) and adult (ad) leaves of control, low-salinity (125 mol m^{-3} NaCl) and high-salinity plants (500 mol m^{-3} NaCl).

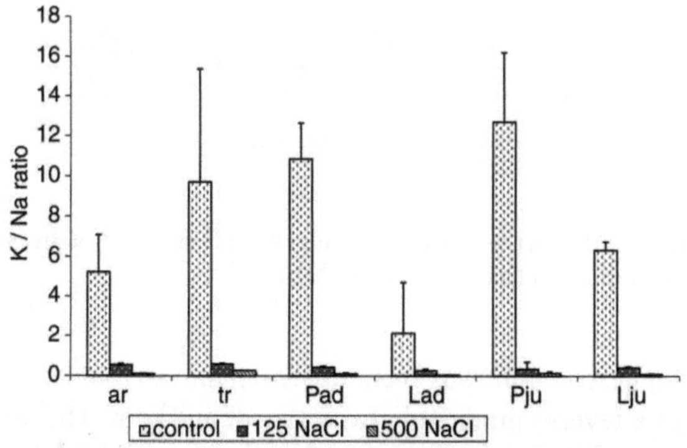

Fig. 5. K/Na ratio in adventitious roots (ar), tap root (tr), petiole (P) and lamina (L) of juvenile (ju) and adult (ad) leaves of controls, low-salinity (125 mol m^{-3} NaCl) and high-salinity plants (500 mol m^{-3} NaCl).

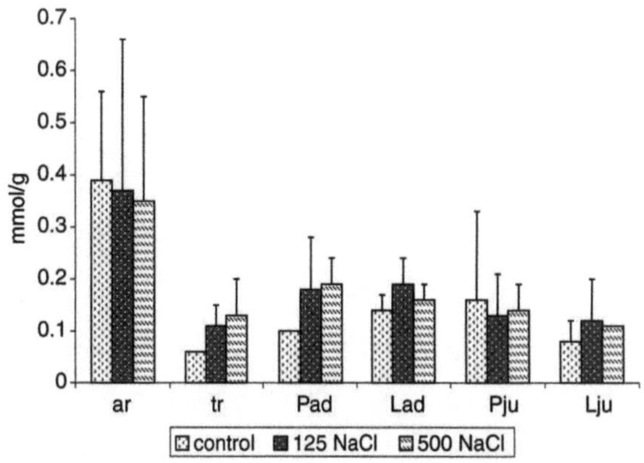

Fig. 6. Ca concentrations (mmol g^{-1}) in adventitious roots (ar), tap root (tr), petiole (P) and lamina (L) of juvenile (ju) and adult (ad) leaves of control, low-salinity (125 mol m^{-3} NaCl) and high-salinity plants (500 mol m^{-3} NaCl).

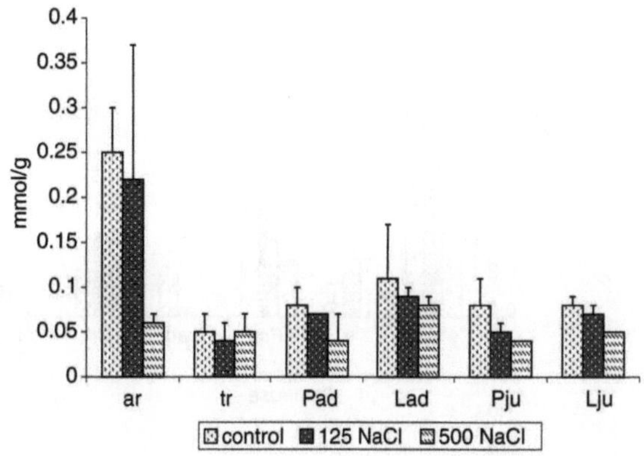

Fig. 7. Mg concentrations (mmol g^{-1}) in adventitious roots (ar), tap root (tr), petiole (P) and lamina (L) of juvenile (ju) and adult (ad) leaves of control, low-salinity (125 mol m^{-3} NaCl) and high-salinity plants (500 mol m^{-3} NaCl).

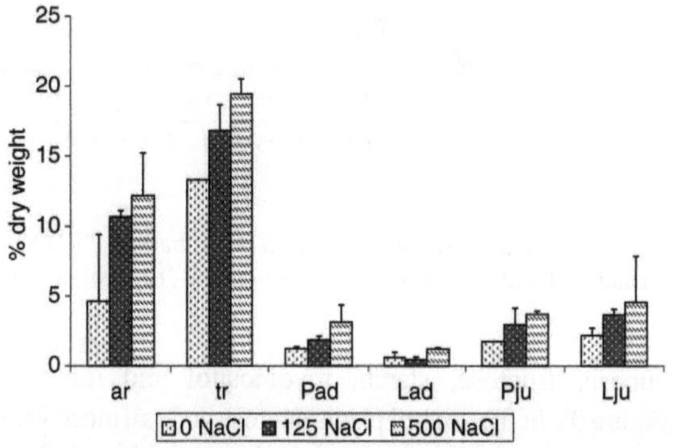

Fig. 8. Total carbohydrates concentration (mg % DW) in adventitious roots (ar), tap root (tr), petiole (P) and lamina (L) of juvenile (ju) and (ad) leaves of controls, low-salinity (125 mol m^{-3} NaCl) and high-salinity (500 mol m^{-3} NaCl) plants.

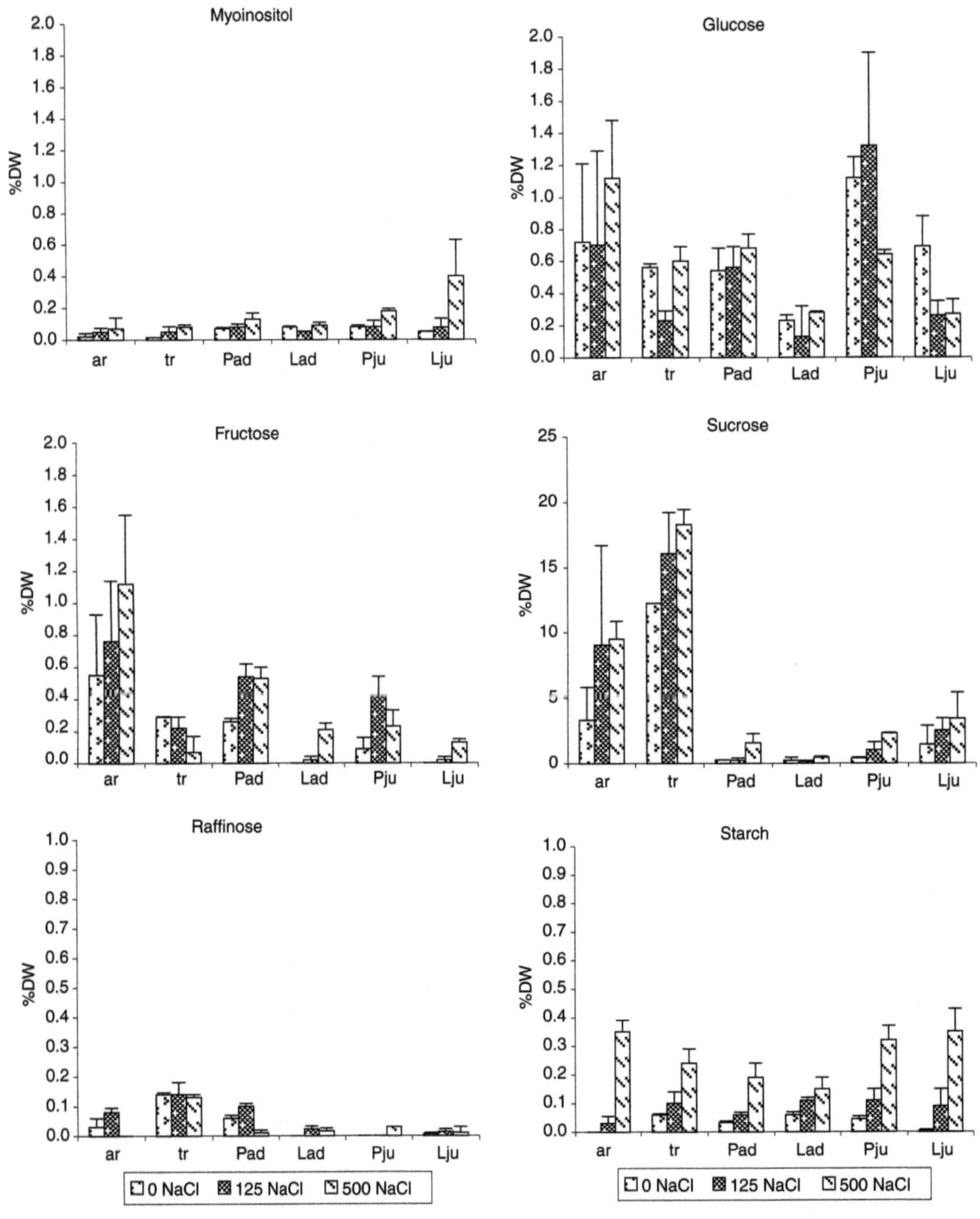

Fig. 9. Soluble carbohydrates and starch concentrations (% DW) in adventitious roots (ar), tap root (tr), petiole (P) and lamina (L) of juvenile (ju) and adult (ad) leaves of control, low-salinity (125 mol m^{-3} NaCl) and high-salinity plants (500 mol m^{-3} NaCl).

glucose, fructose, starch, myoinositol and raffinose (Figure 9). In the control plants sucrose and raffinose were in the tap root higher than in the other parts (Figure 9).

Controls showed higher proline values in leaf lamina than in roots, but salinity caused an increase of proline concentration in all parts of the plant (Figure 10). The highest amount was accumulated in the tap root leading to a reverse gradient between shoot and root. The content of proline in control leaf lamina was about 5 folds higher than in the tap root whereas in 500 mol m^{-3} NaCl, proline content was 3 folds higher in the tap root than in leaf lamina.

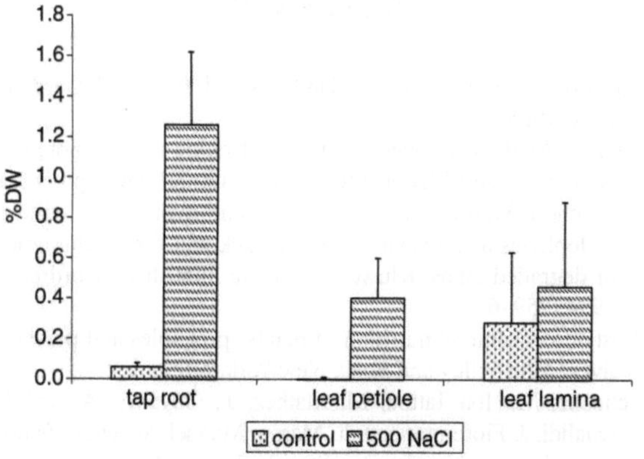

Fig. 10. Proline content (% DW) in tap root, petiole and lamina of leaves, of controls and high-salinity (500 mol m^{-3} NaCl) plants.

In summary the high Na concentrations in the leaves lead mainly to the high osmolality values at high salinity levels while the adjustment in the tap root was reached mainly by carbohydrate and proline accumulation.

DISCUSSION

Growth

The sea beet has the typical features of a halophytic species since plant growth was enhanced in low salinity (125 mol m^{-3} NaCl). This response can be explained by the role of Na as an osmoticum and its effect on cell expansion and water balance of plants (Marschner, 1995; Marschner and Possingham, 1975). However growth was greatly affected by high salinity levels (Kelly et al., 1982; Harrouni et al., 1999; Flowers and Lauchli, 1983) and shoot growth was more reduced by salinity than root growth (Greenway and Munns, 1980; Koyro and Huchzermeyer, 1999a). The reduction of growth at high salinity was reported to be a reaction to ion toxicity in old leaves, to water deficit and to shortage of carbohydrates in the younger leaves (Koyro and Huchzermeyer, 1999b).

Osmotic Adjustment

The increase of NaCl in the culture medium generates a decrease of the water potential and of the osmotic potential in all parts of the sea beet tissue. The difference in osmotic potential between the nutrient solution and the roots and between the roots and the leaves is one of the driving forces for the uptake of water through the soil–plant–atmosphere continuum. The osmotical adjustment to salinity enables the sea beet to delay wilting, to maintain its turgescence and to adjust its shoot osmotic pressure. It is practically reached by the accumulation of high amount of salts in the shoot which led to an increase in the proportion of inorganic matter in the dry weight (Gorham, 1996; Flowers and Lauchli, 1983). Flowers and Yeo (1988) reported that high NaCl concentrations are accumulated in large vacuoles of leaf cells. Minerals and water content were co-ordinated to maintain constant osmotic potential gradient between shoot tissue and the external solution (Gleen and O'Leary, 1984; Terry et al., 1983; Weber, 1995). The osmotic potential was generated in the leaves of the sea beet at high-salinity levels mainly by the accumulation of NaCl in the vacuoles and of proline as osmoprotectant in the cytoplasm. Soluble carbohydrates (especially) sucrose accumulation in tap root has an effect on cellular water relations. It contributes to a decrease in cell osmotic potential (Koyro and Huchzermeyer, 1999b). The stimulation of proline synthesis with salinity in different parts of the plant and especially in the tap root confirms its role as osmoregulatory amino acid as described by several authors (Flowers et al., 1986; Joshi et al., 1996; Koyro, 1997).

Water Content

To compensate the high value of minerals in the growing medium adult leaves accumulated considerable amounts of salt which in turn enabled them to uptake water. Salt accumulation induces leaf succulence which explains the increase of the LMA ratio. Succulence is considered an important buffer mechanism against high changes in leaf water potential under saline conditions (Flowers and Lauchli, 1983). Mineral accumulation in adult leaves prevent the growing ones from the effect of ion toxicity.

Mineral Relations

The specificity of the sea beet for the uptake of K, Mg and Ca was not sufficient to balance the important concentration of Na under high-salinity conditions. Under these conditions high proportions of K, Ca and Mg can be replaced in the leaf tissue by Na (Koyro and

Huchzermeyer, 1999b) which is essential as mineral nutrient for the sea beet by the extent to which it can replace potassium functions and its additional growth stimulation effect (Marschner and Possingham, 1975; Marschner, 1995; Koyro et al., 1999). The distribution of Ca in the plant with increasing salinity differed depending on the considered part of the plant. Gorham (1996) and Marschner (1995) reported that Ca has a role in increasing salt tolerance of plants since it maintains membrane integrity in the presence of high concentration of Na and controls selectivity of ion uptake and transport. On the other hand, high Na concentration in the culture medium induces Ca deficiency in the shoots as described by Gorham (1996) and Marschner (1995). The decrease of Ca in cells at high salinity is reported to be correlated with an inhibition of membrane functions like ATPase activity and an increase in passive Cl and Na transport (Koyro et al., 1993; Koyro, 1997; Koyro et al., 1999).

CONCLUSION

The sea beet is able to survive high salinity by osmotic adjustment and solute and ion compartmentation. The accumulation of high amounts of salt in the shoot vacuole cells provides the sea beet with sufficient water to maintain growth and development. Excess salt ions at high-salinity level induced a toxic effect to the plants, which ended up with a great reduction in biomass production thus putting the limit of salt tolerance (50% biomass production) between 375 and 500 mol m^{-3} NaCl. The quick check system used in this experiment showed to be highly efficient in the screening of plants for salt tolerance.

ACKNOWLEDGEMENTS

This work was carried out in the framework of the Concerted Action Project "Sustainable Utilisation of Halophytes in the Mediterranean and Subtropical Dry Regions". The authors would like to acknowledge the facilities provided by the Justus-Liebig Institute of Giessen where the experiments were undertaken. The help of the Institute's technical staff is fully appreciated. Special thanks are to Dr. B. Huchzermeyer, School of Veterinary Medicine, Hannover for his support and useful discussions.

REFERENCES

Aronson, J. Plants for arid lands. Royal Botanical Gardens, Kew, 1986.

Choukr-Allah, R., Jones, G.W. and Kenny, L. The potential economic use of halophytes for agricultural development of Southern Morocco. In Squires, V.R. and Ayoub, A.T., editors. Halophytes as a resource for livestock and for rehabilitation of degraded lands. Kluwer Academic Publishers, Dordrecht, 1994; 259–61.

Epstein, E. Mineral nutrition of plants: principles and perspectives. John Wiley and Sons, New York, 1972.

Fennane, M., Ibn Tattou, M. Mathez, J., Ouyahya, A. and El Oualidi, J. Flore pratique du Maroc. Manuel de détermination des plantes vasculaires. Pteridophytae, Gymnospermae, Angiospermae (Lauraceae-Neuradaceae), Vol. 1. 1999; 558 pp.

Flowers, T.J., Hajibagheri, M.A. and Clipson, N.J.W. Halophytes. Quaternary Rev Biol 1986; 61(3):313–37.

Flowers, T.J. and Lauchli, A. Sodium versus potassium: substitution and compartmentation. Encyclopaedia of plant physiology. New series, Vol. 15B. 1983; 651–81.

Flowers, T.J. and Yeo, A.R. Ion relations of salt tolerance. In Baker, D.A. and Hall, J.L., editors. Solute transport in plant cells and tissues. Longmann Scientific and Technical, Essex, England, 1988; Chap. 10.

Gleen, E.P. and O'Leary, J. Relationship between salt accumulation and water content of dicotyledonous halophytes. Plant Cell Environ 1984; 7:253–61.

Gonzalez, E., Arrese-Igor, C., Aparicio-Tejo, P.M. and Koyro, H.-W. Osmotic adjustment in different foliar structures of semileafless pea subjected to water stress. Plant Physiol Biochem (Special Issue) 1996; 240.

Gorham, J. Mechanisms of salt tolerance of halophytes. In Choukr-Allah, R., Malcom, C.V., and Hamdy, A., editors. Halophytes and biosaline agriculture. Marcel Dekker, Inc., New York, 1996; 31–53.

Greenway, H. Plant response to saline substrates. I. Growth and ion uptake of several varieties of *Hordeum* during and after sodium chloride treatment. Aust J Biol Sci 1962; 15: 16–38.

Greenway, H. and Munns, R. Mechanism of salt tolerance in non halophytes. Annu Rev Plant Physiol 1980; 31:149–90.

Harrouni, M.C., Daoud, S., El Alami, A., Debbagh, B., Chokr-Allah, R. and Bengaddour, M. Responses of some halophytes to seawater irrigation in Morocco. In Hamdy, A., Lieth, H., Todorovic, M. and Moschenko, M., editors. Halophyte uses in different climates II, 1999; 57–75.

Hoffmann-Thoma, G., Van Bel, A.J.E. and Ehlers, K. Ultrastructure of minor vein phloem and assimilate export in summer and winter leaves of the symplasmically loading evergreens *Ajuga reptans* L., *Aucuba japonica* Thunb., and *Hedera helix* L. Planta. 2000 (accepted for publication).

Joshi, A.J., Bhoite, A.S. and Rejith Kumar, K.S. Effects of seawater on accumulation of organic and inorganic metabolites in *Aeluropus lagopoides* Linn. Physiol Mol Biol Plants 1996; 2:149–52.

Kelly, D.B., Goodin, J.R. and Miller, D.R. Biology of *Atriplex*. In Sen, D.N. and Rajpurohit, K.S., editors. Tasks for vegetation, Vol. 2. 1982; Chap. 5.

Koyro, H.-W. Ultrastructural and physiological changes in root cells of Sorghum plants (*Sorghum bicolor* X *S. sudanensis* cv. Sweet Sioux) induced by NaCl. J Exp Bot 1997; 48:693–706.

Koyro, H.-W. Protocol for the installation of an improved quick check system in the I.A.V. Hassan II (Agadir, Morocco). Third annual report: Sustainable Halophyte Utilisation in the Mediterranean and Subtropical Dry Regions, 1999.

Koyro, H.-W. 2000. In this volume.

Koyro, H.-W. and Huchzermeyer, B. Salt and drought stress effects on metabolic regulation in maize. In Pessarakli, editor. Handbook of plant and crop stress, 1999a.

Koyro, H.-W. and Huchzermeyer, B. Influence of high NaCl-salinity on growth, water and osmotic relations of the halophyte *Beta vulgaris* ssp. *maritima* – development of a quick check. In Hamdy, A., Lieth, H., Todorovic, M. and Moschenko, M., editors. Halophyte uses in different climates I. 1999b; 89–103.

Koyro, H.-W., Stelzer, R. and Huchzermeyer, B. ATPase activities and membrane fine structure of rhizodermal cells from Sorghum and *Spartina* roots grown under mild salt stress. Botanica Acta 1993; 106:110–19.

Koyro, H.-W., Wegmann, L., Lehmann, H. and Lieth, H. Adaptation of the mangrove *Laguncularia racemosa* to high NaCl salinity. In Hamdy, A., Lieth, H., Todorovic, M. and Moschenko, M., editors. Halophyte uses in different climates I. 1999; 43–64.

Le Houerou, H.N. Salt tolerant plants of economic value in the Mediterranean basin. Reclamat Revegetat Res 1986; 5:319–41.

Lieth, H. and Al Masoom, A.A., editors. Towards the rational use of high salinity tolerant plants, Vol. I of the Proc. of the Al Ain conference 1990, T: VS Vol. 27. Kluwer Academic Publishers, Dordrecht, 1993; 521 pp.

Marschner, H. Mineral nutrition of higher plants. Academic Press, London, 1995; 889 pp.

Marschner, H. and Possingham, J.V. Effect of K+ and Na+ on growth of leaf discs of sugar beet and spinach. Z Pflanzenphysiol 1975; 75:6–16.

O'Leary, J.W. A critical analysis of the use of *Atriplex* species as crop plants for irrigation with highly saline water. In Ahmed, R. and San Pietto, A., editors. Prospects of biosaline research. Publ. Botany Dept., Karachi Univ., Pakistan, 1986; 415–32.

Terry, N., Waldron, L.J. and Taylor, S.E. Environmental influences of leaf expansion. In Dale, J.E. and Milthope, F.L., editors. The growth and functioning of leaves. Cambridge University Press, Cambridge, 1983.

Weber, D.J. Mechanisms and reactions of halophytes to water and salt stress. In Ajmal Khan, M. and Irwin, A. Ungar, editors. Biology of salt tolerant plants, 1995.

Salinity tolerance of *Beta vulgaris* ssp. *maritima*. Part II. Physiological and biochemical regulation

S. DAOUD, H.-W. KOYRO, M.C. HARROUNI, A. SCHMIDT AND J. PAPENBROCK

INTRODUCTION

Beta vulgaris ssp. *maritima* (L.) occurs naturally along the Atlantic coasts of western Europe and along the coasts of nearly all Mediterranean countries (Letschert, 1993). To be able to survive in these habitats, the sea beet has developed mechanisms, which enable it to overcome saline stress. Tolerance to salinity is determined by a number of separate but interrelated mechanisms which operate at different levels of tissue structure to control the distribution of salt and other solutes within the plant and to maintain a gradient of water potential through the plant from the soil to the atmosphere (Gorham, 1992). The balance between water and salt uptake could be maintained by reducing transpiration, but at the expense of reduced carbon fixation and reduced growth rate (Gorham, 1996). Under salt stress, the reduction of growth is greater than the decrease in photosynthesis (Cheeseman, 1988). Salinity affects carbon assimilation because of a smaller leaf area rather than a reduced rate of photosynthesis (Klenke et al. in preparation). The reduction of leaf area is among the mechanisms salt includer species use in order to minimise the evaporating surface (Koyro and Huchzermeyer, 1999a) and consequently increase their water content leading to succulence. In the sea beet leaf succulence is a consequence of salt and water accumulation which induce changes in leaf structure, especially the increase of the volume of the palisade and spongy parenchyma cells (Koyro and Huchzermeyer, 1997).

The aim of our study is to investigate the physiological parameters determining salt tolerance of the sea beet. One approach toward understanding the mechanisms of salt tolerance at the hole plant level is to follow the series of physiological transformations that exposure to salinity can generate in water relations, photosynthesis, transpiration and salt and metabolites accumulation (Koyro and Huchzermeyer, 1999a).

MATERIAL AND METHODS

Beta vulgaris ssp. *maritima* plants used for sulfolipids, chlorophyll a and b and gas exchange measurements are the same ones used in the chapter of Daoud et al. (2003 in this book).

The chlorophyll was extracted with 80% acetone from 8 leaf discs (0.78 cm^2) of fresh material (3 replicates per treatment). The chlorophyll a and b contents were determined by a DU-6-spectrophotometer (BECKMAN). Sulfolipids were extracted from plant cells with chloroform/methanol (2:1 v/v) and quantified by thin-layer chromatography, followed by photodensitometric scanning method (modified by Archer et al., 1997).

Stomata frequency was determined by thin coating with transparent "Jade" varnish. The varnish was spread on small leaf surfaces between the secondary veins. Both adaxial and abaxial sides of the leaf were sampled. After drying, the thin layer was stripped of the leaf surfaces and transferred on a glass slide into the beam of a light microscope for the counting of the stomates.

Net photosynthesis, intracellular CO_2, stomatal resistance, stomatal conductance and transpiration, were measured on juvenile and adult leaves using a photosynthesis analyser LI 6200 by enclosing the leaf in a closed chamber.

RESULTS

In juvenile leaves of the sea beet the level of sulfolipids increased with the increase of salinity. In the old

H. Lieth (ed.), Cash Crop Halophytes: Recent Studies, 51–57.

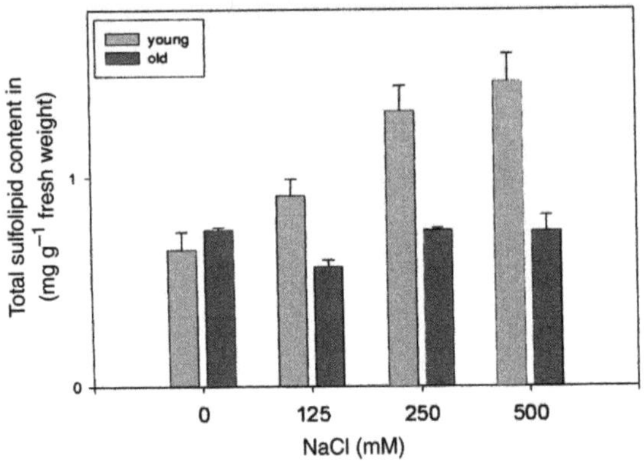

Fig. 1. Sulfolipids content in *Beta vulgaris* ssp. *maritima* watered with different NaCl concentrations.

leaves sulfolipids content remained at higher levels, roughly similar to those in the control young leaves (Figure 1).

Chlorophyll a + b contents of juvenile and adult leaves increased with the increase of salinity in the growing medium, reached a maximum value at 250 mM NaCl, and decreased slightly at high-salinity levels. Nevertheless the rate of chl a + b in juvenile as well as in adult leaves of plants at high-salinity level (500 mM NaCl) remained higher than in the control plants. It was 2.3 times higher in juvenile leaves and 1.4 times higher in adult leaves compared to the control. The ratio chl a/b was similar in all treatments in adult as well as in juvenile leaves (Figure 2).

In both adult and juvenile leaves, the response of net photosynthesis to increasing salinity was such that optimum rates were reached at 250 mM NaCl. Thereafter

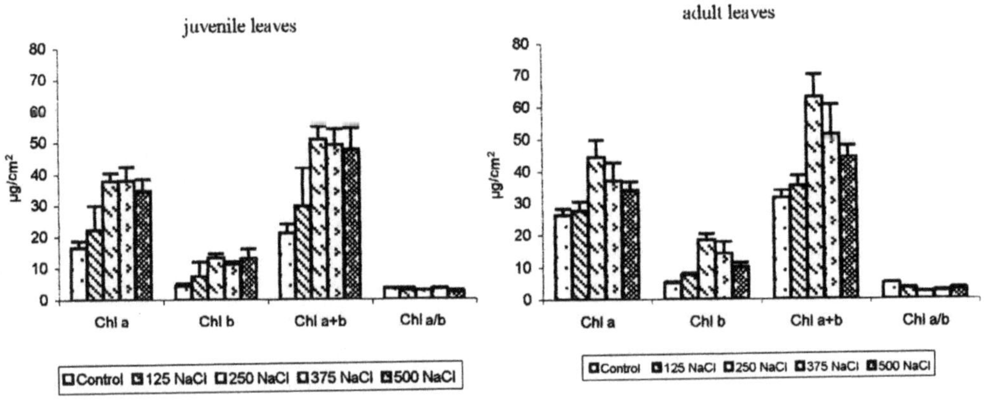

Fig. 2. Chlorophyll a (chl a) and b (chl b) concentrations (µg/cm²), chlorophyll a + b and chl a/b ratio in juvenile and adult leaves of *Beta vulgaris* ssp. *maritima* treated without (control) or with the addition of 500 mM NaCl added to the nutrient solution.

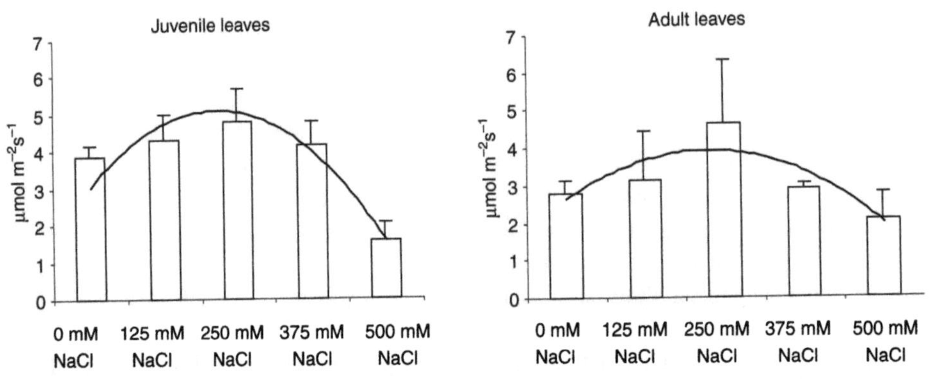

Fig. 3. Net photosynthesis (µmol $m^{-2}s^{-1}$) of juvenile and adult leaves of *Beta vulgaris* ssp. *maritima* of controls and 4 salinity treatments (125, 250, 375 and 500 mM NaCl).

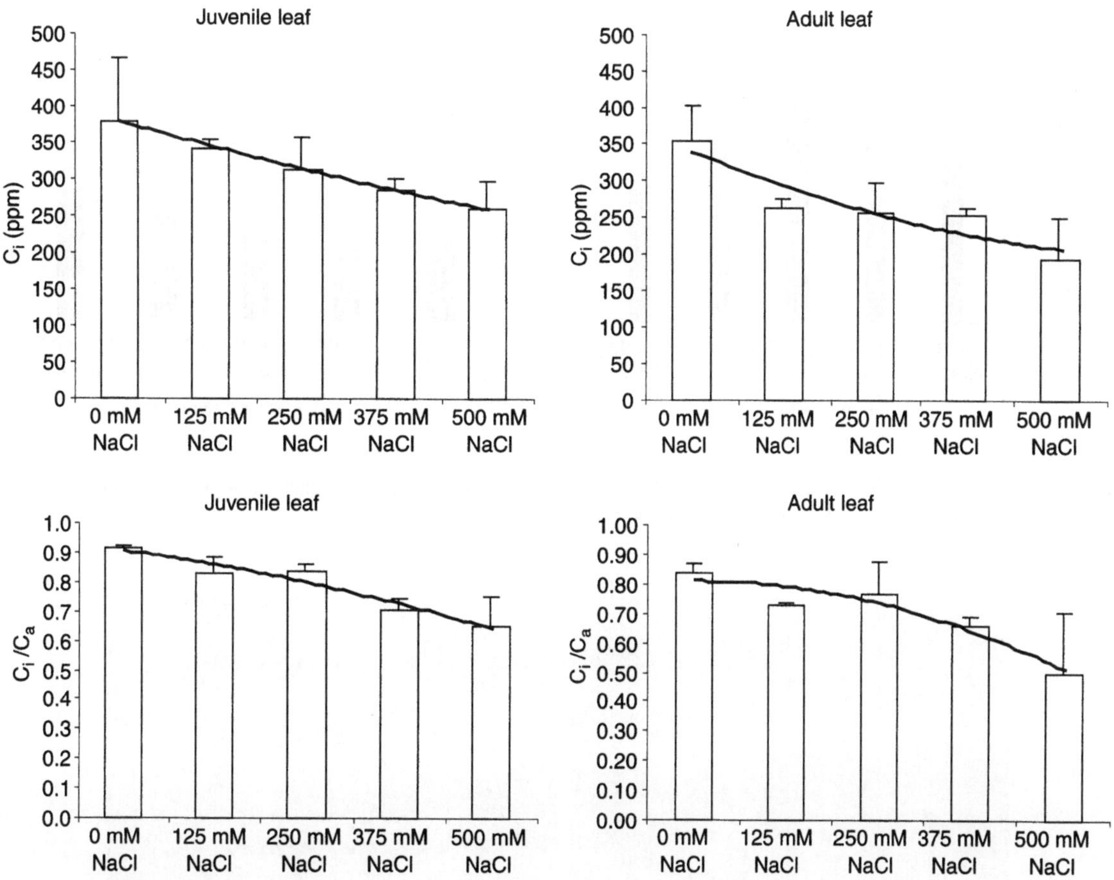

Fig. 4. CO_2 intercellular (C_i) concentrations (ppm) and CO_2 intercellular to atmospheric CO_2 ratio (C_i/C_a) of juvenile and adult leaves of *Beta vulgaris* ssp. *maritima* of control and 4 salinity treatment (125, 250, 375 and 500 mM NaCl).

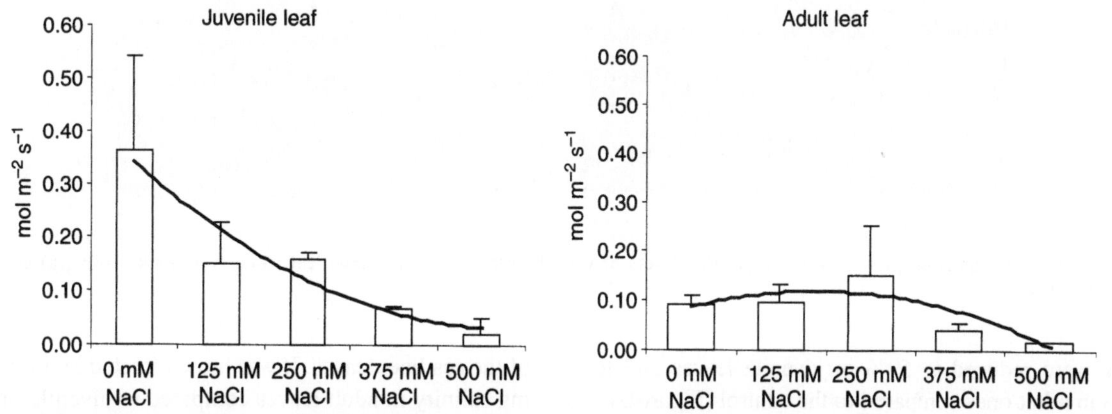

Fig. 5. Stomatal conductance (mol m^{-2} s^{-1}) of juvenile and adult leaves of *Beta vulgaris* ssp. *maritima* of control and 4 salinity treatment (125, 250, 375 and 500 mM NaCl).

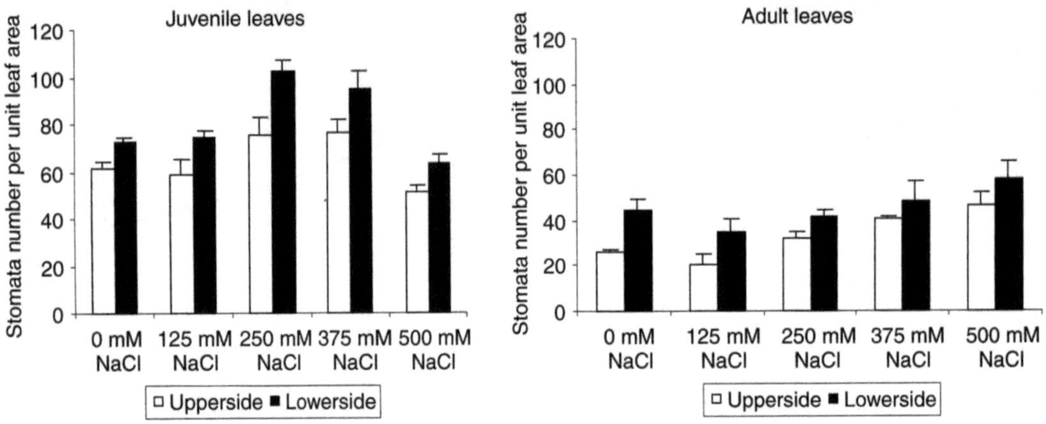

Fig. 6. Stomata frequency (number per unit leaf area) of adaxial and abaxial sides of juvenile and adult leaves of *Beta vulgaris* ssp. *maritima* of control and 4 salinity treatment (125, 250, 375 and 500 mM NaCl).

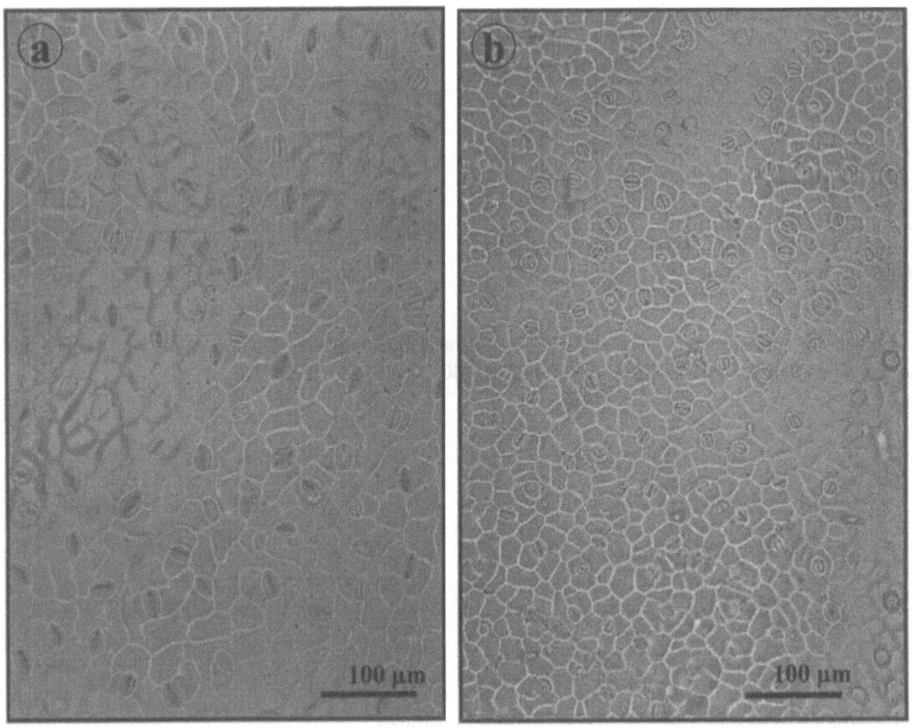

Fig. 7. Typical lightmicroscopical photographs of thin layers of varnish stripped of the adaxial leaf surfaces of adult leaves: (a) control; (b) 500 mM NaCl.

rates were reduced to 59% in juvenile leaves and to 25% in adult ones compared to the control (Figure 3).

The leaf intercellular CO_2 (C_i) and the ratio of leaf intercellular CO_2 to atmospheric CO_2 (C_i/C_a) decreased with increasing salinity in the root medium of the sea beet leaves. It was more affected by increasing salinity in adult leaves compared to juvenile ones. The reduction of (C_i) and (C_i/C_a) was 45% and 40.5% respectively in adult leaves and 32% and 29% respectively in juvenile leaves (Figure 4).

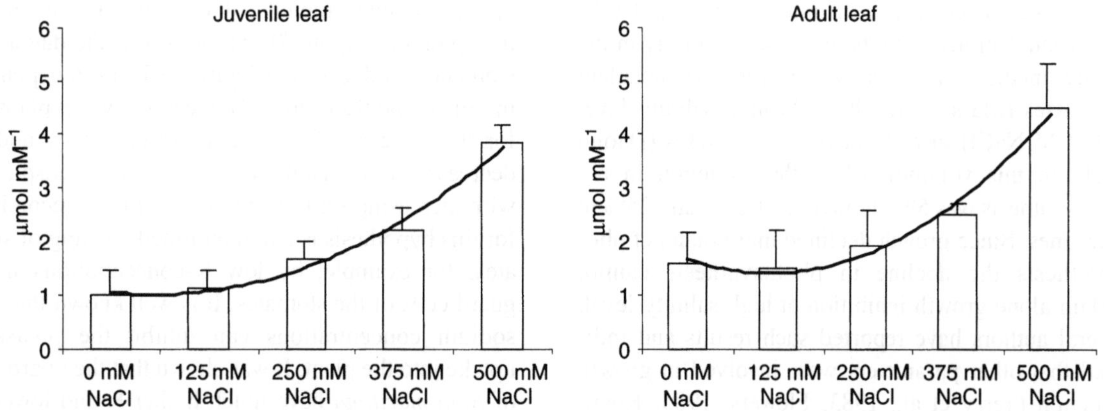

Fig. 8. Water use efficiency (WUE) µmol mM^{-1} of juvenile and adult leaves of *Beta vulgaris* ssp. *maritima* of control and 4 salinity treatment (125, 250, 375 and 500 mM NaCl).

The stomatal conductance of adult leaves increased slightly from the control to optimum growth salinity (250 mM NaCl) and then decreased at higher salinity (Figure 5). The reduction was about 87% compared to the optimum. In juvenile leaves stomatal conductance decreased progressively from the control to high salinity level where the reduction was about 94%. Transpiration of sea beet leaves in response to increasing salinity paralleled the stomatal conductance response (Figure 6).

The stomata frequency was higher in the lower side of the leaf than in the upper side regardless the age (Figure 7). Additionally, the stomata frequency in juvenile leaves was higher in all treatments compared to adult leaves (result not shown). In adult leaves stomata frequency increased with the increase of salinity.

The water use efficiency in both adult and juvenile leaves increased with the increase of salinity as a result of reduced stomatal conductance and transpiration (Figure 8).

DISCUSSION

Sulfolipids and Chl a + b

The effect of increasing salinity by the supply of NaCl in the culture medium of the sea beet resulted in growth stimulation at moderate salinity (125 and 250 mM NaCl) and a reduction of growth at higher salinity levels (Daoud et al., in this volume). Marschner and Possingham (1975) reported that cell expansion of leaf disc of sugar beet and spinach in response to increasing Na-concentration was not simply due to water uptake but involved cell development and maturation since it was combined with increases in dry matter and chloroplast number. They also reported that despite its stimulating effect on fresh and dry matter production and chloroplast formation Na had a depressive effect on chlorophyll formation in sugar beet. In experiment on sea beet Na had not a depressive effect at high salinity levels since the content of chl a + b at high salinity level was 2.3 and 1.4 folds higher in juvenile and adult leaves respectively than in the control plants.

Our results showed an increase of chl a + b, sulfolipid contents and photosyntesis rate of the sea beet at moderate salinity (125 and 250 mM NaCl) which suggests a correlation between these mechanisms. Archer et al. (1997) showed the role of sulfolipids in photosynthesis rate. The high sulfolipid level in young leaves, and the constant level in adult ones may contribute to salt resistance in the sea beet. Archer et al. (1997) reported that sulfolipids have more than one function in the cell, they enter in the structure of thylacoids and chloroplasts. Kuiper and Kuiper (1978) who worked on *Plantago* and Stuiver et al. (1981) who worked on sugar beet reported that high sulfolipid levels in the roots may indicate salt resistance.

Gas Exchange Parameters

Growth and photosynthesis rate of the sea beet varied in the same manner with increasing salinity. Optimal growth was reached at moderate salinity (250 mM NaCl) and declined with the increase of salinity in the culture medium as it is typical for non-succulent halophytes (Marschner, 1995). At high salinity level (500 mM NaCl) growth reduction was 63% (Daoud et al., in this volume), while the reduction in net photosynthesis was 59% in juvenile leaves and 25% in adult ones. Since growth declined more than net photosynthesis the decline in photosynthesis cannot explain alone growth inhibition at high salinity level. Several authors have reported such results and indicated that other parameters were involved in growth reduction (Terry et al., 1983; Flowers, 1985; Koyro and Huchzermeyer, 1999b). Koyro and Huchzermeyer (1999b) and (Daoud et al., in this volume) reported that there is a build up of total carbohydrates in all parts of the sea beet plants with increasing salinity which suggests that photosynthesis does not limit growth in salt stressed sea beet. Kelly et al. (1982) suggested that the growth reduction of *Atriplex halimus* at high salinity without ceasing is an indication that most of the enzymes of photosynthesis and other metabolic pathways were still functional although the degree of activation may have been affected. Flowers et al. (1986) reported that changes in leaf morphology are of greater significance than effects on the metabolic processes of photosynthesis and that salinity affects carbon assimilation per plant via smaller leaf area rather than a reduced rate of photosynthesis. Koyro et al. (1999) and Daoud et al. (in this volume) showed that increasing salinity in the growing medium resulted in an increase of Na-concentration in the leaves of the sea beet together with the loss of K. Terry and Ulrich (1973) reported that in the sugar beet low K-content decreased photosynthesis through an increase of the mesophyll resistance to CO_2, and not through stomatal diffusion resistance.

The increase in stomata frequency (number per leaf area) of the sea beet adult leaves with the increase of salinity was found also in the sugar beet (Hampe and Marschner, 1982) and in *Laguncularia racemosa* (Koyro et al., 1999a). The increase of stomata frequency and the decrease of leaf surface area was explained with a reduction of cell expansion in epidermal cells caused by the low-water potentials in the saline media. The increase of stomata frequency was interpreted from the author as a way to minimise the opening times for gas exchange. At moderate high-water potentials in the air and/or the soil (e.g. after rain) a high stomata frequency enables the sea beet the uptake of high amounts of CO_2 and a reduction of transpiration (Figure 7). At the rest of the day a low-stomatal conductance (Figure 6) helps the plant to minimise the water loss. This can be one explanation for the increase of the water use efficiency and the decrease of stomatal conduction and transpiration with increasing salinity (Figures 6–8). A precondition for this hypothesis is a non inhibited opening of stomates, for example, by low K-concentrations in the guard cells of the stomates. It is well known that high sodium concentrations can inhibit the potassium uptake into the plant. It was shown that the guard cells of *Beta maritima* have much higher K and lower Na concentrations than other epidermal cells (results not shown). This difference between epidermal cells and guard cells give reason to assume that the stomates are functioning even at high-salinity levels.

Since salt transport to the shoots is to some extent related to transpiration rate, an improvement in water use efficiency (the amount of water transpired for a given increase in CO_2 fixed, or a given increase in fresh weight) should help to retard the accumulation of salt in the leaves (and especially in the guard cells; Gorham, 1996) and represent a more conservative and efficient use of water which may contribute to plant survival under long-term exposure to salinity.

Raising the Na-concentration in the root medium increased the leaf mass to area ratio (leaf thickness) and water content of the sea beet by inducing an obvious increase in the volume of the palisade and spongy parenchyma cells (Koyro et al., 1997). Similar effects have been reported previously on the sugar beet (Hampe and Marschner, 1982) and *Atriplex patula* (Longstreth and Nobel, 1979) in which both length and diameter of palisade cells increased with salinity. As a first reaction to salinity, Koyro et al. (1999b) explained that sea beet plants showed a decrease of leaf surface per plant. The reduction in leaf area (in relation to the mass) is one mechanism of salt including species to increase their water content and to minimise the evaporating surface.

CONCLUSION

Salinity tolerance in the sea beet is controlled by several interrelated mechanisms. The compartmentation of NaCl in the vacuoles of shoot cells under saline conditions and low Na-concentrations in the other organelles and in the cytoplasma are a protection for many essential enzymatic activities. Indeed at moderate salinity chlorophyll and sulfolipid contents and consequently photosynthesis were promoted compared to the control plants. At high-salinity levels chlorophyll content and the rate of net photosynthesis were less affected than growth, since there may be a build up of carbohydrates in the tissue levels during salt stress. Growth reduction was mainly due to the reduction in leaf number and area. NaCl-compartmentation was more pronounced in adult leaves which serve as depositories for salts which would otherwise accumulate in juvenile leaves. In adult leaves regulation of transpiration in response to salinity was achieved by the decrease of stomata frequency leading to a decrease in stomatal conductance and an increase in water use efficiency. The reduction of stomatal conductance and transpiration participates to maintain the level of leaf turgescence which was done at the expense of plant growth because the energy needed for growth was lost to maintain high-water level in the plant.

ACKNOWLEDGEMENTS

This work was financially supported by the Concerted Action Project "Sustainable Utilisation of Halophytes in the Mediterranean and Subtropical Dry Regions". The authors are indebted to Prof. Fangmeier for supervising the photosynthetic measurements and to miss Bölke for technical assistance.

REFERENCES

Archer, S.D., McDonald, K.A. and Jackman, A.P. Effect of light irradiance on the production of sulfolopids from Anabaena 7120 in a fed-bach photobioreactor. Appl Biochem Biotech 1997; 67:139–52.

Cheeseman, J.M. Mechanisms of salinity tolerance in plantes. Plant Physiol 1988; 87:547–50.

Daoud, S., Koyro, H.-W. and Harrouni, M.C. 2003. Page 41 this volume.

Flowers, T.J. Physiology of halophytes. Plant Soil 1985; 89:41–56.

Flowers, T.J., Hajibagheri, M.A. and Clipson, N.J.W. Halophytes. Quaternary Rev Biol 1986; 61(3):313–37.

Gorham, J. Grass response to salinity. Desertified grasslands: their biology and management 1992; 165–80.

Gorham, J. Mechanisms of salt tolerance of halophytes. In Choukr-Allah, R., Malcom, C.V. and Hamdy, A., editors. Halophytes and biosaline agriculture. Marcel Dekker, Inc. New York, 1996; 31–53.

Hampe, T. and Marshner, H. Effect of sodium on morphology, water relations and net photosynthesis of sugar beet leaves. Z Pflanzenphysiol Bd 1982; 108:S.151–62.

Kelly, D.B., Goodin, J.R. and Miller, D.R. Biology of *Atriplex*. In Sen, D.N. and Rajpurohit, K.S., editors. Tasks for vegetation, Vol. 2. 1982; Chap. 5.

Klenke, I., Koyro, H.-W. and Huchzermeyer, B. Salinity affects chloroplasts maturation in maize. Plant physiology (in preparation).

Koyro, H.-W. and Huchzermeyer, B. The physiological response of *Beta vulgaris* ssp. *maritima* to sea water irrigation. In Lieth, H., Hamdy, A. and Koyro, H.-W. Water management, salinity and pollution control towards sustainable irrigation in the mediterranean region. Special volume: Salinity problems and halophyte use. Technomack, bari, 1997; 29–50.

Koyro, H.-W. and Huchzermeyer, B. Salt and drought stress effects on metabolic regulation in maize. In Pessarakli, M., editor. Handbook of plant and crop stress 1999a; 843–77.

Koyro, H.-W. and Huchzermeyer, B. Influence of high NaCl-Salinity on growth, water and osmotic relations of the halophyte *Beta vulgaris* ssp. *maritima* – development of a quick check. In Hamdy, A., Lieth, H., Todorovic, M. and Moschenko, M., editors. Halophyte uses in different climates I, 1999b; 89–103.

Koyro, H.-W., Wegmann, L., Lehmann, H. and Lieth, H. Adaptation of the mangrove *Laguncularia racemosa* to high NaCl salinity. In Hamdy, A., Lieth, H., Todorovic, M. and Moschenko, M., editors. Halophyte uses in different climates I, 1999; 43–64.

Kuiper, D. and Kuiper, P.J.C. Lipid composition of the roots of Plantago species: response to alteration of the level of mineral nutrition and ecological significance. Physiol Plant 1978; 44:81–86.

Letschert, J.P.W. *Beta* section *Beta*: biogeographical patterns of variation, and taxonomy. Wageningen Agricultural University. Papers 93-1, 1993.

Longstreth, D.J. and Nobel, P.S. Salinity effects on leaf anatomy, consequences for photosynthesis. Plant Physiol 1979; 63:700–3.

Marschner, H. and Possingham, J.V. Effect of K^+ and Na^+ on growth of leaf discs of sugar beet and spinach. Z Pflanzenphysiol 1975; 75:6–16.

Stuiver, C.E.E., Kuiper, P.J.C., Marschner, H. and Kylin, A. Effects of salinity and replacement of K^+ by Na^+ on lipid composition in two sugar beet inbred lines. Physiol Plant 1981; 52:77–82.

Terry, N. and Ulrich, A. Effects of potassium deficiency on the photosynthesis and respiration of leaves of sugar beet. Plant Physiol 1973; 51:783–6.

Terry, N., Waldron, L.J. and Taylor, S.E. Environmental influences of leaf expansion. In Dale, J.E. and Milthope, F.L., editors. The growth and functioning of leaves. Cambridge University Press, Cambridge, 1983.

Effect of seawater irrigation on biomass production and ion composition of seven halophytic species in Morocco

M.C. HARROUNI, S. DAOUD AND H.-W. KOYRO

INTRODUCTION

Morocco is faced today with an increasing demand for water. With the frequency of drought years, the intensification of food production depends heavily on irrigation and the efficient management of limited water resources. The extension of irrigated agriculture and the intensive use of water resources combined with high evaporative rates, have inevitably given rise to the problems of salinity in the soil and in underground water. It should be noted that salinity problems have been observed in several regions of the country and are liable to spread as irrigation is intensified and irrigated areas are extended (Choukr-Allah and Harrouni, 1996). At the present time, Morocco has more than 800,000 ha of irrigated land. And most irrigated areas show different degrees of salinity. The standard approach to this problem would be to increase salt tolerance of conventional crop plants, but the gain in yield is generally low. The alternative approach to the problem is to make use of the plants that already have the requisite level of salt tolerance, and are highly productive at high-external salinity levels: the halophytes. It is possible to make a reasonable use of them and to expect to find suitable candidates for domestication among that large pool of plants (O'Leary, 1986; Aronson, 1986). Within the past decade, substantial progress has been made to evaluate the potential use of halophytes as crop plants (Aronson, 1986). Halophytes represent an important potential as they can be used for fodder, fuel, oil, wood, pulp and fibre production. They also can be used for land reclamation, dune stabilisation and landscaping (Lieth and Al Massoum, 1993a). Some of 2,500 species of halophytes (gramineae, shrubs and trees), occur in saline coastal environment and inland deserts. Increasing attention has been paid to research and development of halophytes and several authors proposed utilising undiluted seawater on a large scale for irrigation (O'Leary and Glenn, 1984; Aronson, 1986). As a matter of fact two thirds of the world are covered by the oceans, this unlimited source of seawater could be used for the irrigation of halophytes on coastal areas. The good drainage associated with sand allows for a high volume and/or high frequency of irrigation with seawater, thus preventing salt accumulation in the root zone (O'Leary et al., 1985; Goodin et al., 1990). However halophytic species have different degrees of salinity tolerance, and a very limited number of the large pool of halophytes tolerate seawater salinity (Lieth and Al Massoum, 1993a). It therefore is important to select promisable salt tolerant plants with adequate tolerance and yield characteristics (Koyro, 1997).

The aim of this research is to determine the level of salinity tolerance of a number of halophytic species and to study the mechanism by which plants survive high salinity using a quick check system in which plants were grown in coastal sand and irrigated with five seawater dilutions.

MATERIALS AND METHODS

Seven halophytic plants were studied with the aim of estimating their tolerance and responses to salinity. The screening procedure was conducted under greenhouse conditions at IAV Hassan II, Agadir (Morocco), in a "quick check system" (Koyro, 1999). It is an automated irrigation and drainage system where plants were put in 1,001 containers and regularly flooded with the required water (Figure 1). Small pumps were used to recycle water several times per day. The plants were cultivated in plastic containers on coastal sand substrate with four replicates per species. The watering regimes were: tap water (control), 25%; 50%; 75% and

H. Lieth (ed.), Cash Crop Halophytes: Recent Studies, 59–70.

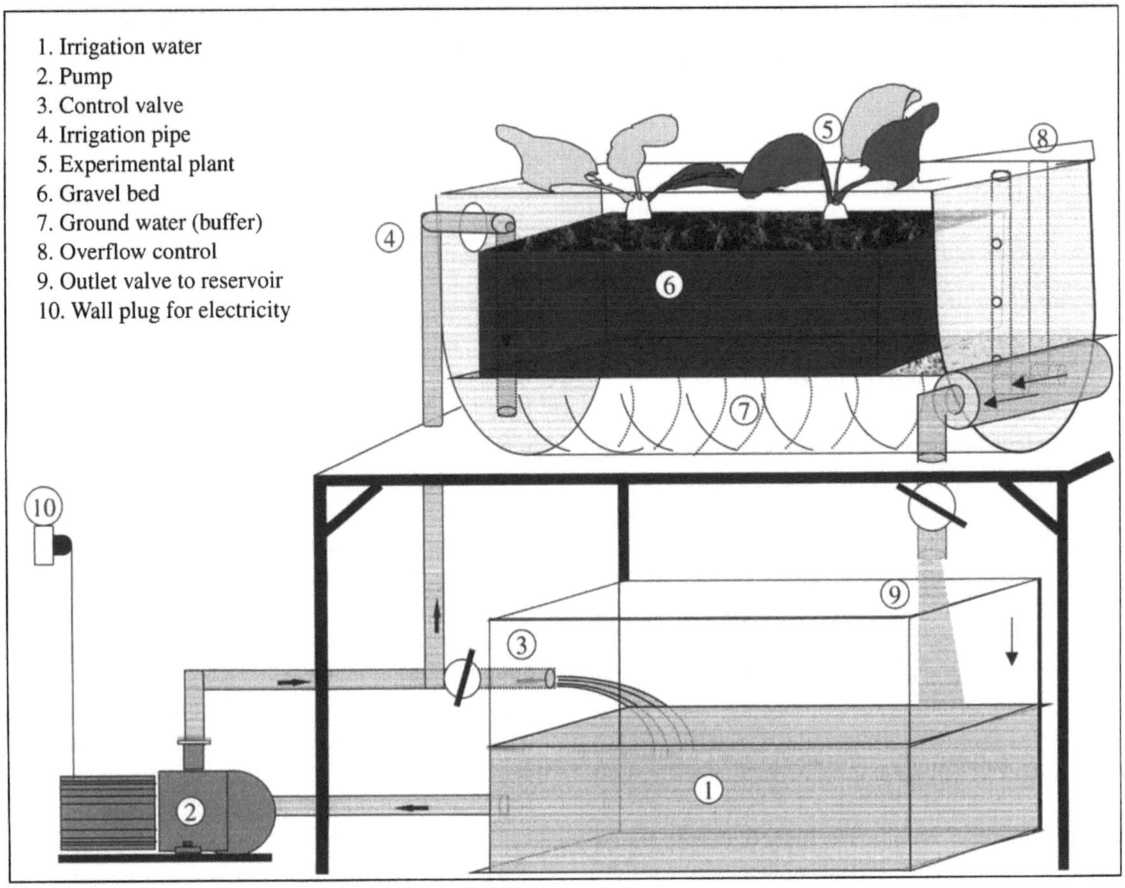

Fig. 1. Scheme of the quick check system. The arrows point to the flow direction of the seawater dilutions.

100% seawater (see Harrouni et al., 1999 for sand and water characteristics). These dilutions were obtained by tap water addition. Seawater was added progressively at the step of 1/8th every 2 days until the required concentration was reached. The dilutions were fertilised with a nutrient solution containing 60 mg/l N, 6.5 mg/l P, 40 mg/l K and traces of trace elements. The experiment with saline treatments started on 8 August and ended on 15 October, 1999.

The plants studied in this trial were: *Aster tripolium, Batis maritima, Halimione portulacoides, Kosteletzkya virginica, Salsola longifolia, Tamarix aphylla* and *Tamarix gallica*. Some plants were propagated by cuttings and others by seeds but all of them were regularly watered with fresh water until the beginning of the experiment. A hygro-thermometer was put in the greenhouse for the monitoring of temperature and air relative humidity.

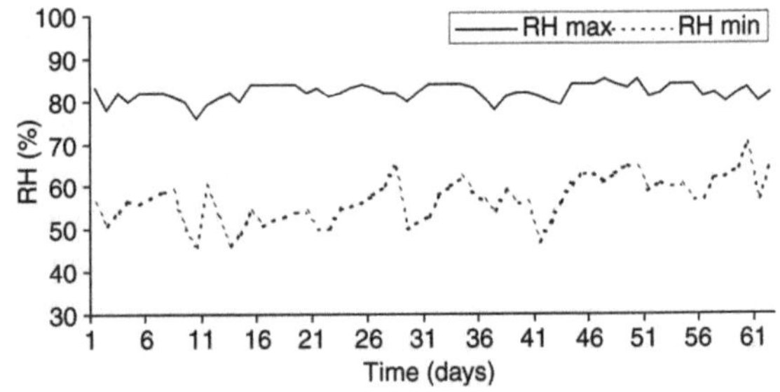

Fig. 2. Maximum and minimum air relative humidity in the greenhouse during the experimental period.

At the end of the experiment the plants were harvested for biomass and other parameters assessment. Root fresh weight (fw) was determined after washing the sand through a fine sieve. Refrigerated (0°C) water was used to avoid any water absorption by roots during

Table 1. Characteristics of tap water and seawater dilutions prior to fertiliser addition

Treatments	pH	EC (dS/m)	Na^+ (mg/l)	Cl^- (mg/l)	K^+ (mg/l)	Ca^{++} (mg/l)	Mg^{++} (mg/l)	$N\text{-}NO_3^-$ (mg/l)	P (mg/l)
Tap water	8.05	00.77	039.91	540.00	003.50	095.59	0011.20	Trace	1.35
25% seawater	7.70	12.10	3722.00	6452.37	116.67	098.53	0065.60	0.60	1.35
50% seawater	7.55	22.00	5201.80	8968.80	183.33	132.35	0984.00	0.30	1.25
75% seawater	7.64	30.80	8071.70	13453.20	194.44	141.18	1064.00	0.16	1.50
100% seawater	7.55	38.50	11211.0	18834.48	377.78	266.18	1976.00	0.21	2.05

Table 2. Plant survival (%) in the five treatments

Plant species	Tap water	25% seawater	50% seawater	75% seawater	100% seawater
A. tripolium	100	100	100	100	100
B. maritima	100	100	100	100	100
H. portulacoides	100	100	100	100	100
K. virginica	100	100	100	100	0
S. longifolia	100	100	100	100	100
T. aphylla	100	100	100	100	100
T. gallica	100	100	100	100	100

Table 3. RGR (gg-1 wk-1) of halophytes grown at five seawater dilutions and the ratio: RGR 25% seawater/RGR control

	0%	25%	50%	75%	100%	RGR 25 : RGR 0%
B. maritime	0.22	0.40	0.45	0.39	0.40	1.78
H. portulacoides	0.28	0.36	0.39	0.33	0.30	1.27
A. tripolium	0.36	0.29	0.25	0.27	0.17	0.79
S. longifolia	0.18	0.23	0.17	0.20	0.17	1.27
T. gallica	0.19	0.12	0.10	0.09	0.09	0.64
K. virginica	0.29	0.16	0.25	0.09	*	0.56
T. aphylla	0.09	0.01	0.04	0.00	0.01	0.14

*Dead plant.

Table 4. Root to shoot ratio of halophytes grown at five seawater dilutions

	0%	25%	50%	75%	100%
B. maritima	0.19	0.10	0.08	0.10	0.11
H. portulacoides	0.14	0.08	0.07	0.09	0.19
A. tripolium	0.39	0.40	0.57	0.42	0.61
T. gallica	0.18	0.27	0.25	0.40	0.41
T. aphylla	0.29	0.63	0.40	0.48	0.65
S. longifolia	0.21	0.13	0.11	0.13	0.08
K. virginica	1.75	2.33	1.74	1.05	*

*Dead plant.

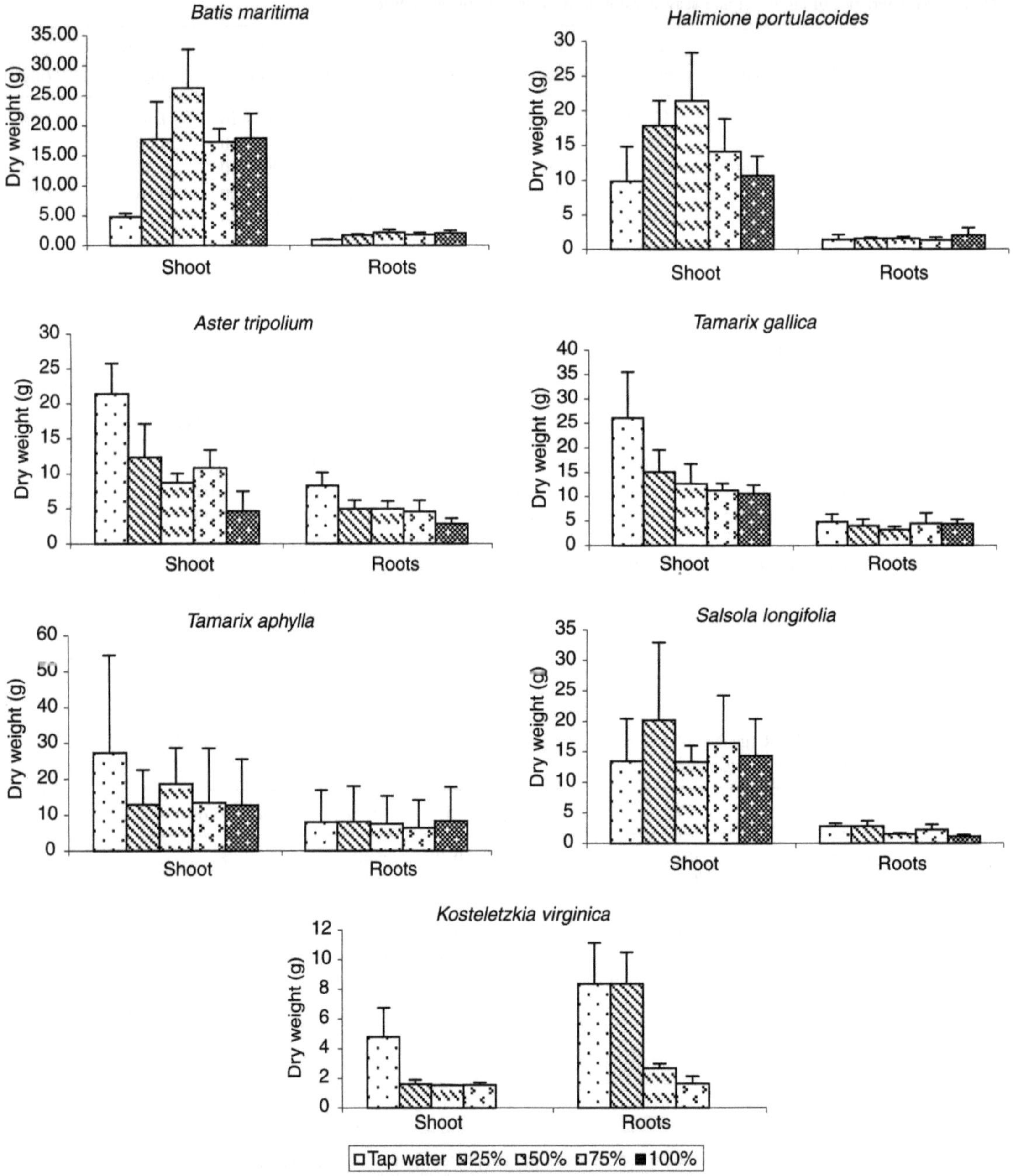

Fig. 3. Dry weight of plants grown in the five seawater treatments.

the procedure which was performed soon after harvest. Shoot and root dry weights (dw) were obtained after oven drying at 65–70°C for at least 48 hours. The water content (ml/g) was determined by the formula: (fw − dw)/dw. Growth was estimated by the relative growth rate calculated on the basis of the increase of dry matter over the experimental period using the formula: RGR = (Ln (final dw) − Ln (initinal dw)/time. Samples of each part of the plants were ground in an electric grinder and 500 mg samples were taken and burnt in a muffle furnace at 485°C. The ash content was obtained by weight difference before and after calcination. Mineral composition of shoots and roots for each species were determined from ash analyses. Na and K contents were determined by flame photometer 410 after dissolving the ash in 2N HCl. Ca and Mg contents were determined with an atomic absorption spectrometer 3110 (Perkin Elmer). Concentrations of these minerals in plant tissues are expressed on a dry weight basis. Tissue osmotic potential was determined in the leaf sap with an osmometer (Osmomat 030) based on the theory of freezing-point-depression. For sap extraction plant material was frozen in Ependorf tubes, thawed at 80°C and crushed using a metal rod. The tubes were centrifuged at 5,000 g for 10 min, the sap was then directly used for osmotic potential determination.

EC of sand was determined with the 1/5 soil/water extract method.

RESULTS

General Conditions

Average maximum and minimum temperatures during the experimental period were 41.5°C and 20.5°C respectively. The absolute maximum and minimum temperatures were 46°C and 16°C respectively. Air relative humidity inside the greenhouse fluctuated quite little during the experimental period (Figure 2). The average maximum RH was 82% with an absolute maximum of 85% and the average minimum RH was 57% with an absolute minimum of 46%. Maximum RH was steadier than minimum RH due to the fluctuation of temperature in daytime during the experimental period. However values of both max and min RH were within reasonable ranges.

Mineral analysis of the seawater dilutions used for irrigation in the different saline treatments in comparison with tap water showed that Cl, Na, Ca and Mg concentrations increased with the increase of seawater ratio. However, Cl had the highest values followed by Na, Mg, K and Ca respectively. Table 1 shows the composition of tap water and the different dilutions of seawater used in the experiment.

Plant Survival and Biomass Production

The plants studied in this experiment survived in all treatments except *K. virginica* which died in full strength seawater (Table 2).

Regarding relative growth rate (RGR), three groups of plants can be distinguished (Table 3). A group (*A. tripolium, K. virginica, T. gallica* and *T. aphylla*) where tap water irrigation enabled the plants to produce maximum growth compared with the saline treatments which reduced the growth rate. The RGR reduction in full strength seawater was more pronounced in *T. aphylla* (89%) than in *T. gallica* and *A. tripolium* (about 53%). *B. maritima* and *H. portulacoides* constitute the second group as RGR was higher in 50% seawater treatment. In these species, tap water irrigation resulted in a low RGR compared especially with 50% treatment. Growth of *S. longifolia* was enhanced in the 25% seawater treatment. Nevertheless RGR was not much affected in more saline treatments.

The ratio of growth in 25% seawater treatment to growth in tap water (Table 3) showed stimulation of growth at moderate salinity for three species. However *B. maritima* had highest value at 50% seawater (its optimum salinity). Among the 4 species which had the ratio below 1, *A. tripolium* showed the highest ratio (0.79); the ratio of *T. aphylla* and *K. virginica* was very low suggesting the sensitivity of these species to low salinity.

Increasing salinity in the root medium showed different responses concerning root/shoot ratio of plants studied (Table 4). However, the root/shoot ratio of *B. maritima* and *H. portulacoides* decreased in treatments 25% and 50% seawater and then started to increase at higher salinity, this result indicates the stimulation of shoot growth of these species at low salinity level and its reduction at high salinity. *A. tripolium, T. gallica* and *T. aphylla* showed an increase of the ratio with salinity suggesting that the shoots are more affected by increasing salinity than the roots. In the other species (*S. longifolia* and *K. virginica*), roots were more affected by increasing salinity than shoots since the root/shoot ratio decreased with salinity (Figure 3).

Mineral Content of Halophytes

The ash produced by the plants in the different treatments is shown in Figure 4. All species had their ash contents increased with the increase of the proportion of seawater in the irrigation solution and shoots accumulated more ash than roots. *B. maritima, S. longifolia, H. portulacoide* and *A. tripolium* accumulated more inorganic matter in their leaves than the other species. Indeed in full strength seawater *B. maritima and S. longifolia* had the highest values of ash in % dry weight (62% and 54%) followed by *H. portulacoide* (43%)

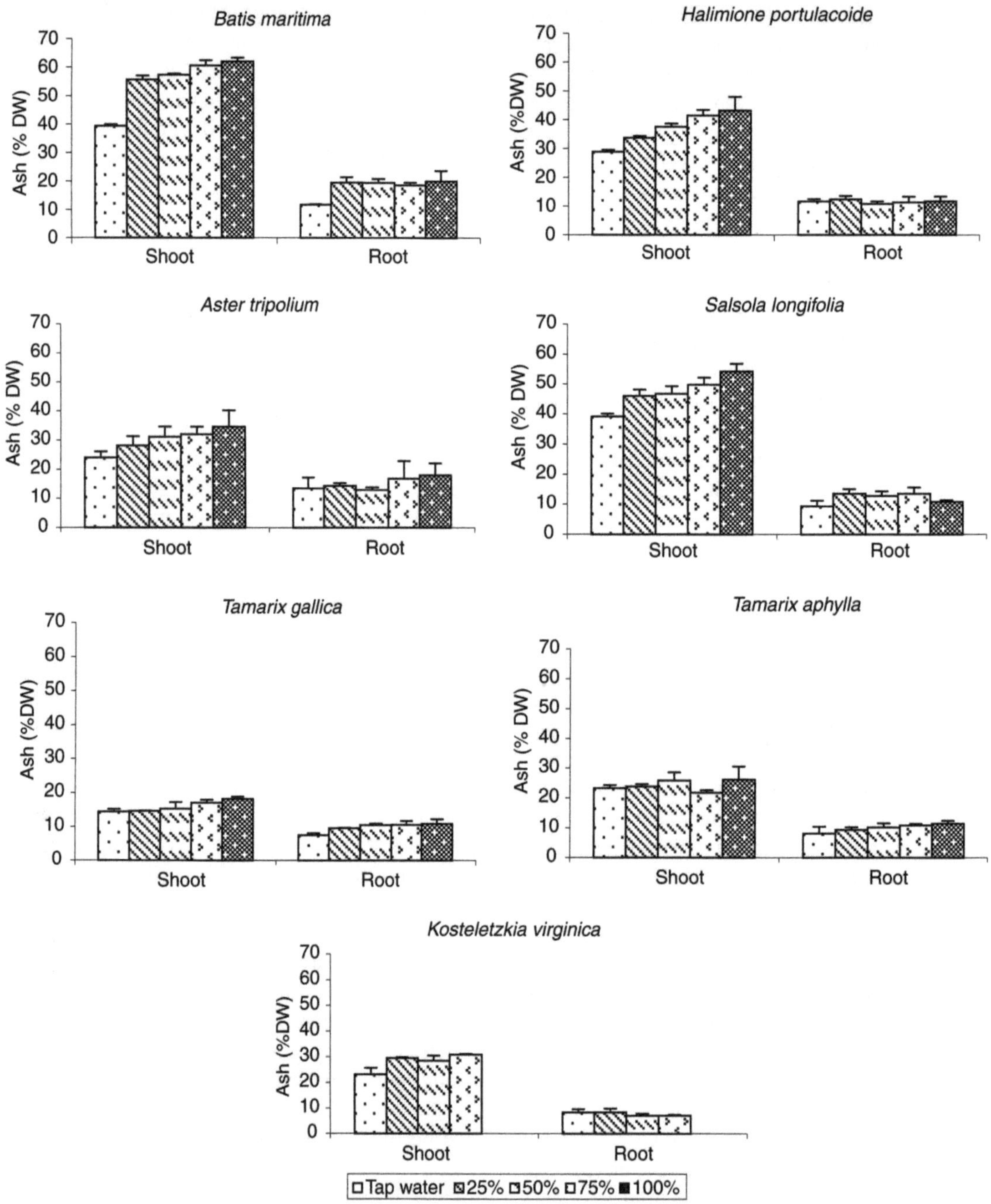

Fig. 4. Ash weight of plants grown in five sea water treatments.

A. tripolium (35%) *K. virginica* (31%) *T. aphylla* (26%) and *T. gallica* (18%). In the roots of most species the ash contents did not show any significant difference between treatments and it was about 11% in high salinity except for *A. tripolium* (18%) and *B. maritima* (20%).

Chemical analysis of the ash showed an accumulation of high amount of Na in the shoots with the increase of salinity in the growing medium, accompanied by a reduction in K content. However some species (*S. longifolia, H. portulacoides, B. maritima* and *A. tripolium*)

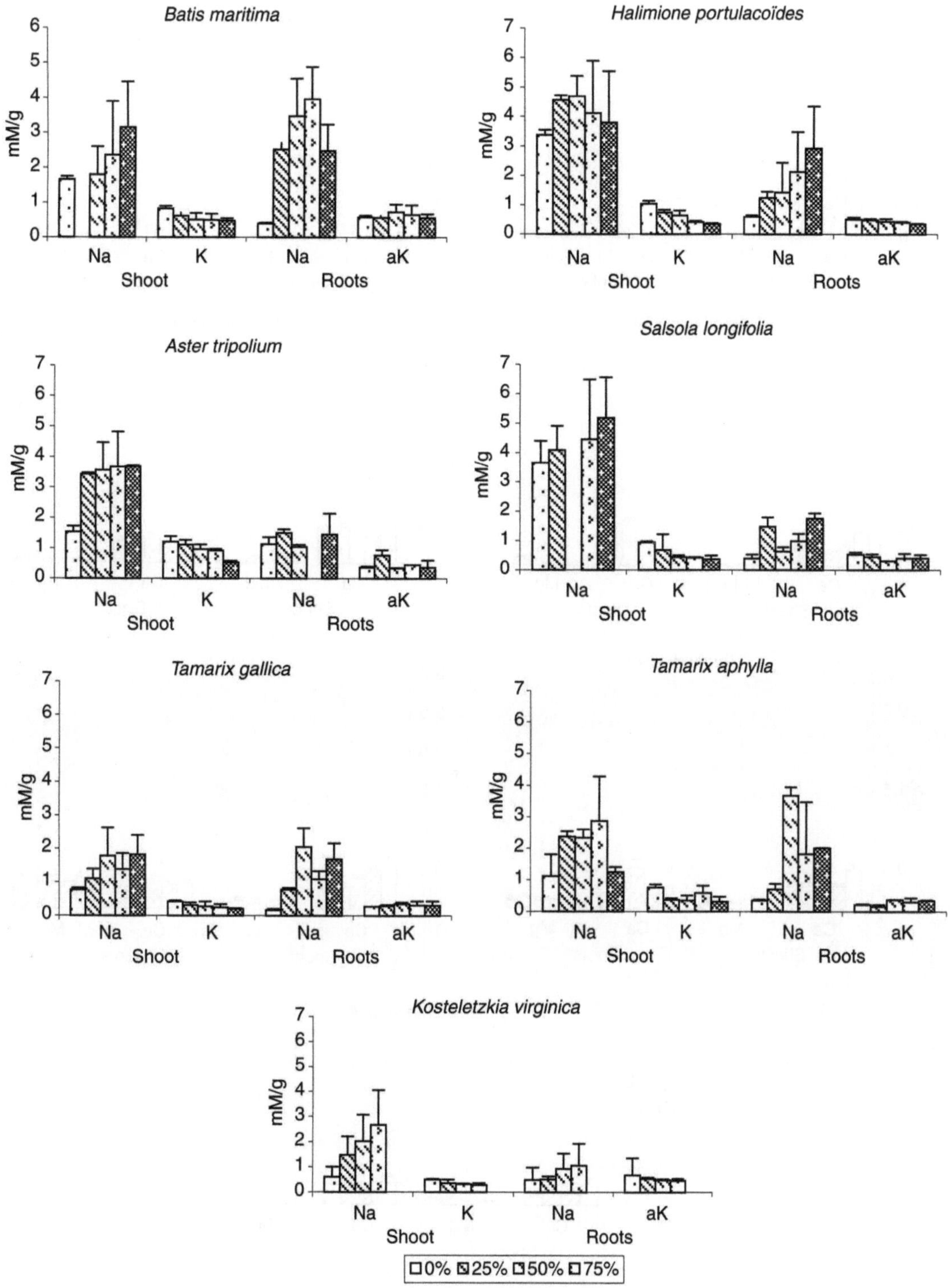

Fig. 5. Na and K content of shoots and roots of plants in five seawater treatments (tap water; 25%; 50%; 75% and 100% seawater).

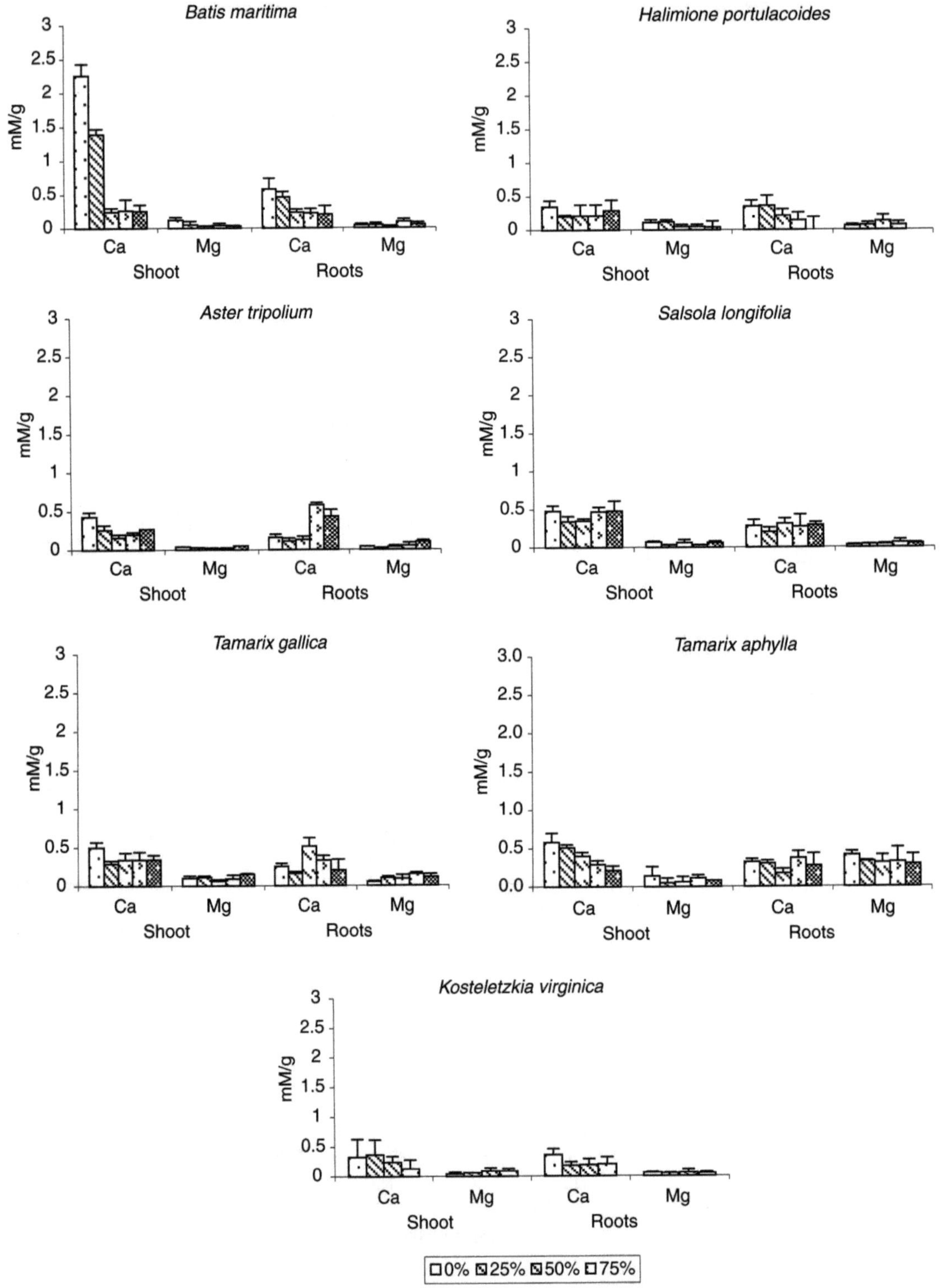

Fig. 6. Ca and Mg content of shoots and roots of plants in five seawater treatments (tap water; 25%; 50%; 75% and 100% seawater).

concentrated high amounts of Na in their shoots as compared to the others (Figure 5). In all species the K content was less than the Na content in the shoots as well as in the roots, and unlike Na its concentration decreased with the increase of salinity.

Ca and Mg were present at much lower levels than Na and K with the exception of *B. maritima* in which the shoot Ca concentration in the control was higher than the Na concentration. In the control and 25% seawater treatments shoots of *B. maritima* had the highest level as compared with the other species, it decreased with the increase of seawater concentration in the root medium, a response which was observed also in *T. aphylla* and *K. virginica*. In the other species the Ca concentration of shoot plants was similar in all saline treatments. In all species the Mg level was low in all parts of the plants compared with Ca and did not show any difference between treatments (Figure 6).

The osmolality of the cell sap of the leaves of *B. maritima, S. longifolia, H. portulacoides* and *A. tripolium* increased with the increase of seawater concentration suggesting a decrease of osmotic potential (due to ion accumulation in plant tissues) in response to the increase of salinity in the root medium (Figure 7).

This regulatory mechanism allows plants to maintain their turgidity by keeping a constant osmotic potential gradient between their shoot tissue and the external solution. Indeed, in these species the water content was high (Figure 8). In *B. maritima* the water content was higher in all saline treatments as compared with the control except in full strength seawater where it was hardly 18.52% lower. *Salsola* had its maximum water content in the treatment with 50% seawater. Plants in 25% and 75% seawater had respectively higher and equal water content as compared with the control. *Halimione* had its maximum water content in the control and 25% seawater treatments. *A. tripolium, T. aphylla* and *T. gallica* showed a decrease in the water content with the increase of seawater concentration.

DISCUSSION

In all species was growth as expressed by dry weight reduced in high seawater concentration treatments; moreover sensitive species showed a greatly reduced biomass. The relative growth rate was differently

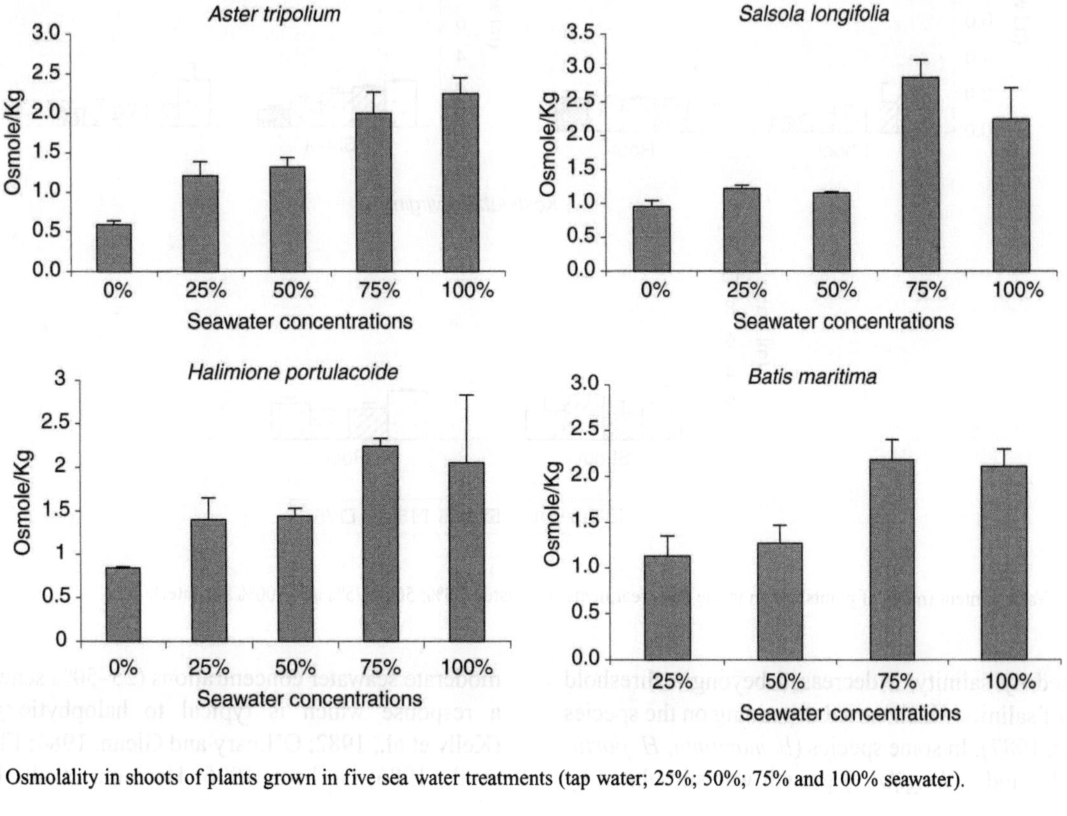

Fig. 7. Osmolality in shoots of plants grown in five sea water treatments (tap water; 25%; 50%; 75% and 100% seawater).

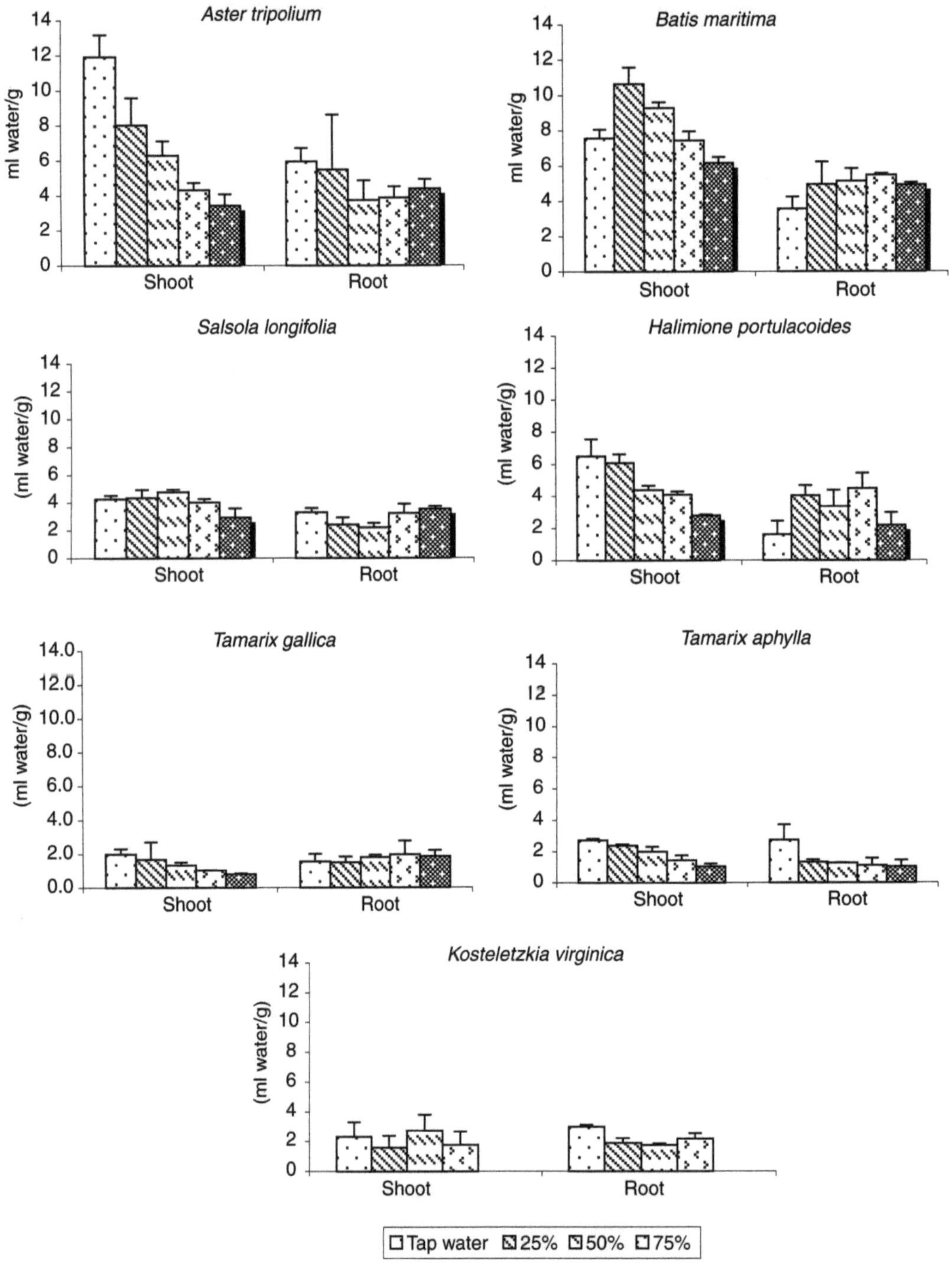

Fig. 8. Water content (ml/g) of plants grown in the five treatments (tap water; 25%; 50%; 75% and 100% seawater).

affected by salinity, it decreased beyong a threshold level of salinity which varied depending on the species (Maas, 1987). In some species (*B. maritima, H. portulacoides* and *S. longifolia*) growth was stimulated by moderate seawater concentrations (25–50% seawater), a response which is typical to halophytic plants (Kelly et al., 1982; O'Leary and Glenn, 1984; Flowers et al., 1986; Gorham, 1996; Harrouni et al., 1999).

This stimulation can be explained by the effect of Na on cell expansion and on water balance (Marschner, 1995). Some species (*A. tripolium, T. gallica, T. aphylla* and *K. virginica*) did not show any increase in growth in the presence of seawater. But in contrast to glycophytes, these species were able to complete their life cycle at high-salt concentrations. Salinity tolerance is related to the ability of different plants to generate osmotic adjustment by regulating their ion and water uptake from culture medium and to avoid toxicity of excess ion accumulation. The necessity to regulate the internal concentration of ions lies in the need to keep tissue water potentials more negative than those in the external medium, in order to maintain the symplastic water content while preventing too great an accumulation of ions, which would be metabolically damaging (Flowers et al., 1986). In species which are stimulated by low-salinity treatments, the increase in weight was largely due to the accumulation of minerals and water which resulted in the decrease of osmotic potential with the increase of seawater concentration (Flowers, 1985). In the highly saline treatments the decreased water content in the plants may be explained by the water stress induced by excess salts in the root zone. At high-salinity growth reduction might be caused by a reduced ability to adjust osmotically, as a result of the saturation of solute uptake (Munns et al., 1983), or nutrient deficiency and ion toxicity (Koyro, 1997; Koyro and Huchzermeyer, 1999). With respect to salt tolerance, plants showed considerable selectivity in ion accumulation. Sodium and chloride are the ions which are most commonly assimilated for osmotic adjustment depending on the species (Jefferies, 1981; Gorham, 1996; Harrouni et al., 1999; Koyro and Huchzermeyer, 1999). The decrease of K content with salinity can be explained by its replacement by Na in its function as osmoticum. Marschner (1995) pointed out the essentiality of the Na for some species in their mineral nutrition by its additional growth enhancement effect. The presence of high amount of Ca (ca. 96 to 266 mg/l, respectively in the control and full strength seawater treatments) in the growing medium may have contributed to the increase of salt tolerance of plants, since Ca plays an important role in maintaining membrane stability and controlling the selectivity of plant ion uptake and transport under saline conditions (Lauchli and Wieneke, 1979; Marschner, 1995). In this experiment's conditions increasing seawater concentration was paralleled with the increase of both Na and Ca, conditions which prevent Na induced Ca deficiency and enhance plant growth. However plant species considerably differed in their sensitivity to Na induced Ca deficiency since RGR of *B. maritima, H. portulacoides* and *S. longifolia* was not much affected by highly saline treatments compared with the other plants.

CONCLUSION

The results reproduce the typical responses of halophytes to saline treatments. What in general should be pointed out is the survival of most species even in full strength seawater and their production of substantial biomass. Salinity stress is composed of osmotic and ionic components, both of which could potentially affect plant performance. The species studied in this experiment are either fodder or fuel wood plants. Their ability to survive and produce biomass under reasonably high salinity shows their potential to be used for salt affected land reclamation with saline irrigation given that appropriate irrigation and drainage system is applied.

ACKNOWLEDGEMENTS

This work was carried out in the framework of the Concerted Action Project "Sustainable Utilisation of Halophytes in the Mediterranean and Subtropical Dry Regions," financed by the EU and co-ordinated by Prof. H. Lieth, University of Osnabrueck, Germany.

REFERENCES

Aronson, J. Plants for arid lands. Royal Botanic Gardens, Kew, 1986.

Choukr-Allah, R. and Harrouni, M.C. The potential use of halophytes under saline irrigation in Morocco. Symposium on the conservation of Mangal Ecosystems, 15–17 December 1996, Al-Ain, United Arab Emirates, 1996.

Clipson, N.J.W., Tomos, A.D., Flowers, T.J. and Wyn Jones, R.G. Salt tolerance in the halophyte Sueda maritima L. Dum. The maintenance of turgor pressure and water-potential gradients. Planta 1985; 165:392–6.

Flowers, T.J., Hajibagheri, M.A. and Clipson, N.J.W. Halophytes. Quaternary Rev Biol 1986; 61(3):313–37.

Flowers, T.J. Physiology of halophytes. Plant Soil 1985; 89: 41–56.

Gleen, E.P. and O'Leary, J. Relationship between salt accumulation and water content of dicotyledonous halophytes. Plant Cell Environ 1984; 7:253–61.

Goodin, J.R., McKell, C.M. and O'Leary, J.W. Introduction. In Saline agriculture, salt-tolerant plants for developing countries. National Academy Press. Washington, DC, 1990; 1–9.

Gorham, J. Mechanisms of salt tolerance of halophytes. In Choukr-Allah, R., Malcom, C.V. and Hamdy, A., editors. Halophytes and biosaline agriculture. Marcel Dekker, Inc. New York, 1996; 31–53.

Harrouni, M.C., Daoud S., El Alami, A., Debbagh B., Chokr-Allah, R. and Bengaddour, R. Responses of some halophytes to seawater irrigation in Morocco. In Hamdy, A., Lieth, H., Todorovic, M. and Moschenko, M., editors. Halophyte uses in different climates II. 1999; 57–75.

Jefferies, R.L. Osmotic adjustment and the response of halophytic plants to salinity. BioScience 1981; 31(1):42–6.

Kelly, D.B., Goodin, J.R. and Miller, D.R. Biology of Atriplex. In Sen, D.N. and Rajpurohit, K.S., editors. Tasks for vegetation, Sc., Vol. 2, 1982; Chap. 5.

Koyro, H.-W. Ultrastructural and physiological changes in root cells of Sorghum plants (*Sorghum bicolor* X *S. sudanensis* cv. Sweet Sioux) induced by NaCl. J Exp Bot 1997; 48:693–706.

Koyro, H.-W. and Huchzermeyer, B. Salt and drought stress effects on metabolic regulation in maize. In Pessarakli, M., editor. Handbook of plant and crop stress, 1999.

Koyro, H.-W and Huchzermeyer, B. Influence of high NaCl-Salinity on growth, water and osmotic relations of the halophyte *Beta vulgaris* ssp. *Maritima* – Development of a quick check. In Hamdy, A., Lieth, H., Todorovic, M. and Moschenko, M., editors. Halophyte uses in different climates I. 1999; 89–103.

Koyro, H.-W. Protocol for the installation of an improved quick check system in the I.A.V. Hassan II (Agadir, Morocco). Third annual report: Sustainable Halophyte Utilisation in the Mediterranean and Subtropical Dry regions, 1999.

Lauchli, A. and Wieneke, J. Studies on growth and distribution of Na, K and Cl in soybean varieties differing in salt tolerance. Z Pflanzenernahrung Bodenkunde 1979; 142:3–13.

Lieth, H. and Al Masoom, A.A., editors. Towards the rational use of high salinity tolerant plants, Vol. I of the Proceedings of the Al Ain conference 1990, T: VS Vol. 27. Kluwer Academic Publishers, Dordrecht, 1993; 521 pp.

Maas, E.V. Salt tolerance of plants. In Christie, B.R., editor. CRC handbook of plant sciences in agriculture, vol. II. CRC Press, Boca Raton, 1987; 57–57.

Marschner, H. Mineral nutrition of higher plants. Academic Press, London, 1995; 889 pp.

Munns, R., Greenway, and Kirst, G.O. Halotolerant eucaryotes. In Lange, O.L., Nobel, P.S., Osmond, C.B. and Ziegler, H., editors. Physiological plant ecology II. 1983; 59–135.

Encyclopedia of Plant Physiology, New Series 12 C. Springer-Verlag, Berlin.

O'Leary, J.W. and Glenn, E.P. New crop opportunities from halophytes. University of Arizona, Annual meeting, 24–29 May 1984, New York, 1984; 1–9.

O'Leary, J.W., Glenn, E.P. and Watson, M.C. Agricultural production of halophytes irrigated with seawater. Plant Soil 1985; 89:311–12.

O'Leary, J.W. A critical analysis of the use of Atriplex species as crop plants for irrigation with highly saline water. In Ahmed, R. and San Pietto A., editors. Prospects of biosaline research. Publ. Botany Dept., Karachi University, Pakistan, 1986; 415–32.

Some physiological and biochemical aspects of salt tolerance of *Sesuvium portulacastrum*

D. MESSEDI, N. SLEIMI AND C. ABDELLY

INTRODUCTION

Drought and salinity are responsible for substantial losses of crop yield, deterioration of plant cover and erosion of soils. Currently, desertification threatens an important part of our planet. Indeed, according to the United Nations, approximately one sixth of the world's population, 70% of all dry lands with a total area of 3.6 billion ha, and a quarter of the total land surface of the earth are endangered by desertification (Jeltsch and Wiegand, 1999). In these regions, soils are often sandy limiting plant development by salinity, drought and mineral deficiencies. The specific plant associations of these zones are essentially psammo-halophyte species. Among these stress tolerant plants, some species could be interesting for soil protection and fixation as well as the rehabilitation of marginal zones. *Limonium delicatulum*, capable to use water efficiently by reducing its water expenditure despite the high-evaporative power of the atmosphere, is used to fix the Mediterranean coastal dunes of Egypt (Batanouny et al., 1992). Other species like *Cymodocea ciliata, Schangenia aegyptiaca, Zygophyllum simplex* and *Suaeda vermiculatas* are very frequent in the UAE coasts (Zahran et al., 1999). In the coastal zones, the fluctuation of water and salinity levels influences strongly plants distribution. Indeed, *Spartina alterniflora* and more particularly *S. patens* dominate in high zone over wash and provide the front line of protection against erosion processes (Courtemanche et al., 1999). *Cakile maritima*, a psammo-halophile species, is the first that colonizes the beach at locations where the lower horizons are humid and salty. Whereas, pioneer species like *Ammophila arenaria* colonize the first mounds of sand (Mezziani, 1984). In addition of its ecological value, *C. maritima* produces oleaginous seeds (El Almi et al., 1999).

All these studies show that halophytes or psammo-halophiles species are potentially interesting for dunes fixation, greenification of salted marsh and valorisation of uncultivated marginal zones. As part of this approach, we have retained a psammophile species, *Sesuvium portulacastrum*. This plant, which belongs to the Aizoaceae family, is a perennial dicotyledonous halophyte, very frequent in the salty marshes of Latin America (Chapman, 1960). This creeping plant seems to present a higher vegetative multiplication potentialities. This property confers on this plant a higher ability to protect and to fix soils (Lonard and Judd, 1997). The authors also showed that *S. portulacastrum* is capable of tolerating salinity, higher temperatures and mineral deficiencies. Some studies recommend utilisation of this halophyte particularly in landscaping (Pasternak and Nerd, 1996). Flowers are often pink purplish and occasionally white (Lonard and Judd, 1997). Like all Aizoaceae plants, *S. portulacastrum* could be used for ornamentation activities.

The general objective of the present work which is included in screening programs for cash crop halophytes, was to determine limits of salt tolerance of *S. portulacastrum* and to identify some physiological and biochemical criteria involved in its higher capacity to tolerance salinity.

METHODS

Young plants of *S. portulacastrum* were obtained by cuttings propagation. Stems of 3 cm length with one node and two opposite leaves were taken from mother plants, disinfected for 5 min in saturated hypochloric calcium solution, rinsed abundantly with distilled water, and then rooted in a sandy soil enriched with organic matter and irrigated daily with tap water

H. Lieth (ed.), Cash Crop Halophytes: Recent Studies, 71–77.

during 7 days and a nutrient solution (Hewitt, 1966) diluted 10 times during 6 days. Rhizogenesis took place after one week. Thereafter, seedlings were transferred to inert sand and divided into eight lots of 10 plants each. They were irrigated with a complete nutrient solution (Hewitt, 1966) added with increasing NaCl concentrations: 0, 100, 200, 300, 400, 600, 800 and 1000 mM. The nutrient solution also contained iron as complex EDTA-K-Fe (Jacobson, 1951) and micronutrients as a mixture of salts: $MnCl_2$; $CuSO_4$, $5H_2O$, $ZnSO_4$, $7H_2O$; $Mo_7O_{24}(NH_4)_6$, $4H_2O$ and H_3BO_3 (Arnon and Hoagland, 1940).

The experiments were carried out in a greenhouse with a 14 h photoperiod. Mean temperature and relative humidity were 30 ± 5°C, 55 ± 5% respectively during the day and 16 ± 2°C, 90 ± 5% at night.

Two harvests were made: the first one at the beginning of the treatment (initial harvest) and the second one after 60 days of salt stress (final harvest). At harvesting, plants were divided into leaves, stems and roots. The measured parameters were fresh and dry matter production, number of leaves, and the contents of water, sodium, chlorides, potassium, calcium, nitrogen and proline in the tissues. Proline was extracted and estimated by the method suggested by Bates et al. (1973) except that the dry matter (and not the fresh material) was used for the extraction of this amino acid.

RESULTS

Growth

Changes of biomass production, dry matter distribution between roots and aerial parts and number of leaves with medium salinity are shown in Figure 1. *S. portulacastrum* grew at all NaCl concentrations (Figure 1a). Dry matter production was highest at a substrate salinity between 100 and 200 mM of NaCl. At these NaCl concentrations, biomass production was twofold with respect to the control plants. *S. portulacastrum* is capable to express a growth activity even in presence of excessive NaCl concentrations of 1000 mM (58.5 g L^{-1}). This dose of salt considerably reduces growth but does not lead to the development of toxicity or mineral deficiency symptoms. Changes of the number of leaves with substrate salinity are similar to those of the whole dry weight. They also show that in condition of moderate saline concentrations, between 0 and 200 mM, NaCl stimulate formation and expansion of young leaves (Figure 1b).

The depressive effect of salt appears only at NaCl concentration exceeding 300 mM. Therefore, the halophile character of this plant is clearly illustrated by dry matter production and morphogenesis variations with medium salinity.

Salt effect on modifications of the dry matter distribution between the aerial parts and roots were evaluated by roots DW/shoots DW ratio variations.

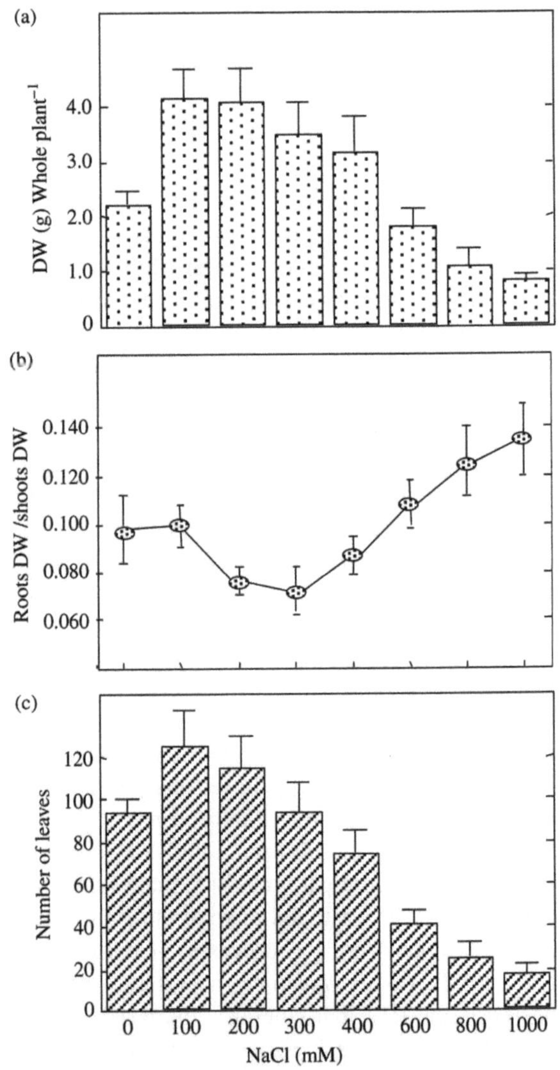

Fig. 1. Changes in *S. portulacastrum* growth with salinity of culture medium. (a) salt effect on whole plant dry matter production; (b) roots/shoots ratio; (c) number of leaves. Parameters of growth were measured after 60 days of salt stress. Means of 10 repetitions. Bars indicate ± standard error ($p = 0.05$).

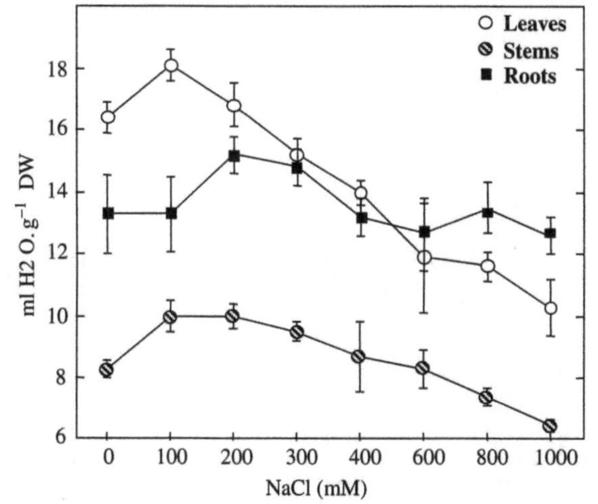

Fig. 2. Changes in shoots, stems and roots water contents with salinity of culture medium. Means of 10 repetitions. Bars indicate ± standard error ($p = 0.05$).

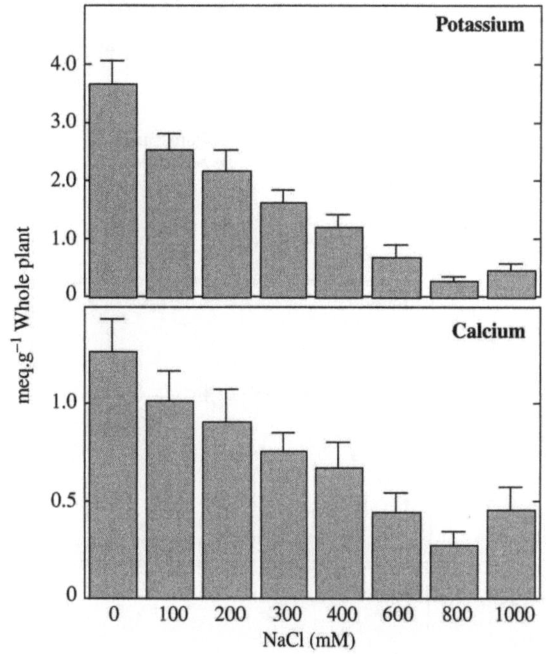

Fig. 3. Changes in potassium and calcium quantities absorbed during treatment with salinity of culture medium. Means of 10 repetitions. Bars indicate ± standard error ($p = 0.05$).

At NaCl concentrations exceeding 300 mM, salt led to a preferential development of roots as shown by the increase of the roots/shoots ratio (Figure 1c). This behaviour is often observed in plants submitted to mineral and/or water deficiencies, suggesting that salt reduces plant supply of some indispensable nutrients to the plant, under condition of excessive salinity of culture medium.

Tissues Water Contents

Salt improved the hydration of the tissues when it was added to the nutrient solution at moderate doses, 100–200 mM NaCl (Figure 2). Changes in tissues water content are similar to those of biomass production. This result suggests a close relationship between plant growth and the capacity to ensure an appropriate supplying of the tissues with water.

Mineral Composition of the Tissues

Potassium and sodium

The cultivation of plants in the presence of salt strongly reduces K^+ absorption as is shown by the decrease of potassium quantities measured in plants submitted to salt (Figure 3). Nevertheless, plant growth was either stimulated or less affected than K^+ absorption. A considerable impoverishment of tissues in K^+ results particularly in the aerial parts. In the roots, K^+ content was less influenced by salinity of culture medium.

Sodium was accumulated especially in the shoots. Leaves constitute a preferential site for Na^+ accumulation. However, high Na^+ content in the leaves was observed in plants cultivated on 200 mM NaCl and did not increase at the above concentrations (Figure 4). This species seems able to regulate Na^+ tissues concentration through a higher dry matter production.

Calcium

Like K^+, Ca^{2+} absorption was systematically inhibited by salt leading to a decrease of tissues content of this nutrient. This impoverishment of Ca^{2+} affects exclusively the leaves that represent an important part of plants (Figure 5). This situation could be caused by a competition between Na^+ and the other cations at the level of absorption sites and by their substitution by Na^+ in the xylem vessels. Decrease of the sap flux induced by salt seems less implied in the restriction of the leaves Ca^{2+} alimentation. Indeed, we noticed that water nutrition of leaves was assured appropriately even in conditions of excessive salinity.

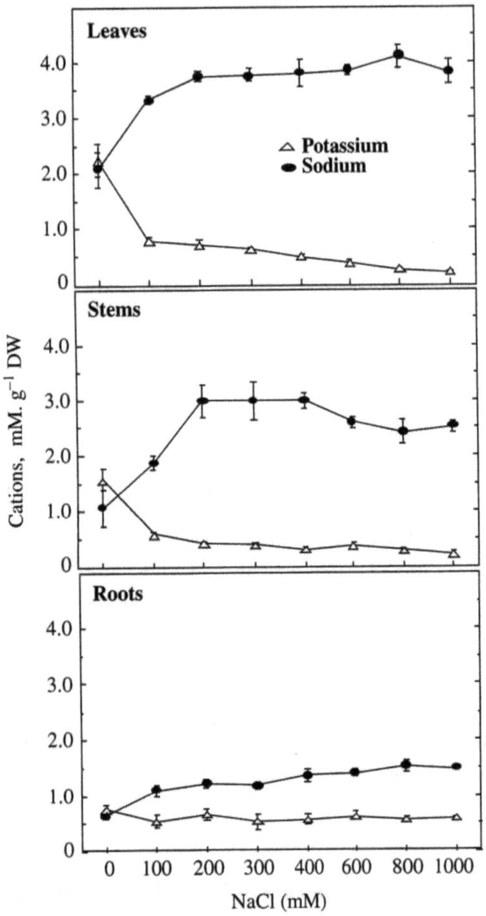

Fig. 4. Changes in leaves, stems and roots Na^+ and K^+ concentrations with salinity of culture medium. Means of 10 repetitions. Bars indicate ± standard error ($p = 0.05$).

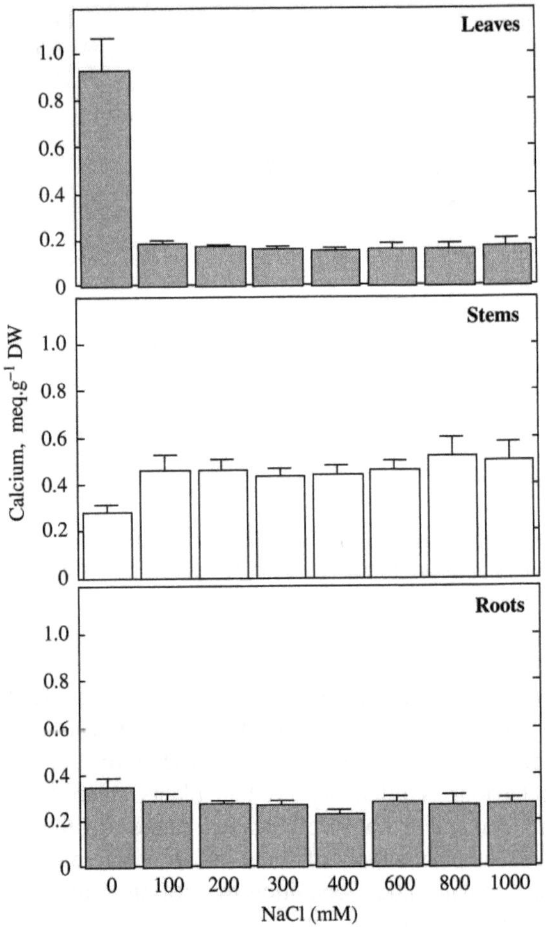

Fig. 5. Salt effects on leaves, stems and roots Ca^{2+} content with medium salinity. Means of 10 repetitions. Bars indicate ± standard error ($p = 0.05$).

Nitrogen tissues content

Cultivation of plants in presence of salt led to an impoverishment of the leaves in reduced nitrogen. This phenomenon appears at a low NaCl concentration (100 mM) and becomes more pronounced when salinity in the nutrient solution increases. Thus, in plants cultivated on 1,000 mM NaCl, reduced nitrogen concentration represents 60% of that of the control ones, grown in salt free medium (Figure 6). In stems and roots, nitrogen content did not depend on medium salinity. As leaves represent the dominant part of plants, it seems that the decrease of nitrogen tissues content indicates a higher salt sensitivity of nitrogen assimilation with respect to growth. It is quite certain that in this species nitrate assimilation takes place in all of its green parts, but not in the roots. Thus, decrease of nitrate assimilation would result from inhibition of photosynthesis and increase of maintenance respiration in plants submitted to salt (Gale and Zeroni, 1985). It seems that in other species, the nitrate uptake by the roots is reduced when photosynthetic products become limited under salt stress; i.e. when only few carbon is available as acceptor for nitrogen (Bouma and De Visser, 1993). As a consequence, this leads to less amino-nitrogen.

Proline accumulation

In control plants, proline was accumulated more in the aerial parts (particularly in the stems) than in the roots. The salinity of the nutrient solution led to an

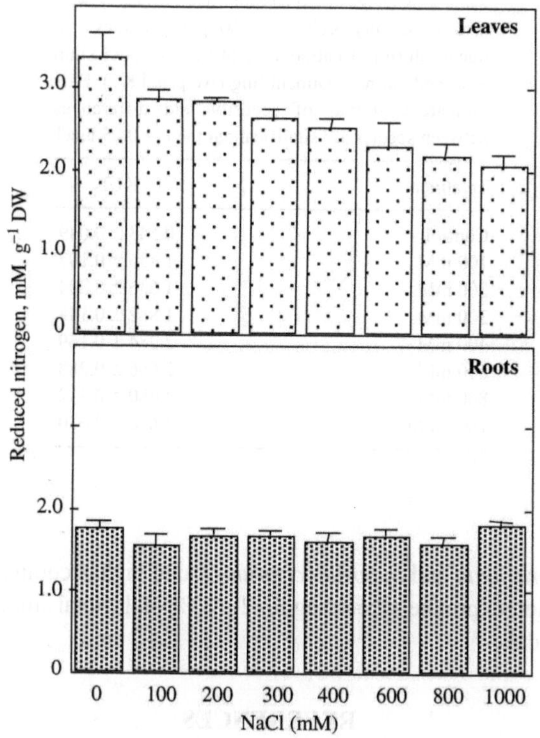

Fig. 6. Salt effect on leaves and roots nitrogen contents. Means of 10 repetitions. Bars indicate ± interval of security and differences between species are significant at the $p = 0.05$ level.

Table 1. Effect of salt on proline tissues contents (μmol g^{-1} DW). Means of 10 repetitions. Bars indicate ± interval of security and differences between species are significant at the $p = 0.05$ level

Treatments	Leaves	Stems	Roots
Control	080.1 ± 15.3	110.2 ± 14.7	045.4 ± 7.7
100 mM	128.7 ± 37.2	202.2 ± 30.1	118.3 ± 26.3
200 mM	320.5 ± 27.5	250.6 ± 32.0	131.8 ± 26.4
300 mM	337.6 ± 89.5	290.0 ± 30.4	160.7 ± 20.4
400 mM	364.6 ± 66.6	324.5 ± 43.8	228.1 ± 31.8
600 mM	379.2 ± 59.1	338.1 ± 48.9	268.8 ± 29.4
800 mM	471.6 ± 83.4	184.9 ± 13.5	254.3 ± 11.7
1000 mM	415.2 ± 35.5	248.9 ± 50.8	375.5 ± 38.8

increase of the content of this compatible compound (Table 1). Proline cytosolic concentration, evaluated by supposing that cytoplasm volume represents 5% of total cell volume (Flowers and Yeo, 1986) increased regularly with salinity of culture medium and reached levels of the vacuolar ionic concentration estimated by 2 ($Na^+ + K^+$).

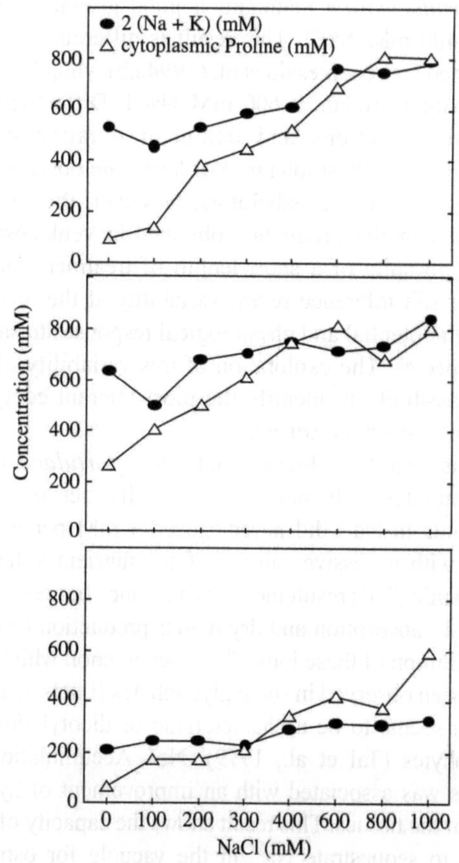

Fig. 7. Changes in cytoplasmic proline concentration with salinity of culture. Proline concentration was evaluated by supposing that cytoplasm volume represents 5% of total cell volume.

These results obtained in the different parts of the plant illustrate the osmotic balance between the two compartments of the cell. The vacuole accumulates mineral ions (particularly Na^+) while the cytoplasm accumulates proline (Figure 7). Our results suggest that proline is the only compatible compound implied in the osmotic balance between the vacuole and the cytoplasm.

CONCLUSION

Our study confirms the halophile character of *S. portulacastrum*: this species requires salt to express its maximal growth potential. The saline concentration ensuring an optimal dry matter production is situated between 100 and 200 mM NaCl. Nevertheless,

S. portulacastrum maintains a good growth activity until 400 mM NaCl. This result is different to those obtained by Venkatesalu et al. (1994a,b) who observed a maximal growth at 600 mM NaCl. Differences in culture conditions and treatment duration could explain the highest tolerance of *Sesuvium* observed by Venkatesalu and collaborators. However, the significant dry matter production obtained in Venkatesalu's study, in spite of a short length of treatment, and a higher salt tolerance reveal variability at the level of growth potential and physiological responses to salt in this species. The exploration of this variability offers the possibility to identify the most tolerant ecotypes within *S. portulacastrum*.

This study shows that *S. portulacastrum* accumulates salt particularly in its aerial parts. Contents in Na^+ did never exceed 4 mM per g DW, even with excessive salinity of the nutrient solution, 1,000 mM. This result indicates a balance between Na^+ and Cl^- absorption and dry matter production leading to a dilution of these ions. This phenomenon which has also been observed in some glycophytes (Lachaal et al., 1996) seems to be a characteristic of dicotyledonous halophytes (Tal et al., 1979). Na^+ Accumulation in leaves was associated with an improvement of hydration of the tissues. This result shows the capacity of this plant to sequestrate Na^+ in the vacuole for osmotic adjustment. The absence of necrosis (specific symptoms of apoplasmic Na^+ accumulation) confirms the efficiency of the compartmentalisation systems. Therefore, it appears that *S. portulacastrum* adopts two mechanisms to protect its tissues against a toxic Na^+ and Cl^- accumulation, a high growth rate combined with an efficient salt sequestration. The aptitude of plants to sequestrate Na^+ in the vacuole instead of potassium explains the increase of the K^+ use efficiency in plants grown in presence of salt (Table 2). However, the inhibition of K^+ and Ca^{2+} uptake by salt seems to limit plant growth under conditions of a highly salted medium. Our results suggest that the restriction of the NO_3^- uptake and transport towards the aerial parts is also responsible for the growth inhibition under excessive saline conditions.

Finally, our results show that proline is the only compatible compound implied in the osmotic balance between the cytoplasm and the vacuole. This result is in agreement with that of Heun et al. (1981) who observed a high proline accumulation in plants of the Aizoaceae family like *Mesembryanthemum cristallinum*. Joshi (1982) while studying *S. nudiflora, Salicornia brachiata* and *S. portulacastrum*, found high accumulation of proline accounting for 20–83% of the total amino acids.

Table 2. Changes of potassium use efficiency with medium salinity. KUE = Dry weight quantities produced during treatment/quantities of potassium absorbed during treatment (mg DW/µmol K^+). Bars indicate ± interval of security and differences between species are significant at the $p = 0.05$ level

Treatment	KUE
Control	0.543 ± 0.049
100 mM	1.435 ± 0.107
200 mM	1.636 ± 0.204
300 mM	1.882 ± 0.148
400 mM	2.284 ± 0.169
600 mM	2.666 ± 0.288
800 mM	3.480 ± 0.312
1000 mM	4.616 ± 0.630

REFERENCES

Arnon, D.I. and Hoagland, D.R. Crop production in artificial solutions and in soils with special reference to factors affecting yields and absorption of inorganic nutrients. Soil Sci 1940; 50:463–84.

Batanouny, K.H., Hassen, A.H. and Fahmi, G.M. Ecophysiological studies on Halophytes in arid and semi-arid zones. Flora 1992; 186:105–16.

Bates, L.S., Waldren, R.P. and Teare, I.D. Rapid determination of the free proline in water stress studies. Plant Soil 1973; 39:205–8.

Bouma, T.J. and De Visser, R. Energy requirements for maintenance of ion concentrations in roots. Physiol Plant 1993; 89:113–42

Chapman, V.J. Salt Marshes and salt deserts of the world. Interscience Publishers, INC., New York. 1960; 263–4

Courtemanche, R.P. Jr., Hester, M.W. and Mendelssohn, I.A. Recovery of a Louisiana barrier island marsh plant community following extensive Hurricane-induced over wash. J Coastal Res 1999; 15:872–83.

EL Almi H., Zarrouk M., Abdelly, C., Smaoui, A., Ben Miled, D. and Chérif, A. Composition lipidique des graines de quelques espèces halophytiques locales. 10èmes Journées Biologiques de l'A.T.S.B., 20–22 Mars 1999, Monastir, Tunisie, 1999; 12.

Flowers, T.J. and Yeo, A.R. Ion relation of plants under drought and salinity. Aust J Plant Physiol 1986; 13:75–91.

Gale J. and Zeroni, M. The cost to plants of different strategies of adaptation to stress and the alleviation of stress by increasing assimilation. Plant Soil 1985; 89:57–67.

Heun, A.M., Gorham, J., Lunge, U. and Jones, R.G. Changes of water-relation characteristics and levels of organic cytoplasmic solute during salinity induced transition of

Mesembryanthemum cristallinum from C3 photosynthesis to Crassulacean acid metabolism. Oecalogia 1981; 50:66–72

Hewitt, E.J. Sand and water culture methods used in the study of plant nutrition. 2nd edn. Commonwealth Bureau of Horticulture tech. Com. 22., 1966.

Jacobson, L. Maintenance of iron supply in nutrient solution by a single addition of ferric-potassium-ehylene-diamine-tetra acetate. Plant Physiol 1951; 26:411–13.

Jeltsch, F. and Wiegand, T. Long-term dynamics in arid and semiarid ecosystems. In Farina, A., editor. Perspectives in ecology. Backhuys Publishers, Leiden, 1999; 79–86.

Joshi, G.V., Jamale, B.B. and Bhosale, L.J. Ion regulation in mangroves. Walsh G.E., Snedaker, S.C. and Teas, H.J., editors. Int. Food Agric. Sci. Univ. Florida, Gainesville, 1975; 595–607.

Lachaâl, M., Abdelly C., Grignon, C., Soltani, A. and Hajji, M. Variation de la sensibilité au sel en fonction du stade de développement chez la lentille (*Lens culinaris*). Agronomie 1996; 16:381–90.

Lonard, R.I. and Judd, F.W. The biological flora of coastal dunes and wetlands. *Sesuvium portulacastrum* (L.) L. J Costal Res 1997; 13:96–104.

Mezziani, K. Phytoecological study and proposals for the management of coastal dunes of the Mostaganem area (Algeria). Thèse Doct. Ing., 1984; 213.

Pasternak, and Nerd. Research and utilization of halophytes in Israel. In Choukr-Allah, R., Malcolm, C.V. and Hamdy, A., editors. Marcel Dekker, Inc, Nex York, 1996; 325–48.

Tal, M., Rosental, I., Abramovitz, R. and Forti, M. Salt tolerance in *Simmondsia chinensis*: Water balance and accumulation of chloride, sodium and proline under low and high salinity, Ann Bot 1979; 43:701–8.

Venkatesalu, V., Raj Kumar, R. and Chellappan, K.P. Growth and mineral distribution of *Sesuvium portulacastrum* L., a salt marsh Halophyte, under sodium chloride stress. Commun Soil Sci Plant Anal 1994a; 25:2797–805.

Venkatesalu, V., Raj Kumar, R. and Chellappan, K.P. Sodium chloride stress on organic constituents of *Sesuvium portulacastrum* L. a salt marsh halophyte, J Plant Nutr 1994b; 17:1635–45.

Zahran, M.A., Al Ansari, F.M. and Abbes, J.A. The ecology of Al-Samaliah Island, Estuarine, Coastal and Shelf Science. Biol Coastal Environ 1999; 49:11–19.

Salt-tolerance strategy of two halophyte species: *Spartina alterniflora* and *Suaeda fruticosa*

N. SLEIMI AND C. ABDELLY

INTRODUCTION

After their absorption from the soil, ions can be accumulated in the root cortex or transported into the xylem, which ensures their transport towards the aerial parts. In the latter, ions follow the water flow up to the transpiring leaves. There, they may be absorbed by leaf cells, or re-exported through the phloem towards non-transpiring organs (Pitman, 1975). Among these ions, Na^+ represents a particular case, at least for two reasons. First, it may be present at very high concentrations in the soil solution (a soil judged moderately salty, or a water of irrigation considered as fairly brackish contain NaCl 25–50 mM. Second, all electrophysiological studies confirm the existence of active exclusion of Na^+ from cytoplasm, by two types of ATPases localised on the plasma membrane and on the tonoplast (Jeschke and Nassery, 1981).

Biochemical studies have shown that protection of cytosolic enzymes against excess Na^+ concentration is necessary for both animals and plants. This is achieved by excretion of cytosolic Na^+ in the vacuole and/or out of the cell (Kramer, 1983). In leaves, the apoplast volume is very small as compared to the cell volume, and does not exceed 0.01 ml g^{-1} FW (Smith and Fox, 1975; Zid, 1983). Therefore, it is indispensable that an exact balance would be maintained between (i) Na^+ import by xylemic flow, (ii) Na^+ absorption by leaf cells associated to Na^+ sequestration into vacuoles, or (iii) Na^+ recycling by the phloem or secretion out of leaf epidermis (Greenway and Munns, 1980).

According to the destiny of the absorbed Na^+, halophytes adopt different strategies to tolerate salt. The dicotyledonous species are capable to compartmentalise Na^+ in their vacuoles (osmotic regulation), or to export it in salt glands at their leaf surface. In the latter species, the included strategy is efficient because it is associated with a high-growth potentialities and sequestration of inorganic ions in the vacuole. Because of toxic effects of higher salinity in the cytoplasm several organic solutes with low-molecular weight are accumulated in this compartment in order to maintain an osmotic balance between this compartment and vacuole (Pollard and Wyn Jones, 1979). The second strategy is adopted by the excluder plants. These plants, which exclude NaCl from their tissues, often display low growth potentialities (Flowers et al., 1977; Greenway and Munns, 1980; Wyn Jones, 1981; Yeo and Flowers, 1983).

In the present study, physiological responses to salt are studied in two fodder halophytes species. *Suaeda fruticosa*, an indigenous Chenopodiaceae in Tunisia, is quite frequent in semi-arid, arid and desertic bioclimatic stages and well appreciated by livestock (Le Houérou, 1996). *Spartina alterniflora*, a Poaceae, is dominant in saline marsh and coastal regions in the East of U.S.A. The physiological responses of these species to NaCl were analysed in the framework of the two salt tolerance strategies described above.

METHODS

Young plants of *S. fruticosa* and *S. alterniflora* were obtained by cuttings propagation. For the first species, stems of 5 cm length with leaves were taken from mother plants at their natural habitat (Soliman Sebkha, near Tunis), sterilised for 5 min in saturated hypochloric calcium solution, rinsed abundantly with distilled water, and then rooted in 1/5 nutritive solution (Hewitt, 1966) and supplied with 35 mM NaCl. Rhizogenesis took place after one month. For *S. alterniflora*, seedling were obtained in the same conditions except that the rooting of rhizome leafed

H. Lieth (ed.), Cash Crop Halophytes: Recent Studies, 79–85.

units (5–10 cm) was achieved in inert sand irrigated with nutrient solution. Rhizogenesis occurred after 45 ± days.

Experiments were carried out in a greenhouse with a 14 h photoperiod. Mean temperature and relative humidity were respectively 25 ± 5°C, 65 ± 5% (day) and 16 ± 5°C, 85 ± 5% (night). After a pre-treatment phase (7 days), platelets were distributed in five lots of 10 plants. They were irrigated with nutritive solution (Hewitt, 1966) added with 0, 100, 300, 400 or 800 mM NaCl. The medium contained also iron as complex EDTA-K-Fe (Jacobson, 1951) and micronutrients as mixture of salts: $MnCl_2$; $CuSO_4$, $5H_2O$, $ZnSO_4$, $7H_2O$; $Mo_7O_{24}(NH_4)_6$, $4H_2O$ and H_3BO_3 (Arnon and Hoagland, 1940).

Two harvests were made: the first one in the beginning of treatment and the second one when clear effects of the treatments were observable: after 45 days for *S. fruticosa* and 100 days for *S. alterniflora*. The plants were divided into shoots and roots. The measured parameters were dry matter production, and tissue content in water, sodium, chlorides, potassium, calcium, magnesium, nitrogen and inorganic phosphorus. Na^+, K^+, Ca^{2+} and Mg^{2+} were determined with an atomic absorption spectrophotometer (Instrumentation Laboratory 151), chloride and nitrogen respectively by coulometrie and Kjeldahl methods (Catlove, 1963).

RESULTS

Growth

Figure 1 represents salt-induced changes in dry matter production. In *S. fruticosa*, maximal dry matter production was obtained at NaCl concentrations comprised between 100 and 300 mM, with a 85 fold increase in dry weight during the 45 day treatment (Figure 1A). *S. alterniflora* expressed maximal growth when irrigated with NaCl free nutrient solution or containing 100 mM NaCl (Figure 1B), its initial dry weight being multiplied by 6 during 100 days of treatment. The comparison of the two species (Figure 1C) shows that *S. fruticosa* was more salt tolerant at moderate NaCl concentrations (300 mM NaCl), but more sensitive at the highest NaCl concentration (800 mM).

For all NaCl concentrations, shoot dry matter amounted to 85% of the whole plant biomass. In *S. fruticosa*, the dry matter allocated to roots was diminished in the presence of salt, resulting in a

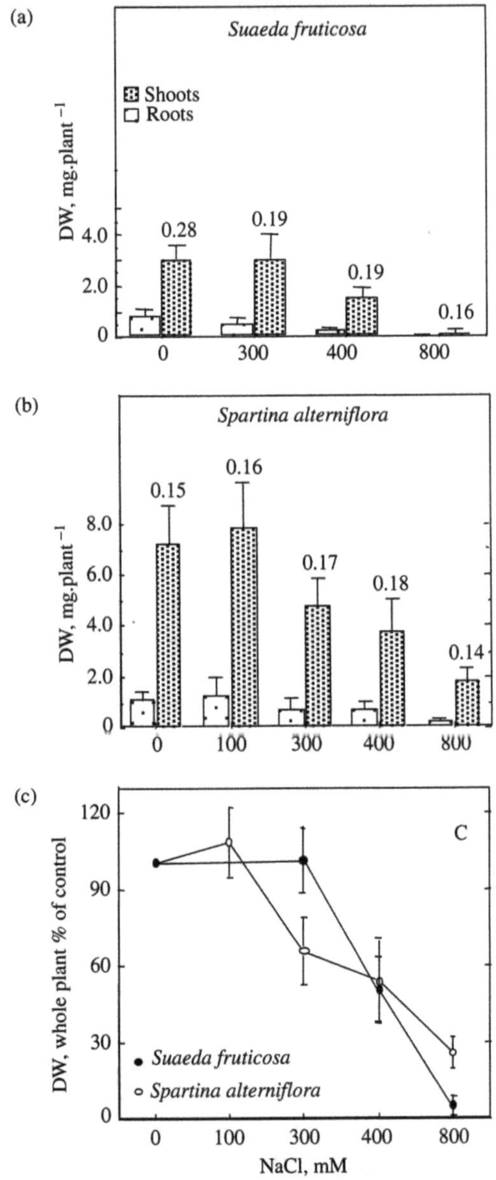

Fig. 1. Effect of medium salinity on *Suaeda fruticosa* and *Spartina alterniflora* growth. (a): *Suaeda fruticosa* dry matter after 45 day culture. (b): *Spartina alterniflora* dry matter after 100 day culture. Numbers on histograms are root to shoot ratio (DW/DW). (c): growth of plant submitted to salinity expressed as % of that of control plants cultivated in the absence of NaCl. Means of 10 repetitions. Bars indicate ± standard error ($p = 0.05$).

smaller root/shoot ratio. In *S. alterniflora*, salt treatment did not significantly modify dry matter distribution between roots and shoots.

Relative growth rate (RGR) expresses dry matter production per unit of time and per unit of dry matter

Table 1. Changes in Relative Growth Rate (RGR) with medium salinity. $RGR = (Ln\ DW_f - Ln\ DW_i)/(T_f - T_i)$; with DW_f: final dry weight measured at the end of treatment, DW_i: initial dry weight measured at the beginning of treatment, Ln: logarithm napierian and $(T_f - T_i)$: treatment duration

	RGR (day^{-1})	
	Suaeda fruticosa	*Spartina alterniflora*
Control (0 NaCl)	0.099	0.017
100 mM NaCl	–	0.018
300 mM NaCl	0.098	0.013
400 mM NaCl	0.082	0.011
800 mM NaCl	0.031	0.003

(Table 1). This parameter eliminates differences in biomass production related to different treatment duration or different initial plants size. RGR values were low as compared with those of other species, particularly annual species. For example, RGR could reach 0.15 day^{-1} in *Medicago* species cultivated under optimal conditions of growth (Abdelly et al., 1995). RGR was strongly decreased in both *S. fruticosa* and *S. alterniflora* when medium NaCl exceeded 300 mM. Nevertheless, *S. fruticosa* showed higher RGR with respect to *S. alterniflora* at all NaCl concentration.

Thus, the higher dry matter of *S. alterniflora* measured at the final harvest and under all NaCl concentrations did not result from a higher growth activity during treatment but from larger initial size and longer treatment duration. The comparison of salt tolerance of the two studied halophytes on the basis of the RGR also confirms that *S. fruticosa* is more salt tolerant than *S. alterniflora* at moderate salinity levels.

Mineral Composition of Tissues

Sodium and chloride

Na^+ and Cl^- shoot content discriminated clearly the two species (Figure 2). *S. fruticosa* showed the highest ion concentrations, which increased with NaCl concentration in the nutrient solution. Inversely, in *S. alterniflora*, Na^+ and Cl^- tissues contents were remained low and did not increase (not exceeding 1.5 mmol g^{-1} DW) in a wide range of NaCl concentrations in the medium (from 300 to 800 mM).

Na^+ and Cl^- root contents were similar in the two halophytes species. Although of weak amplitudes, changes of ions tissues contents with medium salinity were similar to those described in the aerial parts. One explanation for the behaviour of *S. alterniflora* is its ability to protect its organs and particularly photosynthetic ones against an excessive accumulation of Na^+ and Cl^- by secreting these ions by leaves as confirmed by the abundance of salt crystals on the surface of shoots in plant grown on salty medium. This avoidance mechanism has been described in several plants, *Agrostis stolonifera* (Ahmad and Wainwright, 1976), *Leymus sabulosus* and *Elytrigia juncea* (Gorham et al., 1985).

Potassium, calcium and magnesium

The culture of the two species in the presence of salt led to a decrease of shoot and root in both K^+ and Ca^{2+} (Figure 3). This effect was more pronounced in *S. fruticosa* than in *S. alterniflora*. Mg^{2+} concentration was diminished only in *Sueda* roots. Since these changes in K^+, Ca^{2+} and Mg^{2+} tissue concentrations were associated with a visible decrease of their uptake rate in (mmoles per plant; Figure 4) and a reduction of their allocation to the aerials parts (not shown) they resulted from inhibition of ion uptake rather than dilution of tissue ion contents by growth. This situation could be caused by a competition between Na^+ and the other cations at the level of absorption sites and by their substitution by Na^+ in the xylem vessels.

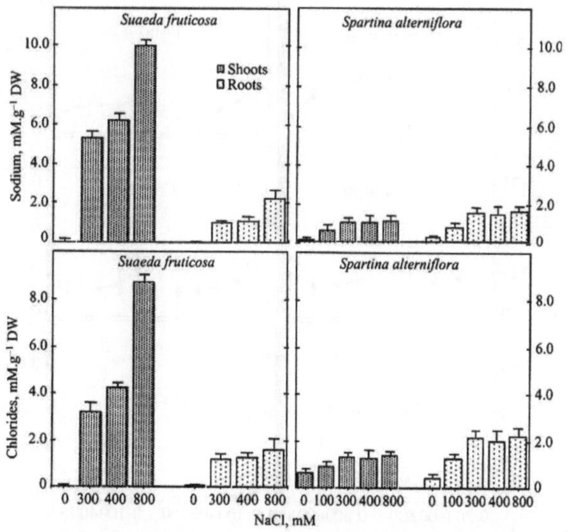

Fig. 2. Changes in shoot and root Na^+ and Cl^- concentrations with NaCl concentrations in the medium. Means of 10 repetitions. Bars indicate $\pm$ standard errors ($p = 0.05$).

Nitrogen and phosphorus

The nitrogen fraction measured in this study corresponds to reduced nitrogen, which represents the dominant nitrogen fraction in plants tissues, generally exceeding 90% of the total nitrogen (Mengel et Kirkby, 1982). In *S. fruticosa*, NaCl treatment led to a significant decrease of shoot content in reduced nitrogen, particularly when plants were grown at 400–800 mM NaCl (Figure 5). The accumulation of nitrogen was little modified in roots.

As the aerial parts represent more than 85% of the total plant biomass, these results indicate that nitrogen assimilation was more diminished by salt than plant growth. This situation which happened exclusively at higher NaCl concentrations could be caused by reduction in NO_3^- absorption (Cl^-/NO_3^- antagonism) or by specific inhibition of nitrate reductase by Na^+ or Cl^- accumulated in the cytoplasm because of a failing of systems involved in ion compartmentalization in conditions of excessive salinity (Papadopoulos and Rendig, 1983; Soltani et al., 1990).

In *S. alterniflora*, nitrogen tissue content did not vary with medium salinity in the aerial parts and in roots (Figure 5). This result suggests a close relationship between dry matter production and nitrogen assimilation. The capacity of *S. alterniflora* to secrete Na^+ and Cl^- by the aerial parts (and probably to exclude them by roots) insures protection of cytoplasm against excessive accumulation of Na^+ and Cl^-. This could favour the maintenance of metabolic and vacuolar NO_3^- pools at higher level, and to ensure optimal conditions to nitrate reductase activity.

The phosphorus fraction measured in this work corresponds to the inorganic phosphate pool (Pi). In *S. fruticosa*, medium salinity decreased Pi tissue concentration, particularly in the roots. Inversely, the Pi concentration was unaffected in *S. alterniflora* (Figure 5). One may hypothesise that phosphorus absorption would be limited in *Suaeda* plants grown in presence of salt, resulting probably in a higher utilization of cell phosphorus reserves to sustain plant growth. In *S. alterniflora*, phosphorus absorption would not be inhibited by salt, and Pi storage not solicited.

Relationship between Ionic Accumulation and Water Tissue Content

A positive relationship appeared between tissue water and Na^+ contents in the shoots of *S. fruticosa* (Figure 6), grown at moderate NaCl concentrations (not exceeding

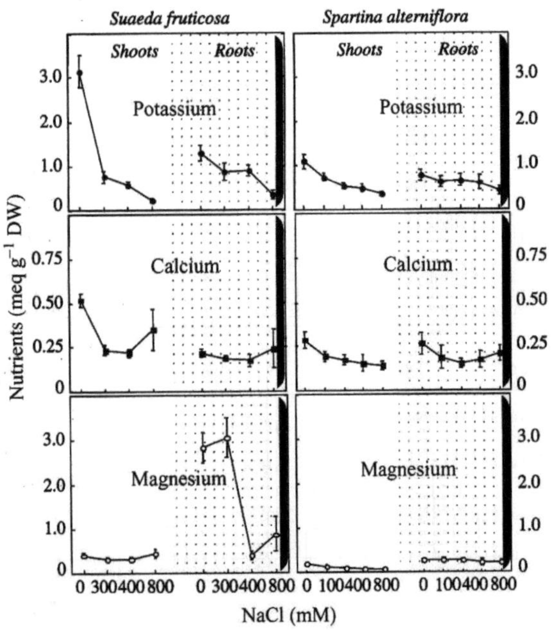

Fig. 3. Changes in shoot and root K^+, Ca^{2+} and Mg^{2+} concentrations with NaCl concentrations. Means of 10 repetitions. Bars indicate ± standard errors (p = 0.05).

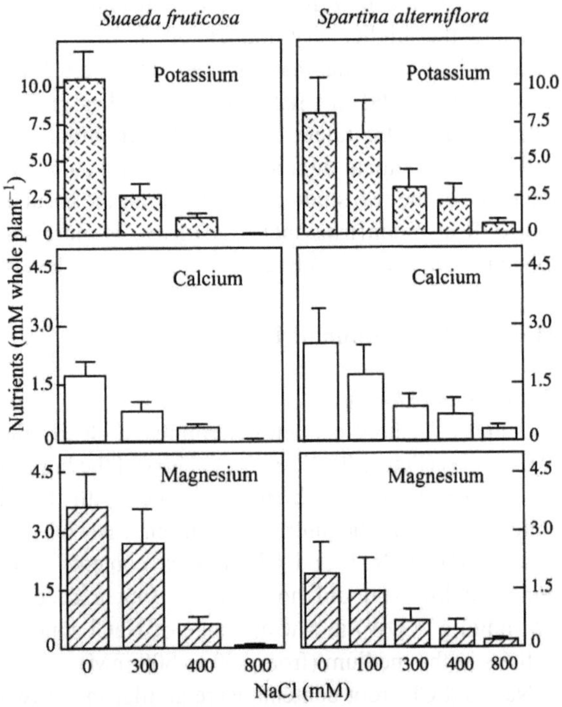

Fig. 4. Effect of NaCl concentration in the medium on potassium, calcium and magnesium quantities absorbed by whole plants. Means of 10 repetitions. Bars indicate ± standard errors (p = 0.05).

400 mM). For NaCl concentrations exceeding 400 mM NaCl, the higher leaf Na^+ content was associated with a decrease in water content. The relation between leaf hydration and K^+ concentration was less strict. This result suggests that the osmotic adjustment was ensured mainly by Na^+ whereas, K^+ being perhaps reserved for more specific functions. At the highest NaCl concentrations in the medium, the systems responsible for Na^+ uptake and compartmentalisation in cells would be overflow, and excess accumulation of this cation in the apoplast would result in tissue dehydration.

In *S. alterniflora* aerial parts, hydration is related positively to K^+ content and negatively with Na^+ content (Figure 6). This suggests that the osmotic adjustment is ensured by K^+ and not by Na^+ in this species. This behaviour, which implies inaptitude of *Spartina* leaf cells to absorb and compartmentalize Na^+ in vacuoles could lead to an accumulation of Na^+ in the apoplast and to a loss of the cellular water. However, secretion of Na^+ by shoot salt glands probably prevents excess Na^+ accumulation in the apoplast, as suggested by the maintenance of leaf hydration.

Potassium Use Efficiency

Potassium use efficiency (KUE) corresponds to dry matter production by unit of K^+ consumed (mg DW/μmol K^+). In condition of appropriate potassium nutrition (no NaCl), this parameter was significantly higher in *S. alterniflora* than in *S. fruticosa* (Table 2). Addition of NaCl to the nutrient solution led to an increase of KUE. At 300 mM NaCl concentration assuring an important biomass production, KUE was increased four times in *S. fruticosa* and only two times in *S. alterniflora*. This specific difference was also maintained in conditions of excessive salinity. The improvement of the KUE reveals the capacity of plants to use Na^+ instead of K^+ for the osmotic adjustment and to preserve K^+ for specific functions like maintenance of an appropriate ionic environment

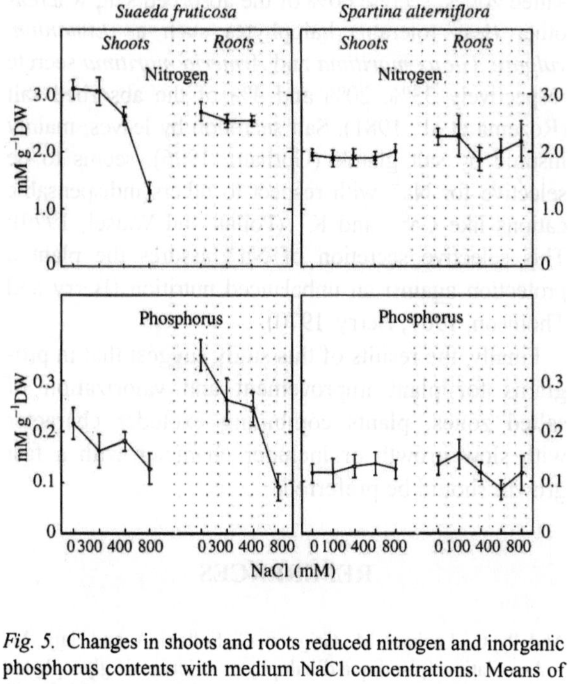

Fig. 5. Changes in shoots and roots reduced nitrogen and inorganic phosphorus contents with medium NaCl concentrations. Means of 10 repetitions. Bars indicate ± standard error ($p = 0.05$).

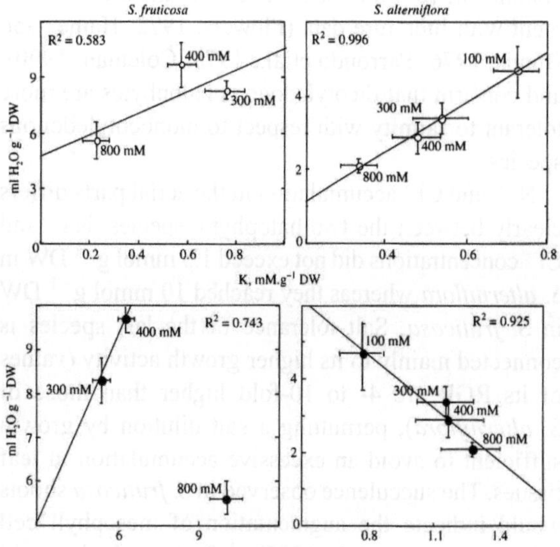

Fig. 6. Relationship between shoot hydration and contents in potassium and sodium. Means of 10 repetitions. Bars indicate ± standard error ($p = 0.05$).

Table 2. Changes of potassium use efficiency (KUE) with medium salinity. KUE = Dry weight quantity produced during treatment/quantity of potassium absorbed during treatment (mg DW/μmol K^+). Bars indicate ± standard error ($p = 0.05$)

	S. fruticosa	*S. alterniflora*
Control	0.37 ± 0.04	0.99 ± 0.11
100 mM, NaCl		1.43 ± 0.15
300 mM, NaCl	1.38 ± 0.33	1.86 ± 0.23
400 mM, NaCl	1.61 ± 0.17	2.04 ± 0.34
800 mM, NaCl	4.23 ± 0.84	3.00 ± 0.43

for the metabolic process in the cytoplasm (Leigh et Wyn Jones, 1984). Our results suggest that the KUE improvement in *Suaeda fruticosa* mainly resulted from the capacity of this species to replace K^+ by Na^+ for the osmotic adjustment. *S. alterniflora* would be rather able to sustain young organs growth by a higher capacity of K^+ mobilization from oldest organs.

CONCLUSION

Our results show that *S. fruticosa*, a dicotyledonous halophyte, express its maximal growth potentialities at 300 mM NaCl. In *S. alterniflora*, a monocotyledonous halophyte, the maximal biomass production was obtained at 100 mM NaCl. These results are in agreement with literature data (Flowers, 1972; Haines and Dunns, 1976; Parrondo et al., 1978; Coleman, 1990), and confirm that dicotyledonous halophytes are more tolerant to salinity with respect to monocotyledonous species.

Na^+ and Cl^- accumulation in the aerial parts differs clearly between the two halophytic species. Na^+ and Cl^- concentrations did not exceed 1.5 mmol g^{-1} DW in *S. alterniflora* whereas they reached 10 mmol g^{-1} DW in *S. fruticosa*. Salt tolerance of the last species is connected mainly to its higher growth activity (values of its RGR are 4- to 10-fold higher than those of *S. alterniflora*), permitting a salt dilution by growth sufficient to avoid an excessive accumulation in leaf tissues. The succulence observed in *S. fruticosa* shoots would indicate the augmentation of mesophyll cell volume (Munns et al., 1983), a feature which would favour salt dilution. As it is based on ion sequestration in vacuoles in order to maintain hypertony with respect of the outside medium, this strategy is called osmotic. Our results suggest that nitrogen nutrition is a fundamental component of this strategy. Indeed, we observed that the failing of this strategy in plants grown on medium containing more than 400 mM NaCl was associated with an inhibition of nitrogen absorption and reduction.

The *S. alterniflora* strategy is based on its capacity to control the NaCl accumulation in its tissues. Two mechanisms could be involved in this strategy, namely Na^+ and Cl^- secretion by leaves and exclusion by roots. We have evaluated that about 90% of Na^+ and Cl^- absorbed by plants were secreted at the level of leaves (Figure 7). This protects tissues against apoplastic accumulation of Na^+ and Cl^-. This avoidance strategy of *Spartina* seems

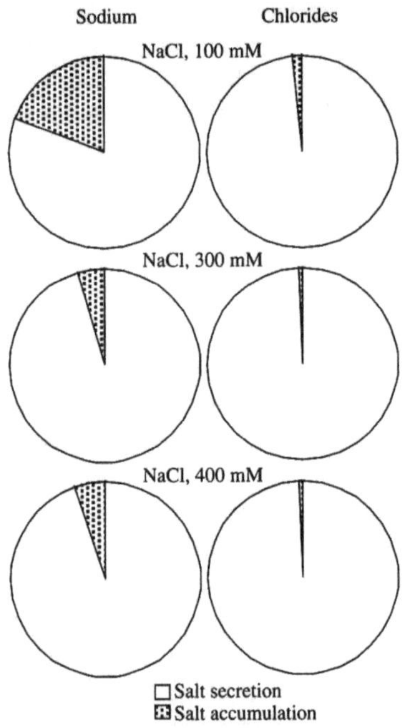

Fig. 7. Salt secretion by leaves of *S. alterniflora* expressed as % of total Na^+ and Cl^- uptake.

more effective than the osmotic strategy of *Suaeda* particularly in higher salinity conditions (800 mM NaCl). The role of salt secretion in tolerance has been described in several studies. Thus, *S. anglica*, native of strongly salted zones, secrete 60% of the absorbed salt, whereas other (less tolerant) halophytes such as *Limonium vulgare, Glaux maritima* and *Armeria maritima* secrete respectively 33%, 20% and 4% of the absorbed salt (Rozema et al., 1981). Salt secretion by leaves, mainly insured by salt glands (Gorham, 1996), seems to be selective for Na^+ with respect to others indispensable cations like Ca^{2+} and K^+ (Pollak and Waisel, 1979). This selective secretion of Na^+ assures the plant a protection against an unbalanced nutrition (Berry and Thomson, 1967; Berry, 1970).

Finally, the results of this study suggest that in programs for plant improvement and valorization of salted zones, plants combining excluder character with slow growth or includer character with a fast growth should be preferred.

REFERENCES

Abdelly, C., Lachaal, M., Grignon, C., Soltani, A. and Hajji, M. Association épisodique d'halophytes stricts et de glycophytes dans un écosystème hydromorphe salé en zone semi – aride. Agronomie 1995; 15:557–68.

Ahmad, I. and Wainwright, S.J. Ecotype differences in leaf surface properties of *Agrostis stolonifera* from salt marsh, spray zone and inland habitats. New Phytol 1976; 76:361–6.

Arnon, D.I. and Hoagland, D.R. Crop production in artificial solutions and in soils with special reference to factors affecting yields and absorption of inorganic nutrients. Soil Sci 1940; 50:463–84.

Berry, W.L. Characteristics of salts secreted by *Tamarix aphylla*. Am J Bot 1970; 57:1226–30.

Berry, W.L. and Thomson, W.W. Composition of salt secreted by salt glands of *Tamarix aphylla*. Can J Bot 1967; 45:1774–5.

Coleman, C.H. An experimental investigation of pore water salinity as a factor in *Spartina alterniflora* dieback in Charleston Harbor. Thesis, University of South California, 1990.

Flowers, T.J. Salt tolerance in *Suaeda maritima* (L.). Dum. J Exp Bot 1972; 23:310–21.

Flowers, T.J., Troke, P.F. and Yeo, A.R. The mechanism of salt tolerance in halophytes. Ann Rev Plant Physiol 1977; 28:89–121.

Gorham, J. Mechanisms of salt tolerance of halophytes. In R. Choukrallah, R., Malcolm, C.V. and Hamdy, A., editors. Halophytes and biosaline agriculture, M. Dekker Inc., New York, 1996; 31–54.

Gorham, J., Wynjones, R.G. and Mc Donnel, E. Some mechanism of salt tolerance in crop plants. Plant Soil 1985; 89:15–40.

Greenway, H. and Munns, R. Mechanism of salt tolerance in non-halophytes. Ann Rev Plant Physiol 1980; 31:149–90.

Haines, B.L. and Dunn, E.L. Growth and resource allocation responses of *Spartina alterniflora* Loisel. To three levels of NH_4-N, Fe and NaCl in solution culture. Bot Gazette 1976; 137:224–30.

Hewitt, E.J. Sand and water culture methods used in the study of plant nutrition, 2nd edn. Commonwealth Bureau of Horticulture tech. Com. 22, 1966.

Jacobson, L. Maintenance of iron supply in nutrient solution by a single addition of ferric-potassium-ethylene-diamine-tetra acetate. Plant Physiol 1951; 26:411–13.

Jeschke, W.D. and Nassery, H. K^+–Na^+ selectivity in roots of *Triticum, Helianthus* and *Allium*. Physiol Plant 1981; 52:217–24.

Kramer, D. The possible role of transfer cells in the adaptation of plants to salinity. Physiol Plant 1983; 58:549–55.

Le Houérou, H.N. Forage halophytes in the mediterranean bassin. In Choukrallah, R., Malcolm, C.V. and Hamdy, A., editors. Halophytes and Biosaline Agriculture. M. Dekker Inc., New York, 1996; 115–36.

Leigh, R.A. and Wyn Jones, R.G. A hypothesis relating critical potassium concentrations for growth to the distribution and functions for this ion in the plant cell. New Phytol 1984; 97:1–13.

Mengel, K.K. and Kikby, E.A. Nitrogen. In Mengel, K. and Kirkby, E.A., editors. Principles of plant nutrition. International Potash Institute, P.O. Box, CH-3048 Worblaufen – Bern/Switzerland, 1982; 335–68.

Munns, R., Greenway, H. and Kirst, G.O. Halotolerant eukaryotes. In Lange, O.L., Nobel, P.S., Osmond, C.B. and Ziegler, H., editors. Encyclopedia of plant Physiol, New series, Vol. 12C. Springer, Berlin, 1983; 59–135.

Papadopoulos, I. and Rendig V.V. Interactive effects of salinity and nitrogen on growth and yield of tomato plants. Plant Soil 1983; 73:47–57.

Parrondo, R.T., Gosselink, J.G. and Hopkinson, C.S. Effects of salinity and drainage on growth of three salt marsh grasses. Bot Gazette 1978; 139:102–7.

Pitman, M.G. Whole plants. In: Baker, D.A. and Hall, J.L., editors. Ion transport in plant cells and tissues. North Holland, 1975; 267–308.

Pollak, G. and Waisel, Y. Ecophysiology of salt excretion in *Aeluropus litoralis* (Gramineae). Physiol Plant 1979; 47:177–84.

Pollard, A. and Wyn Jones, R.G. Enzyme activities in concentrated solution of glycinebetaine and other solutes. Planta 1979; 144:291–8.

Rozema, J., Gude, H. and Pollak, G. An ecophysiological study of the salt secretion of four halophytes. New Phytologist 1981; 89:201–17.

Smith, F.A. and Fox, A.L. The free space of Citrus leaf slices. Aust J Plant Physiol 1975; 2:441–6.

Soltani A., Hajji M., Grignon C. Recherche de facteurs limitant la nutrition minérale de l'orge en milieu salé. Agronomie 1990; 10:857–66.

Waisel, Y. Ecological studies on *Tamarix aphylla* (L.) Karst. III. The salt economy. Plant Soil 1961; 13:356–64.

Wyn Jones, R.G. Salt tolerance. In Johson, C.B. editor. Physiological processes limiting plant productivity. Butterworths, London, 1981; 271–92.

Yeo, A.R. and Flowers, T.J. Varietal difference in the toxicity of sodium ions in rice leaves. Physiol Plant 1983; 59:189–95.

Zid, E. (1983) Mécanismes de la nutrition minérale de la feuille de Citrus et de son agression par le sodium. Thèse de Doct. D'Etat, Tunis, 1983; 419.

Salt-avoidance mechanisms in the halophyte *Distichlis spicata* as a promising source for improved salt resistance in crop plants

AMNON BUSTAN, MICHÈLE ZACCAI, YARON SITRIT, RACHEL DAVIDOVICI AND DOV PASTERNAK

INTRODUCTION

Salinity is an escalating problem in agriculture worldwide (Szabolics, 1989; Hamdy, 1996; Choukr-Allah, 1996). The use of recycled and saline waters to replace fresh water in agriculture contributes, among other factors, to the aggravation of salinity problems. Most crop plants are sensitive to salinity and poorly equipped to cope with salt stress without suffering from impaired growth and development, and even injury (Läuchli and Epstein, 1990; Maas, 1990). The adverse effects of salinity cause considerable reduction in yields, far below the economic threshold of most crops in present use. It can be stated quite confidently that the prevalent future environment of agricultural crops will be much more saline than today. Therefore, improving salt resistance of crop plants is of major interest in agricultural research. Halophytes are the ultimate candidates to serve as a genetic source for this purpose.

SALT STRESS IN PLANTS

Saline environment causes plant responses through several effects, among which osmotic effects, ion toxicity, and mineral nutrition are of principal importance (Greenway and Munns, 1980; Läuchli and Epstein, 1990; Munns, 1993; Gorham, 1996; Jacoby, 1999). Initially, salinity causes water deficit resulting in osmotic stress. When exposed to salinity for long periods or to high levels of salt, most plant species are unable to regulate ion fluxes into interior domains. The resulting salt buildup in the plant leads to disturbed ion homeostasis and ion toxicity.

Osmotic adjustment is usually mediated by the synthesis of intracellular compatible solutes (osmolytes) and by maintenance of ion homeostasis in the cytosol (McCue and Hanson, 1990). Ion homeostasis is controlled by selective uptake mechanisms of ions through the plasma membrane and by ion fluxes through the tonoplast (Maathuis, 1996). In plants, these processes are predominantly mediated by cytosolic concentrations of K^+ (Wyn Jones, 1996). Under saline conditions, when Na^+/K^+ ratio in the rhizosphere is high, selectivity of ion uptake in root plasma membrane is disturbed and sodium leaks into the cytosol (Läuchli, 1996). It has been concluded that the pathways for Na^+ uptake in plants may involve "illicit" transfer by different transporters which together constitute that "cation leak" (Serrano et al., 1999). According to Amtmann and Sanders (1998), a voltage-independent non-selective channel comprises the dominant element of Na^+ currents in high salt conditions in higher plants.

STRATEGIES OF SALT RESISTANCE IN PLANTS

Salt-tolerant metabolic pathways were reported for some algae species but not for higher plants (Munns, 1993). Moreover, no differences were found between *in vitro* salt-sensitivity of key enzymes from halophytes and glycophytes (Greenway and Osmond, 1972). While high intracellular salt concentrations (above 0.4 M) inhibit enzymes activity through a direct damage to protein structure (Wyn Jones and Pollard, 1983), toxic effects on cells occur at much lower concentrations (about 0.1 M) (Jacoby, 1999), pointing to specific toxicity targets (Serrano, 1996). Salt-resistant plants must, therefore, either protect sensitive metabolic pathways or avoid hazardous levels of toxic ions in the cytosol. The production of

H. Lieth (ed.), Cash Crop Halophytes: Recent Studies, 87–91.

compatible compounds in response to salt stress has received much attention as a mechanism of salt tolerance in plants. Compatible compounds such as proline, glycine-betaine and sugar alcohols contribute to the maintenance of osmotic potential in the cytosol (osmolites). More recent information relates compatible compounds to the insulation of enzymes and proteins from toxic ions and free radicals during salt stress.

Nevertheless, salt avoidance, as opposed to salt tolerance, is the major strategy of plants, including halophytes, to prevent ion toxicity. In order to maintain cellular ion homeostasis under conditions of increasing Na^+ leakage into the cytosol, salt-resistant plants should employ efficient Na^+ efflux mechanisms (Flowers et al., 1977; Cheeseman, 1988; Bohnert et al., 1995; Hasegawa et al., 2000). A general description of hypothetical Na^+ influx/efflux mechanisms in plant cells is shown in Figure 1.

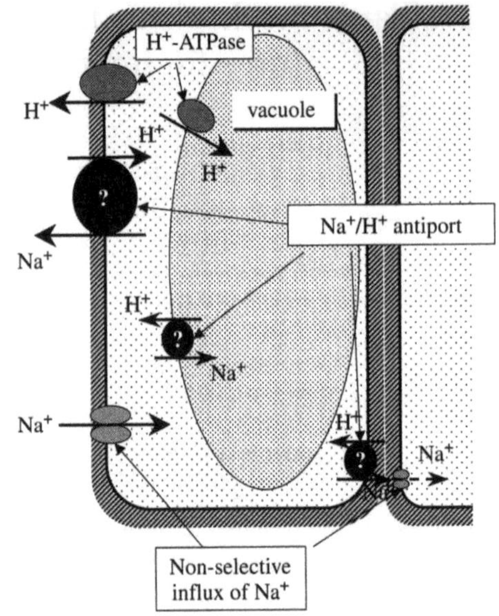

Fig. 1. Hypothetical Na^+ influx/efflux mechanisms in root epidermal cells. When rhizosphere Na^+ concentration increases, Na^+ ions leak into the cytosol through non-selective paths at the plasma membrane. H^+-ATPases are activated at the plasma membrane and tonoplast to extrude protons from the cytosol, thus providing secondary ion transport. To avoid ion toxic effects and to maintain ion homeostasis in the cytosol, Na^+ ions are extruded to the vacuole via Na^+/H^+ antiport activity at the tonoplast. Ion accumulation in the vacuole also contributes to osmotic adjustment. When cytosolic Na^+ concentration further increases, the vacuolar capacity to accumulate Na^+ ions is insufficient to maintain ion homeostasis and Na^+/H^+ antiport is required at the plasma membrane. Na^+ ions excreted to the intercellular domain may leak into adjacent cell layers and find ways to the xylem and above ground plant organs. Alternatively, Na^+ ions may be secreted back to the rhizosphere. The ratio between Na^+ leakage into plant tissues and Na^+ secretion from the roots may determine the difference between salt-resistant and salt-sensitive plants. It is suggested that this latter Na^+/H^+ antiport mechanism is highly efficient in root epidermal cells among halophytes. Differences between halophytic and glycophytic Na^+/H^+ antiport at the level of gene sequence or gene expression, if and when revealed, may bring about opportunities to improve salt resistance among crop species.

SALT AVOIDANCE

It has been proposed that in salt-resistant plants, compartmentation of Na^+ into vacuoles provide an efficient mechanism to avert the deleterious effects of Na^+ in the cytosol. Furthermore, ion accumulation in vacuoles contribute to the maintenance of osmotic balance required to drive water into cells (Flowers, 1985; Glenn et al., 1999). Although enhanced ion compartmentation has been demonstrated recently (Apse et al., 1999) using molecular approach, the role of vacuolar cation accumulation in salt resistance may be limited (Hasegawa et al., 2000). Ion concentration in the vacuole and the size and number of vacuoles necessarily restrict the capacity of salt compartmentation in plants.

In addition to the vacuolar compartmentation, deleterious salt concentration in the cytosol can be avoided through salt buildup in cell walls, salt distribution to older organs or tissues, and through salt glands (Läuchli and Epstein, 1990). This type of mechanisms may be found in many plant species, including glycophytes, but their expression, efficiency, and complexity tend to be much higher in halophytes. The conjoint basis of these mechanisms is their requirement for ion efflux machinery in the plasma membrane.

MEMBRANE Na^+/H^+ ANTIPORTERS

Although the importance of sodium extrusion from the cytosol was emphasized as a pivotal process in salt resistance in plants (Munns, 1993; Gorham, 1996) the evidence remained until recent years at the level of physiological activity. Salt stress increased ATP- and PP_i-dependent H^+-transport activity in sunflower (Ballesteros et al., 1996) and in *Spartina patens* (Wu and Seliskar, 1998) and was shown to induce Na^+/H^+ antiport activity in sunflower (Ballesteros et al., 1997). Na^+/H^+ antiporter activity has been detected in vesicles derived from the plasma membrane of the halophyte *Atriplex* (Hassidim et al., 1990). The use of molecular approaches has been proven necessary for deeper insight into ion transport in plants (Serrano et al., 1999; Hasegawa et al., 2000).

Among several families of Na^+ transport proteins, the Na^+/H^+ antiporters seem as yet to be the most promising group. In *Schizosaccharomyces pombe* the major sodium extrusion system is an electro-neutral Na^+/H^+ antiporter encoded by the *sod2* gene (Dibrov et al., 1997). This type of antiporter is also present in *S. serevisiae* as the product of the *NHA1* gene (Prior et al., 1996). Sod2/Nha1 antiporters exhibit some sequence similarity to the bacterial NhaA antiporter (Dibrov and Fliegel, 1998). The Nhx1/Nha2 antiporter homologues that are present in bacteria and in animal cells (Andre, 1995) have been proposed to encode a vacuolar antiporter (Nass et al., 1997). Based on homologous sequences, a gene (*AtNHX1*) encoding for a vacuolar Na^+/H^+ antiporter was identified and characterized in *Arabidopsis* (Gaxiola et al., 1999).

Over-expression of *AtNHX1* in *Arabidopsis* resulted in increased Na^+ accumulation in the vacuoles and improved salt resistance of the plants (Apse et al., 1999). These results should be interpreted with caution, since differences in the subcellular localization between plant and yeast were found in few cases (Gaxiola et al., 1999; Apse et al., 1999). Compartmentation of Na^+ into the vacuole may contribute to ion homeostasis and water balance in plant cells, nevertheless, it may well be that the major Na^+ efflux mechanism in salt-resistant plants is at the plasma membrane.

The homeostasis of cations intracellular concentrations was probably of primordial importance for most of the primitive organisms if the assumption that life evolved in NaCl-containing seas were correct. Hence, the ancestors of Na^+ pumps have originated very early during evolution (Serrano et al., 1999). This hypothesis is strongly supported by sequence conservation found in Na^+ pumps among organisms across all living systems (Andre, 1995; Nass et al., 1997). While terrestrial animal cells preserved the ancestral Na^+ chemiosmotic circuit, the capability of Na^+ transport progressively lost importance in plants that had acclimated to the Na^+-poor land and fresh water ecosystems (Serrano et al., 1999). However, it is attempting to assume, that the molecular basis encoding for Na^+ transport mechanisms is still included in the DNA of most glycophyte species in various degrees of incompleteness. This hypothesis is supported by the recent identification and characterization of the *AtSOS1* gene that encodes for a putative Na^+/H^+ antiporter in *Arabidopsis* plasma membrane (Shi et al., 2000). Over-expression of this gene brought about significant increase in the salt resistance of this non-halophyte species.

Moreover, it is logical to assume that in halophytes, preserved ancestral Na^+ pumps would largely contribute to their prominent salt resistance. Such genes are expected to be expressed mainly at the boundary between the saline environment and the intracellular domain of the plant – the plasma membrane of root cells. Physiological and anatomical findings in roots of halophytes (e.g. invagination of plasma membrane, and formation of "transfer cells") may indicate for an acute role of plasma membrane in salt resistance (Kramer, 1983; Koyro and Stelzer, 1988; Koyro et al., 1993; Echeverria, 2000). We believe that mechanisms as such displayed by halophytes are more efficient and puissant. Furthermore, the existence of some active parts of the transport mechanism or its control may ease the introduction and successful expression of genes transferred from halophytes to glycophytes.

SALTGRASS (*DISTICHLIS SPICATA*) AS A SOURCE OF GENES

Saltgrass (*Distichlis spicata*, Gramineae) is an extremely salt-resistant plant species, native to coastal and inland saline environments of America (Hitchcock, 1951; Ungar, 1974), distributed from Canada to Patagonia. Saltgrass occupies the edges of tidal marine marshes, where salinity may exceed 40 g NaCl/L (Gallagher, 1985). Saltgrass is found also in inland saline depressions, where it is known as desert saltgrass. In field experiments under high salinity levels (EC_I up to 12 dS m^{-1}), saltgrass had considerable biomass production comparing with other halophytes and salt-resistant species (Pasternak et al., 1993) such as *Paspalum vaginatum, Chloris gayana* and *Cynodon dactylon*. This may demonstrate the ability of the species to grow well at salinity range much higher than the agricultural threshold.

Unlike among some other halophytic plant families (e.g. Chenopodiceae), saltgrass plants do not accumulate high levels of salt in leaves and stems, even when grown under high-saline conditions. Pasternak et al. (1993) found that ash content in saltgrass hey was somewhat lower than that of alfalfa grown under

Table 1. Estimated and measured salt uptake and secretion through salt glands in saltgrass plants grown in water culture at 30 dS m^{-1}. Actual salt uptake was 85% less than expected at the absence of salt avoidance mechanisms in the roots

	Transpiration rate	Theoretical NaCl uptake	Accumulation rate of NaCl in plants	NaCl secretion rate through salt glands	Estimates salt avoidance in the roots
kg·kgFW^{-1}·week^{-1}	0.0030	0.0450	0.0025	0.0040	0.0385
kg·plant^{-1}·week^{-1}	0.0006	0.0090	0.0005	0.0007	0.0078

similar high saline conditions, although alfalfa growth was strongly hampered by salinity. In another study, the degree of salt resistance among different species of the Chloridoidae grass family, including saltgrass, was found to be negatively correlated with the accumulation of salt in leaves (Marcum, 1999). Saltgrass was the most resistant among the studied species, with the lowest sodium concentration in leaf sap. Salt glands activity has been suggested as the major mechanism of salt disposal in saltgrass (Hansen et al., 1976; Marcum, 1999). Nevertheless, calculations show that this might not be the case. Indeed, salt glands may play an important role as an alternative route to regulate salt levels in the leaves. Compartmentation and distribution of Na^+ within the roots should also be considered among salt defense mechanisms of saltgrass. Yet, it is most conceivable that salt uptake is largely prevented by the roots through exclusion and excretion mechanisms. As already demonstrated for *Spartina alterniflora* (Bradley and Morris, 1991), a plant species that shares the same ecological niche with seashore saltgrass, more than 85% of the theoretical amount of salt that should have passed through the plant at the absence of salt exclusion/extrusion in the roots – remain in the growing medium (Table 1).

The fact that saltgrass plants do not absorb other cations and heavy metals (Prodgers and Inskeep, 1991) supports the probability that highly efficient cation efflux systems are active in saltgrass root membrane. These unique characteristics make saltgrass a suitable model plant to study the physiological and molecular basis of salt-avoidance mechanisms in plants. Regardless of salt glands, saltgrass plants have not developed other morphological or anatomical means to cope with salt stress.

The relative simplicity of salt-avoidance mechanism may give rise to the approach that only few genetic modifications will be required to improve salt resistance in glycophytes. It seems realistic, therefore, to assume that some of the molecular factors regulating saltgrass exceptional resistance to salt can be characterized and transferred to other plant species.

At present, we already have good indications that members of the Na^+/H^+ antiporter gene family are active in saltgrass roots. As already mentioned, alignment analysis of Na^+/H^+ antiporters at the protein level shows high conservation among distant organisms through kingdoms. Degenerated primers were designed from conserved regions within these genes. Using the RT-PCR approach we were able to isolate several cDNA fragments that were ranked together with known Na^+/H^+ antiporter genes. Isolated genes will be characterized and their role in salt resistance will be evaluated in saltgrass and in heterologous systems.

ACKNOWLEDGEMENT

This study is supported by the Fohs Foundation of Oregon, USA.

REFERENCES

Amtmann, A. and D. Sanders, D. Mechanisms of Na^+ uptake by plant cells. Adv Bot Res 1998; 29:75–112.

Andre, B. An overview of membrane transport proteins in *Saccharomyces cerevisiae*. Yeast 11:1575–611.

Apse, M.P., Aharon, G.S., Snedden, W.A. and Blumwald, E. Salt tolerance conferred by overexpression of a vacuolar Na^+/H^+ antiporter in arabidopsis. Science 1999; 285:1256–8.

Ballesteros E., Donaire, J.P. and Belver, A. Effects of salt stress on H^+-ATPase and H^+-PPase activities of tonoplast-enriched vesicles isolated from sunflower roots. Physiol Plant 1996; 97:259–68.

Ballesteros E., Blumwald, E., Donaire, J.P. and Belver, A. Na^+/H^+ antiport activity in tonoplast vesicles isolated from sunflower roots induced by NaCl stress. Physiol. Plant. 1997; 99:328–34.

Bohnert, H.J., Nelson, D.E. and Jensen, R.G. Adaptations to environmental stresses. The Plant Cell 1995; 7:1099–111.

Bradley, P.M. and Morris, J.T. Relative importance of ion exclusion, secretion and accumulation in *Spartina alterniflora* Loisel. J Exp Bot 1991; 42:1525–32.

Cheeseman, J.M. Mechanisms of salinity tolerance in plants. Plant Physiol 1988; 87:547–50.

Choukr-Allah, R. The potential of halophytes in the development and rehabilitation of arid and semi-arid zones. In

Choukr-Allah, R., editor. Halophytes and biosaline agriculture. Marcel Dekker Inc., New York, 1996; 3–13.

Dibrov, P. and Fliegel, L. Comparative molecular analysis of Na^+/H^+ exchangers. FEBS Lett 1998; 424:1–5.

Dibrov, P., Smith, J.J., Young, P.G. and Fliegel, L. Identification and localization of the sod2 gene product in fission yeast. FEBS Lett 1997; 405:119–124.

Echeverria, E. Vesicle-mediated solute transport between the vacuole and the plasma membrane. Plant Physiol 2000; 123:1217–26.

Flowers, T.J. Physiology of halophytes. Plant Soil 1985; 89:41–56.

Flowers, T.J., Troke, P.F. and Yeo, A.R. The mechanisms of salt tolerance in halophytes. Annu Rev Plant Physiol 1977; 28: 89–121.

Gallagher, J.L. Halophytic crops for cultivation at seawater salinity. In Pasternak, D. and San Pietro, A., editors. Biosalinity in Action, Martinus Nijhoff, Dordrecht, 1985; 323–36.

Gaxiola, R.A., Rao, R., Sherman, A., Grisafi, P., Alper, S.L. and Fink, G.R. Proc Natl Acad Sci USA 1999; 96:1480–5.

Glenn, E., Brown, J.J. and Blumwald, E. Salt tolerance and crop potential of halophytes. Crit Rev Plant Sci 1999; 18:255–77.

Gorham, J. Mechanisms of salt tolerance in halophytes. In Choukr-Allah, R., editor. Halophytes and biosaline agriculture. Marcel Dekker Inc., New York, 1996; 31–53.

Greenway H. and Munns, R. Mechanisms of salt tolerance in non-halophytes. Annu Rev Plant Physiol 1980; 31:149–90.

Greenway H. and Osmond, C.B. Salt responses of enzymes from species differing in salt tolerance. Plant Physiol 1972; 49:256–9.

Hamdy, A. Saline irrigation: assessment and management techniques. In R. Choukr-Allah, R., editor. Halophytes and biosaline agriculture. Marcel Dekker Inc., New York, 1996; 147–80.

Hansen, D.J., Dayanadan, P., Kaufman, P.B. and Brotherson, J.D. Ecological adaptation of salt marsh grass, *Distichlis spicata* (Gramineae) and environmental factors affecting its growth and distribution. Am J Bot 1976; 63:635–50.

Hasegawa, P.M., Bressan, R.A., Zhu, J.K. and Bohnert, H.J. Plant cellular and molecular response to high salinity. Annu Rev Plant Physiol Plant Mol Biol 2000; 51:463–99.

Hassidim, M., Braun, Y., Lerner, H.R. and Reinhold, L. Na^+/H^+ and K^+/H^+ antiport in root membrane vesicles isolated from the halophyte *Atriplex* and the glycophyte cotton. Plant Physiol 1990; 94:1795–801.

Hitchcock, A.S. Manual of the grasses of the United States. USDA Miscellaneous Publication No. 200, Washington DC, 1951.

Jacoby, B. In Pessarakli, M., editor. Handbook of plant and crop stress. Dekker, New York, 1999; 97–123.

Koyro, H.-W., Stelzer, R. and Huchzermeyer, B. ATPase activities and membrane fine structure of rhizodermal cells from *Sorghum* and *Spartina* roots grown under mild salt stress. Botanica Acta 1993; 106:110–19.

Koyro, H.-W. and Stelzer, R. Ion concentrations in the cytoplasm and vacuoles of rhizodermal cells from NaCl treated *Sorghum, Spartina*, and *Puccinallia* plants. J Plant Physiol 1988; 133:441–6.

Kramer, D. The possible role of transfer cells in the adaptation of plants to salinity. Physiologia Plantarum 1983; 58:549–55.

Läuchli, A. Salinity – potassium interaction in crop plants. In Oosterhuis, D.M. and Berkowitz, G.A., editors. Frontiers in potassium nutrition: New perspectives on the effects of potassium on physiology of plants. Proceeding of a Symposium, Potash&Phosphate Institute of Canada, 1996; 71–6.

Läuchli, A. and Epstein, E. Plant responses to saline and sodic conditions. In Tanji, K.K., editor. Salinity assessment and management. Amer. Soc. Civil Eng., New York, 1990; 113–37.

Maas, E.V. Crop salt tolerance. In Tanji, K.K., editor. Agricultural salinity assessment and management. ASCE manual No. 71, New York, 1990; 262–304.

Marcum, K.B. Salinity tolerance mechanisms of grasses in the subfamily Chloridoideae. Crop Sci 1999; 39:1153–60.

McCue, K.F. and Hanson, A.D. Drought and salt tolerance: towards understanding and application. Trends Biotechnol 1990; 8:358–62.

Munns, R. Physiological processes limiting plant growth in saline soils: some dogmas and hypotheses. Plant Cell Environ 1993; 16:15–24.

Nass, R., Cunningham, K.W. and Rao, R. Intracellular sequesteration of sodium by a novel Na^+/H^+ exchanger in yeast is enhanced by mutations in the plasma membrane H^+-ATPase. J Biol Chem 1997; 272:26145–52.

Pasternak, D., Nerd, A. and Y. DeMalach, Y. Irrigation with brackish water under desert conditions IX. The salt tolerance of six forage crops. Agric Water Manage 1993; 29:321–34.

Prior, C., Prior, S., Souciet, J.-L. and H. Sychrova, H. Characterization of the NAH1 gene encoding a Na^+/H^+ antiporter of the yeast *Saccharomyces cerevisiae*. FEBS Lett 1996; 387:89–93.

Prodgers, R.A. and Inskeep, W.P. Heavy metal tolerance of inland saltgrass (*Distichlis spicata*). Great Basin Naturalist 1991; 51(3):271–8.

Serrano, R. Salt tolerance in plants and microorganisms: toxicity targets and defense responses. Int Rev Cytol 1996; 165:1–52.

Serrano, R., Mulet, J.M., Rios, G., Marquez, J.A., de Larrinoa, I.F., Leube, M.P., Mendizabal, I., Pascual-Ahuir, A., Proft, M., Ros, R. and Montesinos, C. A glimpse of the mechanisms of ion homeostasis during salt stress. J Exp Bot 1999; 50:1023–36.

Shi, H., Ishitani, M., Kim, C. and Zhu, J.K. The Arabidopsis thaliana salt resistance gene SOS1 encodes a putative Na^+/H^+ antiporter. Proc Natl Acad Sci USA 2000; 97:6896–901.

Szabolics, I. Salt Affected Soils. CRC Press, Boca Raton, FL, 1989.

Ungar, I. In W.H. Queen, W.H. and Reimold, R.J., editors. Ecology of halophytes. Academic Press Inc., New York, 1974.

Wu, J. and Seliskar, D.M. Salinity adaptation of plasma membrane H^+-ATPase in the salt marsh plant Spartina patens: ATP hydrolysis and enzyme kinetics. J Exp Bot 1988; 49:1005–13.

Wyn Jones, R.G. Cytoplasmic potassium homeostasis: review of the evidence and its implications. In Oosterhuis, D.M. and Berkowitz, G.A., editors. Frontiers in potassium nutrition: New perspectives on the effects of potassium on physiology of plants. Proceeding of a symposium, Potash&Phosphate Institute of Canada, 1996, 13–22.

Wyn Jones, R.G. and Pollard, A. Proteins, enzymes, and inorganic ions. In Läuchli, A. and Pirson, A. editors. Encyclopedia of plant physiology, New Series, Vol. 15B. Springer, Berlin, 1983; 528–62.

Part II

Chemical Contents

Introduction to Part II: Chemical content of halophytes

H. LIETH

The knowledge of the chemical composition of plants is important for many reasons. For their own structure and function as well as for their utilisation within the ecosystem as for human exploitation.

Halophytes require especially analytical efforts about their biochemistry and inorganic chemistry because much less is known about them. They contain some inorganic chemical compounds in higher amounts than glycophytes. This is not only true for sodium chloride but also for a variety of other chemicals which are either pollutants or potential resources for later industrial utilisation.

The chemical composition of the plant matter used for feed and food requires analytical investigations quite different from the one which should explain the salinity tolerance of the species or the tolerance and uptake of pollutants.

This chapter contains analytical examples for both aspects. Future analytical efforts must be encouraged to uncover the potential of halophytes to accumulate certain inorganic elements for industrial exploitation to produce various organic compounds for consumption by animals and men and for industrial or medical uses. Several aspects of these analytical efforts were documented in the books previously reported by the members of the EU CA and published by Backhuys Publishers as listed on page 127.

In some case decides the chemical analysis about the usability of a potential crop species, at which time, which part can be harvested and used and in which way can it be utilised.

As an example we include in Table 1 the chemical analysis of leaves and fruits of *Avicennia marina*, a mangrove species which can tolerate highly saline water (seawater). The adult foliage of this plant shows a high salinity which reduces its value as single feed source for animals while the fruits have a low salinity which renders it usable as vegetable if other chemical constituents keep it edible. This is the case as shown in Table 1.

Table 1. Comparative chemical analyses of some halophytes with regards to their value for food and feed[1]

		Plant material with point of origin					
		Fruits				Leaves	
		Avicennia marina				*Avicennia marina*	*Aster tripolium*
Chemical parameter	Units	ADH	UAQ	AJM	RAK	ADH	Osnabrueck
Dry matter	%	46.50	46.00	44.20	40.20	33.70	12.50
Water content	%	53.50	54.00	55.80	59.80	66.30	87.50
Ash	%	2.10	1.95	2.35	5.35	8.22	1.86
Raw protein N × 6.25	%	5.54	4.87	4.26	4.09	3.42	1.22
Fat	%	0.45	0.36	0.45	0.34	0.50	0.45
Carbohydrates	%	**	**	**	**	**	**
Starch	%	**	**	**	**	**	**
Raw fiber	%	12.20	11.00	13.70	10.50	14.50	5.30
Saccharose	%	< 0.20	< 0.20	< 0.20	< 0.20	**	**
Glucose	%	0.37	0.57	0.54	0.40	**	**

H. Lieth (ed.), *Cash Crop Halophytes: Recent Studies*, 95–96.

Table 1. (Cont.)

		Plant material with point of origin					
		Fruits				Leaves	
		Avicennia marina				*Avicennia marina*	*Aster tripolium*
Chemical parameter	Units	ADH	UAQ	AJM	RAK	ADH	Osnabrueck
Fructose	%	0.49	0.73	0.63	0.55	**	**
Maltose	%	< 0.30	< 0.30	< 0.30	< 0.30	**	**
Sorbit	%	< 0.30	< 0.30	< 0.30	< 0.30	**	**
Citric acid	%	**	**	**	**	**	**
Tartaric acid	%	**	**	**	**	**	**
Maleic acid	%	**	**	**	**	**	**
Ascorbic acid	%	**	**	**	**	**	**
Sodium	mg/100g	256.00	238.00	364.00	1350.0	2214.00	301.00
Potassium	mg/100g	626.00	559.00	564.00	692.00	838.00	373.00
Magnesium	mg/100g	35.50	30.00	53.80	181.00	168.00	28.20
Iron (total)	mg/100g	1.59	1.39	4.33	5.27	1.34	1.52
Phosphorus (total)	mg/100g	86.00	611.00	73.00	72.00	66.00	30.00
Chloride	mg/100g	410.00	370.00	610.00	2550.0	3500.00	710.00
Zinc	mg/100g	1.01	0.77	0.82	0.59	2.47	0.48
Salt content	mg/100g	651.00	605.00	925.00	3430.00	5630.00	765.00

ADH = Abu Dhabi; UAQ = Umm Al Quawain; AJM = Ajman; RAK = Ras al Kaima; AVM = Abu Dhabi (seedling with young leaves); ATR = *Aster tripolium* leaves grown in 3% salinity from Osnabrueck/Germany; ** = not determined; < = below detection limit

[1] provided by Deutsches Institut für Lebensmitteltechnik e.V. Quakenbrueck.

Chemical composition variation during root decomposition in Tagus estuary salt marshes

ANA LUISA COSTA AND ISABEL CAÇADOR

INTRODUCTION

In salt marshes vascular plants constitute the most important source of available organic matter. In Tagus estuary salt marshes 50% of the root biomass becomes detritus (Caçador et al., unpublished data). Vascular plant-derived detritus have been considered the major source material from which carbon and energy flow (Odum and de la Cruz, 1967; Moran et al., 1989).

Vascular plant material is highly heterogeneous composed of structural, lignocellulosic components, as well as non-lignocellulosic compounds such as sugars, free amino acids, lipids and proteins. Whereas most non-lignocellulosic material are leached and degraded by the microorganisms within days following deposition of plant material, lignocelluloses components are degraded only about 10–20% in the first month after submergence (Benner et al., 1985; Benner and Hodson, 1985).

The percentages of lignocellulose components (lignin, cellulose and hemicellulose) differ among species and in tissues within the same plant. Lignin form close associations with the cell-wall polysaccharides cellulose and hemicellulose, reducing accessibility of these cell-wall constituents to microbial attack. The inhibition of decay rates by lignin has been confirmed in studies of halophytes decomposition (Hemminga and Buth, 1991; Wilson et al., 1986; Buth, 1987; Moran et al., 1989; Moran and Hodson, 1989).

In this paper the results of the litterbag studies have been used to determine the effect of chemical composition variation on decomposition of root litter of two dominant plant communities in two salt marshes in Tagus estuary (Figure 1), with different abiotic conditions. In Pancas temperature values range between 14.7°C and 22.6°C and in Corroios between 16.3°C and 24.7°C. Redox potential values, in Pancas, range between 144 and 507 mV and in Corroios between 17 and 338 mV. These values were determined at 12:00 h, in low tide during the study period.

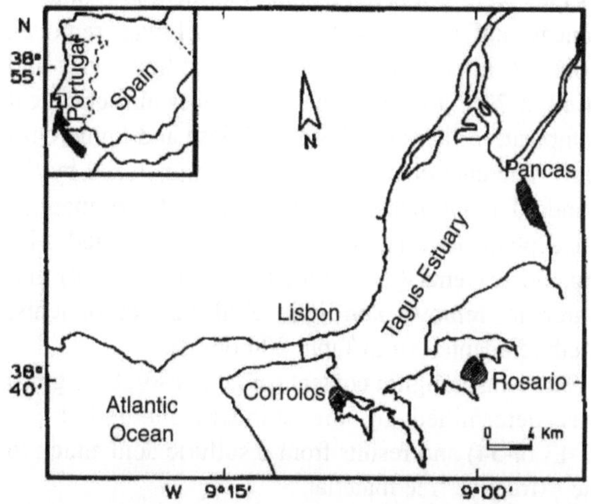

Fig. 1. The Tagus estuary showing the location of the sampling sites: Pancas and Corroios.

The Tagus estuary is one of the largest estuaries on the Atlantic coast of Europe, covering an area of 300 km^2 at low tide and 340 km^2 at extreme high tide. The southern and eastern parts of the estuary contain extensive intertidal mud-flat areas with the presence of *Spartina maritima* (Poales: Poaceae), *Halimione portulacoides* (Caryophyllales: Chenopodiaceae) and *Arthrocnemum fruticosum* (Caryophyllales: Chenopodiaceae).

MATERIALS AND METHODS

Belowground plant material of *S. maritima* and *H. portulacoides* was collected in February 1997, with a coring device 6.7 cm in diameter in Tagus estuary

H. Lieth (ed.), Cash Crop Halophytes: Recent Studies, 97–102.

salt marshes, Pancas and Corroios. The samples were washed and approximately 5 g of root material were placed in litterbags with 450 μm mesh. The bags were prepared for each type of root material and were placed in an undisturbed area below the marsh surface, 10 cm depth, between roots of each species. Additionally three samples of each type of root material were used as reference material.

This study was conducted from March 1997 to September 1997. Three or more litterbags of each species were collected at intervals of one month in Pancas and Corroios. Plant material was removed from the litterbags, washed with distilled water and dried at 70°C to determine weight loss and chemical composition (hemicellulose, cellulose and lignin content). The chemical composition was analyzed by the standard Tappi methods. For chemical determinations dried plant material was successively extracted with organic solvents (dichloromethane and ethanol) and water to remove non-lignocellulosic components. Method adapted from Tappi 204 om-88.

The Klason lignin content (or acid insoluble lignin) was determined by the standard method Tappi (T 13 m-54) and results from a sulfuric acid attack to the extractive free material.

After Klason lignin determination the obtained filtrate was used to analyze the soluble lignin by filtrate absorption in the ultra-violet region 200–208 ηm. Soluble lignin content was determined by the Tappi TUM 250-91 method.

Structural polysaccharides composition (cellulose and hemicellulose) were determined by respective monomeric composition in a high performance liquid chromatography (HPLC), method adapted from Tappi (T249 cm-85).

Percent weight loss of individual components of *S. maritima* and *H. portulacoides* was calculated by multiplying the concentration of each component by the fraction of original weight remaining and expressing this as the percent of the original content of the component.

The Levene's Test was used for homogeneity of variances. The mass loss results were statistically analyzed in a three-way analysis of variance (ANOVA). Significant differences were determined by the Tukey HSD Post Hoc test.

After data transformation the chemical composition results still had heterogeneous variances, so it was used in the Kruskal-Wallis method. The data was statistically analyzed using the Program Statistica '99 edition.

RESULTS AND DISCUSSION

Initial Chemical Composition

The initial chemical composition of radicular system of *H. portulacoides* and *S. maritima* is summarized in Table 1. The total non-lignocellulosic components (extracted material) were 0.122 g g^{-1} and 0.120 g g^{-1} in *H. portulacoides* and in *S. maritima*, respectively. The initial non-lignocellulosic components soluble in dichloromethane, ethanol or water had a similar distribution pattern, in that ordering of the components was dichloromethane<ethanol<water for both species and sites (Figures 2–4). In Corroios, *H. portulacoides* had the lowest values for the three components.

Total lignocellulose content and the individual compounds varied between species. Total lignocellulose content ranged from 76% to 86% dry weight. The

Table 1. Initial chemical composition of root system (g g^{-1} dry weight) for each species

	Lignin	Cellulose	Hemicellulose	Total lignocellulose	Total non-lignocellulosic components
Pancas					
H. portulacoides	0.353 (0.000)	0.198 (0.002)	0.208 (0.008)	0.757	0.134
Corroios					
H. portulacoides	0.387 (0.003)	0.180 (0.006)	0.193 (0.007)	0.760	0.107
S. maritima	0.332 (0.000)	0.252 (0.010)	0.273 (0.008)	0.857	0.120

Standard deviation is in parentheses. Total non-lignocellulosic components had no replicates.

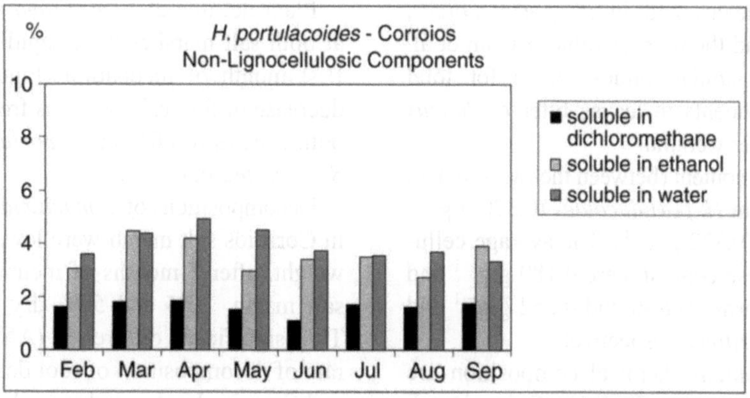

Fig. 2. Percent non-lignocellulosic components of degrading root material of *Halimione portulacoides* in Corroios. The first month values corresponding to percent non-lignocellulosic components in reference material.

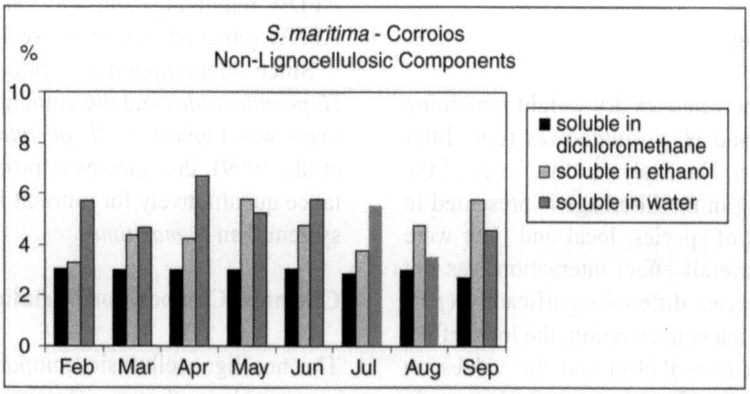

Fig. 3. Percent non-lignocellulosic components of degrading root material of *Halimione portulacoides* in Pancas. The first month values corresponding to percent non-lignocellulosic components in reference material.

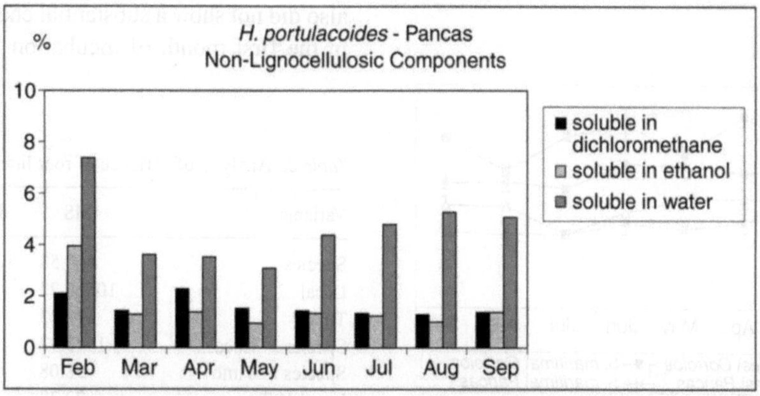

Fig. 4. Percent non-lignocellulosic components of degrading root material of *Spartina maritima* in Corroios. The first month values corresponding to percent non-lignocellulosic components in reference material.

lowest value was determined in *H. portulacoides*, while *S. maritima* had the highest value. Kuehn et al. (2000) have found similar values (78%) for total lignocellulose components in leaves litter of *Juncus effusus* in a freshwater wetland.

The average lignin content (between the two sites), in contrast, was higher in *H. portulacoides* (0.370 g g^{-1}) than in *S. maritima* (0.332 g g^{-1}). The average cellulose and hemicellulose content were 0.189 g g^{-1} and 0.201 g g^{-1} in *H. portulacoides* and 0.252 g g^{-1} and 0.273 g g^{-1} in *S. maritima*, respectively.

These results of initial chemical composition are similar to those determined by Buth (1987) in a Dutch salt marsh: lignin 0.412 g g^{-1} and cellulose 0.251 g g^{-1} in *Halimione portulacoides* and 0.371 g g^{-1} and 0.228 g g^{-1} for lignin and cellulose in other species of *Spartina*, respectively.

Decomposition Rates

Figure 5 shows the percentages dry weight remaining of the *S. maritima* and *H. portulacoides* roots litter during decomposition. The analysis of variance of the dry weight remaining in the litterbags is presented in Table 2. The effects of species, local and time were significant but the overall effect interaction was not significant. Each species differed significantly ($p \leq 0.01$) in litter remaining concentration, the local effect was also significant ($p \leq 0.001$) and the values on most dates differed significantly ($p \leq 0.01$) too. In same date there were an apparently gained in weight. This feature has been observed in several other decomposition studies (Buth, 1987; Wilson et al., 1986). Their suggestion is a flush of colonizing microorganisms after some weeks of incubation.

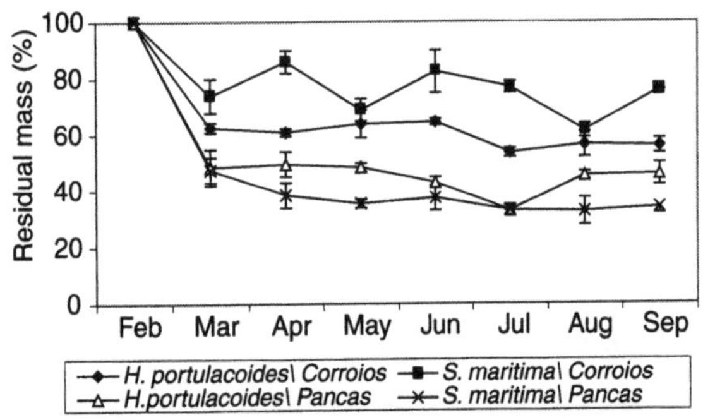

Fig. 5. Weight loss of plant material from litterbags in Pancas and Corroios. Bars give the standard deviation.

Plant detritus of *S. maritima* and *H. portulacoides* at both salt marshes were rapidly leached during the first month of incubation (Figure 5) resulting in a decrease of the residual mass from 50% in Pancas, in both species to 60% in *H. portulacoides* and 75% in *S. maritima* in Corroios.

Decompositions of *S. maritima* and *H. portulacoides* in Corroios salt marsh were lower, 60% and 80% dry weight, after 7 months of incubation, than in Pancas salt marsh, 30% and 50% dry weight, respectively. This significant difference (ANOVA, $p \leq 0.001$) in rate of decomposition of root detritus between Pancas and Corroios has been observed previously (Costa and Caçador, unpublished data).

This range of belowground decay rates were greater than the rates determined by Buth (1987) for root material of *Halimione portulacoides* (about 75% AFDW remaining) and *Spartina anglica* (about 85% AFDW remaining) after 30 weeks.

Since decomposition rates were highest in *H. portulacoides* and the amount of belowground biomass was highest in *H. portulacoides* too (Caçador et al., 1999), this species is probably of more importance quantitatively for nutrient fluxes in the estuarine system than *S. maritima*.

Chemical Composition Variation

The non-lignocellulosic components did not change appreciably with time (Figures 2–4). However the non-lignocellulosic components, in *H. portulacoides* in Pancas decreased substantially during the first 3 months of incubation, reflecting its greater overall decomposition rate (Figure 5).

The total lignocellulose content of the two species also did not show a substantial change with time except in the first month of incubation for *H. portulacoides*

Table 2. Analysis of variance of root litter dry weight

Variable	MS	d.f.	F	*p*-level
Species	265.58	1	11.1649	**
Local	10556.22	1	443.7803	***
Time	148.43	6	6.2401	***
Species × Local	1945.97	1	81.8082	***
Species × Time	58.08	6	2.4418	*
Local × Time	72.38	6	3.0430	*
Species × Local × Time	47.54	6	1.9984	n.s.

*$p \leq 0.05$; **$p \leq 0.01$; ***$p \leq 0.001$; n.s., not significant.

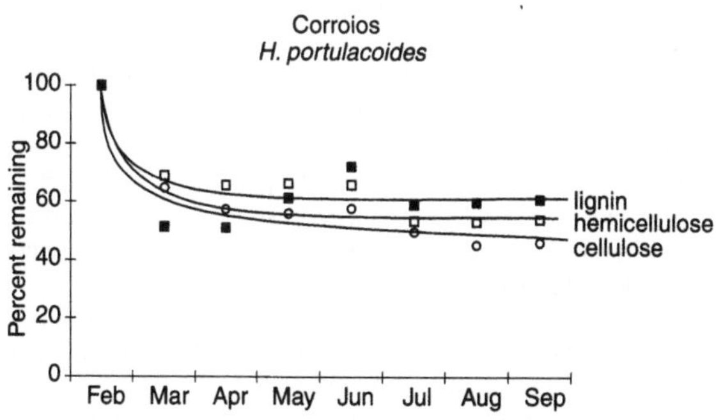

Fig. 6. Loss of lignocellulosic components (cellulose, hemicellulose and lignin) for *Halimione portulacoides* in Corroios.

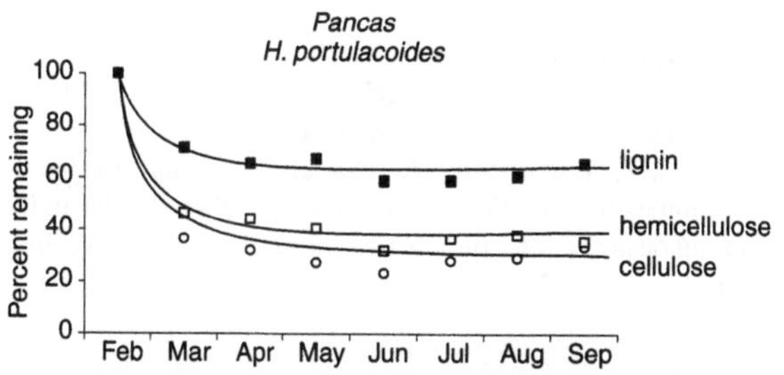

Fig. 7. Loss of lignocellulosic components (cellulose, hemicellulose and lignin) for *Halimione portulacoides* in Pancas.

(Figures 6 and 7). Lignocellulose content of *S. maritima*, incubated in Corroios, did not change appreciably during this time, probably reflecting its much slower overall decomposition rate. Kuehn et al. (2000) also did not found great changes in the structural plant polymers in a decaying macrophyte leaves during the study period. However, they found small differences for water-soluble and dichloromethane-soluble components.

Lignin content had the highest values in the residual litter of *H. portulacoides* (Figures 6–8) and probably reflects the discrimination against the lignin component of root litter during microbial decomposition (Buth, 1987; Moran and Hodson, 1989).

The two polysaccharides components of lignocellulose, cellulose and hemicellulose, decreased as decomposition progressed. The most pronounced loss was in cellulose (about 70% of the initial value) for *H. portulacoides* in Pancas (Figure 7). This preferential utilization of the cellulose fraction of lignocellulose

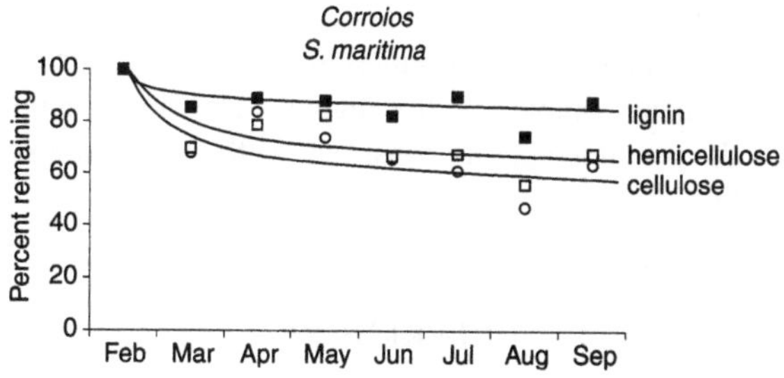

Fig. 8. Loss of lignocellulosic components (cellulose, hemicellulose and lignin) for *Spartina maritima* in Corroios.

relative to the other structural components (hemicellulose and lignin) has been found previously for aquatic vascular plant material degrading in the same type of ecosystem (Moran and Hodson, 1989).

CONCLUSIONS

The significant difference (ANOVA, $p \leq 0.001$) in rate of decomposition of root detritus between Pancas and Corroios has been observed previously (Costa and Caçador, unpublished data) and is likely due to environmental differences between the two salt marshes, rather than to plant-specific compositional differences.

Since decomposition rates were highest in *H. portulacoides* and the amount of belowground biomass was highest in *H. portulacoides* too (Caçador et al., 1999), this species is probably of more importance quantitatively for nutrient fluxes in the estuarine system than *S. maritima*.

Decomposition losses are expected to be greater for plant species lower in initial lignin content (Buth, 1987; Moran and Hodson, 1989; Foote and Reynolds, 1997). According to this criteria, of the two plant species chosen for this study *H. portulacoides*, having the highest initial lignin content, would be expected to degrade more slowly. Probably, more favorable environmental conditions in *H. portulacoides* sediments, such as the highest aeration of these sediments (Costa and Caçador, unpublished data), compensated for decay resistance of *H. portulacoides* material. This decay resistance overcome by changes in environmental conditions was also described by Buth (1987) for *H. portulacoides* and *Spartina* communities and by Godshalk and Wetzel in aquatic macrophytes.

Moran and Hodson (1989) considered that decomposition rates are greater in plants species which are high in non-lignocellulosic components; *H. portulacoides* had in fact the highest values in non-lignocellulosic components (Table 1): the higher decay rate found for *H. portulacoides* is consistent with these authors findings.

REFERENCES

Benner, R., Maccubbin, A.E. and Hodson, R.E. Anaerobic biodegradation of the lignin and polysaccharide components of the lignin and polysaccharide components of lignocellulose and synthetic lignin by sediment microflora. Appl Environ Microbiol 1984; 47:998–1004.

Buth, G.J.C. 1987. Decomposition of roots of three plant communities in a Dutch salt marsh. Aquat Bot 1987; 29:123–38.

Caçador, I., Mascarenhas, I. and Mascarenhas, P. Biomass of *Spartina maritima, Halimione portulacoides* and *Artrocnemum fruticosum* in Tagus estuary Salt Marshes. In Lieth, H., editors. Halophyte uses in different climates I. Ecological and ecophysiological studies. Prog Biometeorol 1999; 13:33–41.

Foote, A.L. and Reynolds, K.A. 1997. Decomposition of Saltmeadow Cordgrass (*Spartina patens*) in Louisiana Coastal Marshes. Estuaries 1997; 20(3):579–88.

Godshalk, G.L. and Wetzel, R.G. Decomposition of aquatic angiosperms. II Particulate components. Aquat Bot 5:301–27.

Hemminga, M.A. and Buth, G.J.C. Decomposition in salt marsh ecosystems of the SW Netherlands: The effects of biotic and abiotic factors. Vegetatio 1991; 92:73–83.

Hodson, R.E., Christian, R.R. and Maccubbin, A.E. Lignocellulose and lignin in the salt marsh grass *Spartina alterniflora*: initial concentrations and short-term, post-depositional changes in detritus matter. Mar Biol 81:1–7.

Kuehn, K.A., Lemke, M.J., Suberkropp, K. and Wetzel, R.G. Microbial biomass and production associated with decaying leaf litter of the emergent macrophyte *Juncus effusus*. Limnol Oceanogr 2000; 45(4):862–70.

Moran, M.A., Benner, R. and Hodson, R.E. Kinetics of microbial degradation of vascular plant material in two wetland ecosystems. Oecologia 1989; 79:158–67.

Moran, M.A. and Hodson, R.E. Bacterial secondary production on vascular plant detritus: relationships to detritus composition and degradation rate. Appl Environ Microbiol 1989; 55(9):2178–89.

TAP Test Methods. Official test methods and provisional test methods. Technical Association Pulp Ana Paper Industry, TAPPI Press, Atlanta, USA, 1994–95.

TAPPI Test Useful Methods. Official test methods and provisional test methods. Technical Association Pulp ana Paper Industry, TAPPI Press, Atlanta, USA, 1991.

Wilson, J.O., Buchsbaum, R., Valiela, I.E and Swain, T. Decomposition in salt marsh ecosystems: phenolic dynamics during decay of litter of *Spartina alterniflora*. Mar Ecol Prog Ser 1986; 29:177–87.

Metal partition in Tagus estuary salt marshes: a case study

ISABEL CAÇADOR AND CARLOS VALE

INTRODUCTION

Salt marshes are among the most productive ecosystems in the world (Adam, 1990). Cities and industrialised areas were often installed in the proximity of salt marshes in estuaries and coastal lagoons. In these cases, tidal flooding transports large quantities of contaminants in both dissolved and suspended particulate forms to the salt marshes. Salt marsh vegetation influences the dynamics of the estuarine ecosystem and retains efficiently anthropogenic metals discharged to the system. Complex interactions between root and sediments result in the redistribution of metals in the sediment-root system (Tinker and Barraclough, 1988). In many marshes it has been shown that plants have a dominant role on the metal cycling (Vale, 1990; Caçador et al., 1996). Metals are taken up by the roots, translocated to the above ground parts of the plants and, when they die, metals return to the sediments when organic matter is oxidised. Since roots accumulate larger proportions of metals (Caçador et al., 2000; Otte, 1991) the interactions between below-ground biomass and sediments are extremely active and may have a strong influence on the form and concentrations of metals in the rhizosphere (Caçador et al., 1996).

However, a number of factors may cause post-depositional mobilisation of metals from industrial and urban origin. Diagenetic reactions related to oxidation of organic matter that usually occurs in high levels in salt marshes and oxidation of metal sulphides in the rooting zone (Madureira et al., 1997) lead to the mobilisation of metals. This chapter reviews the trace-metal distribution in the Tagus salt marshes and reports on the amount of metal recycled annually between sediment and plant.

THE TAGUS ESTUARY SALT MARSHES

The Tagus estuary is one of the largest estuaries on the Atlantic coast of Europe, covering an area of 300 km^2 at low tide and 340 km^2 at extreme high tide. The southern and eastern parts of the estuary contain extensive inter-tidal mud-flat areas with the presence of *Spartina maritima* (Poales: Poaceae), *Halimione portulacoides* (Caryophyllales: Chenopodiaceae) and *Arthrocnemum fruticosum* (Caryophyllales: Chenopodiaceae). Contrary to many cases in Europe where pollutants from industrial regions are discharged into rivers and brought to the estuaries via the rivers, in the Tagus most pollutants are discharged directly into the estuary. The estuary receives effluents from about 2.5 million inhabitants living in the Greater Lisbon area, together with the discharges from industries (chemicals, steelmaking and shipbuilding). Previous studies showed that Tagus salt marshes incorporate large quantities of anthropogenic metals into the sediments. Concentrations of copper (Cu), zinc (Zn), cadmiun (Cd) and lead (Pb) in the upper sediment increased 3–12 times relative to pre-industrial levels, which indicates that a substantial quantity of anthropogenic metals is incorporated into the sediment (Caçador et al., 1996).

MATERIAL AND METHODS

Three pure stands of *Spartina maritima, Halimione portulacoides* and *Arthrocnemum fruticosum* were studied at Rosário, a Tagus estuary salt marsh. The sampling was done in August 1998 and January 1999, corresponding to periods of maximum and minimum plant growth. From each stand above-ground biomass

H. Lieth (ed.), Cash Crop Halophytes: Recent Studies, 103–109.

was determined by clipping the vegetation at ground level from three squares with 0.09 m^2 (0.3 × 0.3 m^2). After cutting the plant material in each quadrate, one core from each quadrate was taken using a stainless steel coring tube with 6.8 cm inner diameter and 60 cm depth. The samples were transported to the laboratory in plastic bags, and the above-ground materials were washed with demineralized water, separated into stems, leaves, flowers and standing dead material, and dried to constant weight at 80°C. Dead or partially decayed material, present as standing dead stems or leaves and surface litter, was identified by the yellowish or brownish coloration and separated from the living material. In the upper layer (0–25 cm), where the majority of below-ground components are typically concentrated (Catarino and Caçador 1981; Gross et al. 1991), below-ground parts were separated from soil particles using a 212 μm mesh sieve. The remaining plant material was dried at 80°C for 48 h and weighed.

Three 60-cm long sediment cores were collected from inter-tidal non-vegetated zones and one from each of the areas on the salt marsh dominated by *Halimione portulacoides, Spartina maritima* and *Arthrocnemum fruticosum*. Sediment cores were sliced immediately at the following layers: 0–5; 5–15; 15–25; 25–35; 35–45 and 45–55 cm of depth. Redox potential (Eh) and pH of the sediment layers were measured using a Criso pH/mV meter. The pH was measured after a short period of stabilization and the Eh electrode was allowed to reach equilibrium for 15 min. Both pH and Eh measurements at each layer were repeated four times. In the laboratory, sediment samples were dried, and, after roots had been removed, they were homogenised. The particle-size distribution was determined by the pipette method (Gee and Bauder, 1986). After destroying the organic matter and the carbonates of the sediment, particle size was determined according to Stoke's law, in a 1000-ml measuring cylinder with distilled water. The organic matter content of the sediment was determined by loss of ignition (LOI) at 600°C for 2 h. Two grams of sediment were extracted with 10 ml of HNO_3/HCl (3:1 v/v) twice at 130°C (Otte, 1991). The plants were washed with demineralized water to remove the soil from the roots and, after separating the roots, shoots and leaves, the fractions were dried and homogenised. For analysis of Zn, Pb and Cu in the three parts of the plants 10 ml $HNO_3/HClO_4$ were added to 100 mg of material according to the methodology described in Otte (1991). The concentrations of Zn, Pb and Cu in the solutions were determined by atomic absorption spectrophotometry (Perkin-Elmer 4000). Standard additions and sludge reference materials (EC standards CRM 145 and 146) plus vegetal reference material were used for sediment and plant analysis, respectively. The analysis of nitrogen and carbon was made using a CHN/O elemental analyser (Fisons, AE 1108).

RESULTS AND DISCUSSION

Sediment Characteristics

All sediment samples were made up of clay and silt containing less than 3% sand. Grain size composition was relatively uniform with the depth, and non-vegetated and vegetated sediments colonised by the three plants showed similar-fine fraction (<63 μm) percentage, considered as the most important fraction in concentrating metals that will be eventually available to be incorporated by the plants (Otte, 1991).

Loss on ignition, an estimation of the organic matter content, was higher in sediments colonised by the three plants than in non-vegetated sediments (Figure 1). Sediments between roots of *Spartina maritima* contained more organic matter than sediments around the roots of *Halimione portulacoides* and *Arthrocnemum fruticosum*. Higher organic matter content in rooting sediment layers is derived mainly from root material, the breakdown of which results in the formation of

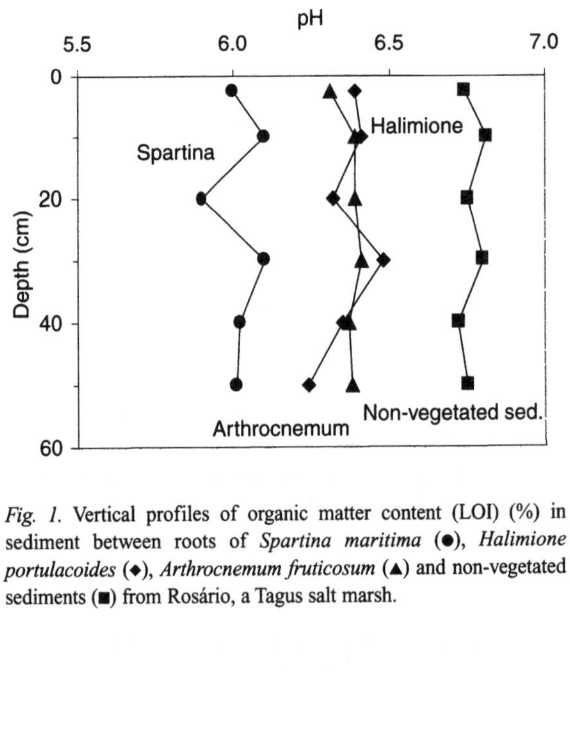

Fig. 1. Vertical profiles of organic matter content (LOI) (%) in sediment between roots of *Spartina maritima* (●), *Halimione portulacoides* (◆), *Arthrocnemum fruticosum* (▲) and non-vegetated sediments (■) from Rosário, a Tagus salt marsh.

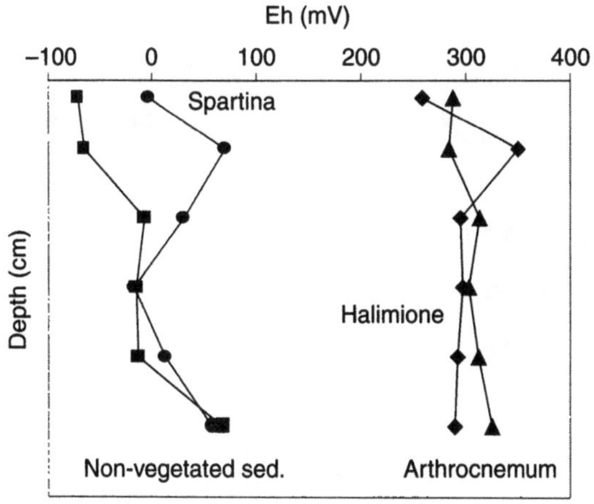

Fig. 2. Vertical profiles of total C (%) in sediment between roots of *Spartina maritima* (●), *Halimione portulacoides* (◆), *Arthrocnemum fruticosum* (▲) and non-vegetated sediments (■) from Rosário, a Tagus salt marsh.

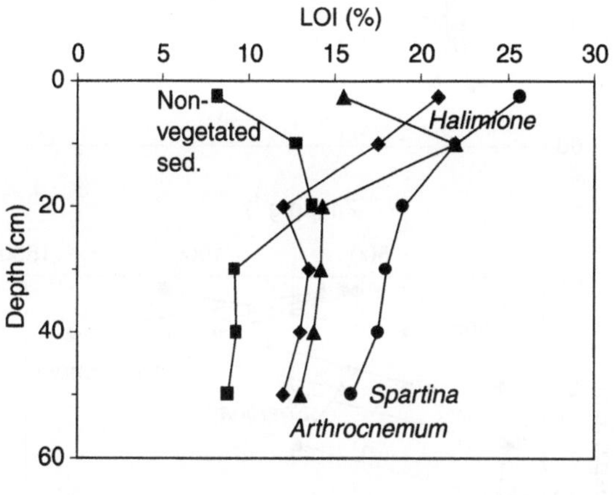

Fig. 3. Vertical profiles of total N (%) in sediment between roots of *Spartina maritima* (●), *Halimione portulacoides* (◆), *Arthrocnemum fruticosum* (▲) and non-vegetated sediments (■) from Rosário, a Tagus salt marsh.

sediment organic matter (Tinker and Barraclough, 1988). Different LOI among the sediment colonised by the three species may be attributed to differences in organic matter degradation rates (Orson et al., 1992; Caçador et al., in prep.). Carbon and nitrogen contents were also higher in sediment colonised by the three salt-marsh plants than in non-vegetated sediment (Figures 2 and 3). However, the C/N ratio was lower in vegetated areas (Figure 4), which suggests that nitrogen is more effectively retained in colonised sediments than carbon. The contribution of the microbial

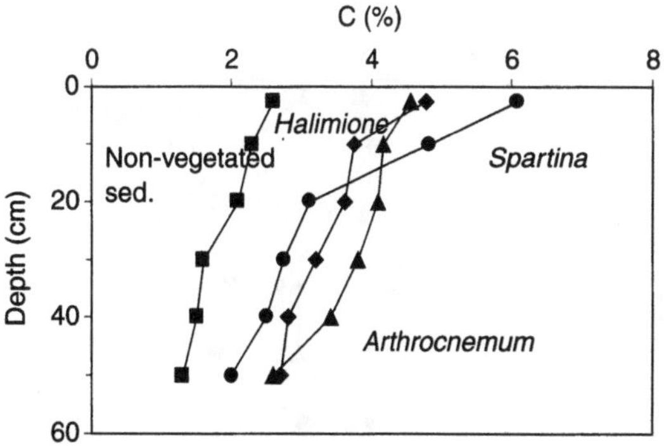

Fig. 4. Ratio (C/N) in sediment between roots of *Spartina maritima* (●), *Halimione portulacoides* (◆), *Arthrocnemum fruticosum* (▲) and non-vegetated sediments (■) from Rosário, a Tagus salt marsh.

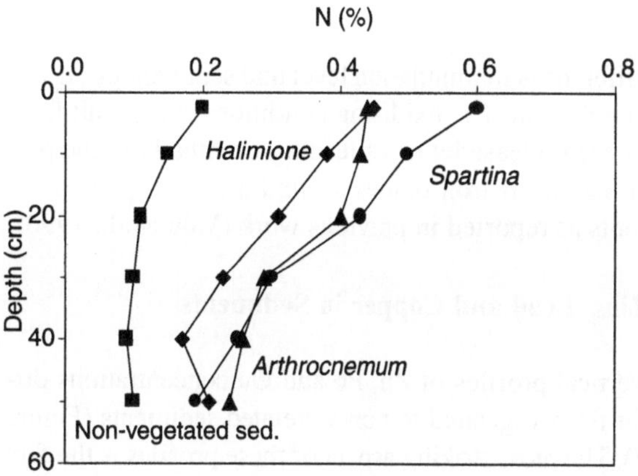

Fig. 5. Vertical profiles of pH in sediment between roots of *Spartina maritima* (●), *Halimione portulacoides* (◆), *Arthrocnemum fruticosum* (▲) and non-vegetated sediments (■) from Rosário, a Tagus salt marsh.

population on C and N budgets should also be taken into account. Accordingly, pH (Figure 5) and redox potential (Figure 6) in sediments between roots of *S. maritima* were lower in bulk sediment, presumably due to the microbial breakdown of organic matter with release of CO_2 (Tinker and Barraclough, 1988). The redox potential of sediment layers from non-vegetated sites and below roots (>30 cm depth) showed lower values than upper sediment layers where larger amounts of roots are present. This indicates that the rhizosphere is a more oxidative and acidic environment than the surrounding sediments. These differences are attributed to plant activity rather than to

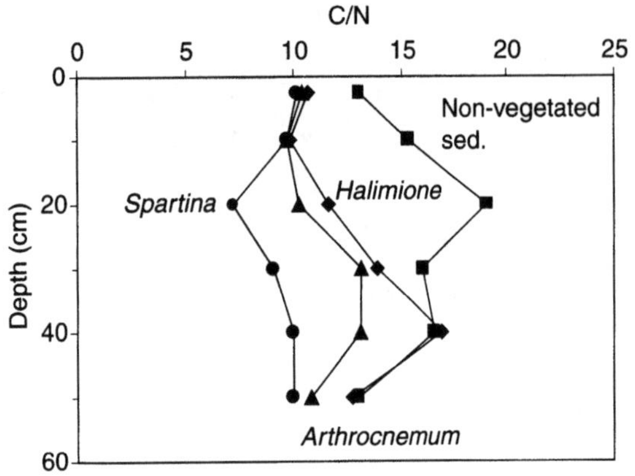

Fig. 6. Vertical profiles of redox potential (Eh) in sediment between roots of *Spartina maritima* (●), *Halimione portulacoides* (◆), *Arthrocnemum fruticosum* (▲) and non-vegetated sediments (■) from Rosário, a Tagus salt marsh.

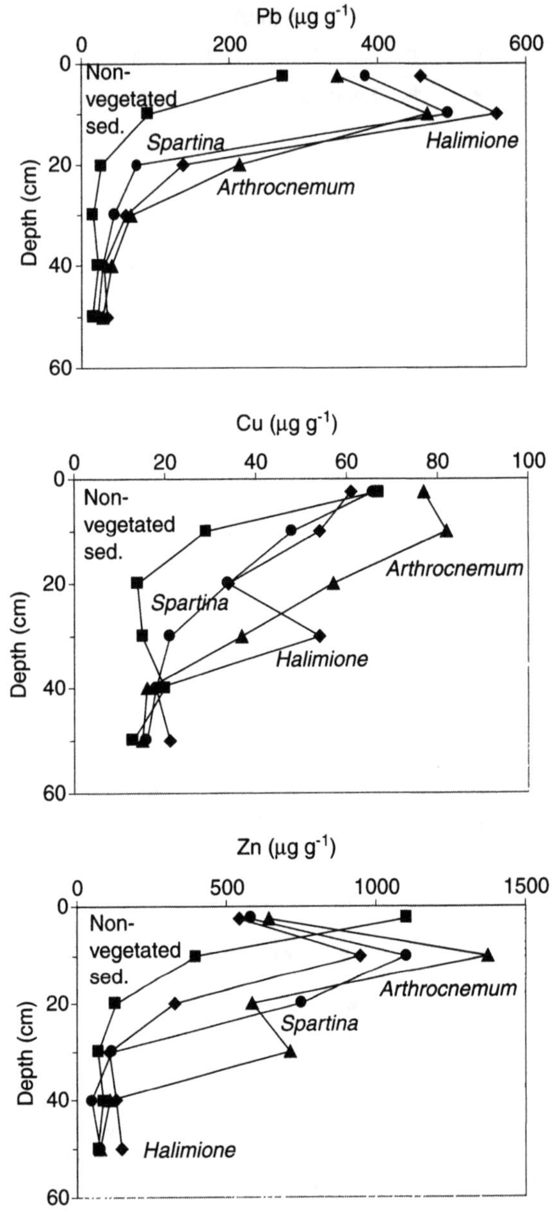

Fig. 7. Vertical profiles of total Zn, Pb and Cu (μg g^{-1}) in sediment between roots of *Spartina maritima* (●), *Halimione portulacoides* (◆), *Arthrocnemum fruticosum* (▲) and non-vegetated sediments (■) from Rosário, a Tagus salt marsh.

differences in inundation level and sediment grain size distribution. The oxidising conditions may result from oxygen release by roots leading eventually to the precipitation of iron oxides around a certain part of the roots as reported in previous work (Vale et al., 1990).

Zinc, Lead and Copper in Sediments

Vertical profiles of Zn, Pb and Cu concentrations differ from vegetated to non-vegetated sediments (Figure 7). The most striking aspect of these profiles is the fact that vegetated sediments contained higher metal contents than non-vegetated sediments. Furthermore, concentration increases at depths of higher root density, in contrast to the metal enrichment of the topmost layers in non-vegetated sediments (Caçador et al., 1993). The shape of metal concentration profiles in vegetated sediments is, thus, not attributed to contamination, rather being related to the presence and activity of roots.

Several researchers have investigated the association between organic matter accumulation in salt marsh sediments and metal concentrations (Ernst, 1990). In some cases, they are directly related and that was interpreted as organic material being a vehicle for the transport of metals from plants. Apparently that was not the case in the Tagus salt marsh studied, since sediments organically enriched and colonised by *S. maritima* showed lower Zn, Pb and Cu concentrations. Other works found also little or no correlation between organic matter content and metal concentrations (Orson et al., 1992). This lack of correlation indicates that organic matter content is not the dominant factor explaining the metal distribution in Tagus salt marsh sediments (Caçador et al., 1999). In fact, available metals extracted by DTPA (expressed in % of total metal concentration) and LOI in sediments between roots are inversely related, which means that metals are preferentially available in sediments impoverished in organic matter (Caçador et al., 1996, 1999). Another possible explanation for metal incorporation in rooting sediments are the complex interactions between roots and sediments. For example, formations of iron oxides enriched in trace metals (Vale et al., 1990), metals incorporated in fresh root debris that forms the sedimentary organic matter, and eventually metals associated with exudates (Morel et al., 1986; Mench and Martin, 1991).

Zinc, Lead and Copper in Plants

Below-ground biomass of *S. maritima, H. portulacoides* and *A. fruticosum* are substantially higher when compared to the above-ground values (Table 1). This is commonly observed in salt marsh plants (Gross et al., 1991) and emphasises the importance of organic matter regeneration at the interior of salt marsh sediments. Roots, stems and leaves of the studied plants contained different levels of zinc, lead and copper (Table 2). Clearly, the higher metal concentrations were found in the root system, and only small fractions were recorded in the above-ground parts. The partition of lead among the analysed parts of the plants was very pronounced, indicating that lead is efficiently stored in roots. Otherwise, zinc and copper are more uniformly distributed, meaning they are partially translocated from the roots to the upper parts of the plant.

Table 1. Minimum and maximum Zn, Pb and Cu concentration (mg/kg) in root system, stem and leaves of *S. maritima, H. portulacoides* and *A. fruticosum*

	Spartina		Halimione		Arthrocnemum	
	Min	Max	Min	Max	Min	Max
	(mg/kg)		(mg/kg)		(mg/kg)	
Zn						
Root system	308	530	485	670	395	430
stems	40	89	39	80	20	67
leaves	40	63	48	90	–	–
Pb						
Root system	208	396	475	580	283	343
stems	2	6	–	4	3	16
leaves	4	7	–	7	–	–
Cu						
Root system	65	82	108	171	32	52
stems	12	18	6	14	15	30
leaves	5	12	4	15	–	–

The halophytes recycle metals in their interior and exchange them with the surrounding sediments. This function appears to be an important vector helping to reduce the effects of metal contamination in estuarine areas. It is well documented that transition elements are accumulated in salt marsh plants during the growing season (Alberts et al., 1990) and the incorporation is more effective in the root system (Rozema et al., 1990). The metal uptake by the roots involves complex processes due to the nature of the root-sediment interactions (Tinker and Barraclough, 1988; Otte, 1991). The chemistry of the rhizosphere is very different from the surrounding sediment environment (Ernst, 1990). As plants colonise the sediment, oxygen is delivered by the roots, sulphide forms are oxidised and the micro-environment around the roots becomes more oxidative and acidic (Madureira et al., 1997; Caçador et al., 1996). In certain parts of the roots iron plaques may be formed (Crowder et al., 1987; Otte et al., 1989). In the Tagus they are degenerated in thick concretions, called rhizoconcretions (Vale et al., 1990). With such a variety of heterogeneous processes, metal uptake by salt marsh plants is still poorly understood. While the transfer of metals, from sediment to root is not well clarified, the Tagus salt marsh plants accumulate considerable quantities of metals. For example, cadmium concentration in roots of *S. maritima, H. portulacoides* and *A. fruticosum* is 2–4 times the levels existing in the bulk sediment (Caçador et al., 2000). The overall effect of accumulation in roots becomes particularly relevant in this ecosystem because the underground biomass is high, and the root:sediment proportion reaches a maximum of 1:4 (weight:weight) at the end of the growing season (Caçador et al., 2000). The roots are, thus, an important component of the salt marsh sediments interfering actively with the metal cycling.

Table 2. Minimum and maximum biomass (kg/m^2) values of root system, stems and leaves of *S. maritima, H. portulacoides* and *A. fruticosum*

	Spartina		Halimione		Arthrocnemum	
	Min	Max	Min	Max	Min	Max
	(kg/m^{-2})		(kg/m^{-2})		(kg/m^{-2})	
Root system	4.10	7.10	4.50	7.70	3.20	5.20
stems	0.10	0.32	0.61	1.05	1.23	2.43
leaves	0.35	0.71	0.12	0.37	–	–
total	4.55	8.13	5.23	9.12	4.43	7.63

As plants die, roots are incorporated into the sediment as organic debris and metals return to the sediment. Only a small fraction of metals transferred to the sediment is rapidly available to plants. Apparently, this fraction barely exceeds 20% and this limit does not vary a great deal with the degree of metal contamination (except for Zn). Metals appear thus to be immobilised as a consequence of root activity and this may prevent the toxic action of these elements in the estuarine ecosystem.

Seasonal Variation of Metal Concentrations in the Root-sediment System

Metal concentrations in roots and sediments between roots in Tagus estuary salt marshes vary with the season (Caçador et al., 2000). The highest metal concentrations occurred in the roots, with lowest levels in January and increasing levels during the growth periods. Zinc, copper and lead in sediments exhibited a corresponding change in concentrations, reaching maximum in January and subsequently decreasing in spring. These results indicate that metals incorporated in the sediments are taken up by the vegetation and accumulated in the root tissues, mainly during the growing season.

Calculation of Excess of Metals in Rooting Sediments

In order to assess the fraction of metals that is recycled between the root and the surrounding sediment, mass balance calculations were done between the two extreme periods of the growing season (August and January).

Considering a hypothetical cylinder of 1 m^2 surface and 35 cm height, the mass of metal (Mme) that is lost by the roots of each plant between the two months (1 and 2) may be estimated by the expression:

$$\text{Mme} = (\text{biomass}_1 \times [\text{Me}]_1) - (\text{biomass}_2 \times [\text{Me}]_2)$$

The quantities of metals obtained for each plant were:

Plant species	Zn (g)	Cu (g)	Pb (g)
Spartina	2.5	0.32	2.0
Halimione	3.0	0.83	2.2
Arthrocnemum	1.0	0.17	0.9

Considering the same sediment volume, the mass of metal (Mep) incorporated in the sediment between roots due to the plant activity may be estimated by the sum of metal accumulated in each sediment slice considered:

$$\text{Mep} = \Sigma\ (\text{mass of dry sediment/m}^2) \times [\text{Me}]_{\text{sed}}$$

The mass of metal incorporated in the sediment (MMe) between the two periods is the difference:

$$\text{MMe} = (\text{Mep}_1) - (\text{Mep}_2)$$

The quantities of metals obtained for the areas colonised by each plant were:

Sediment by	colonised Zn (g)	Cu (g)	Pb (g)
Spartina	40	31	21
Halimione	24	45	30
Arthrocnemum	91	58	29

On the basis of these values one may estimate the percentage of the accumulated metals in roots to the metal variation in surrounding sediments:

Contribution of roots	Zn (%)	Cu (%)	Pb (%)
Spartina	6	1	10
Halimione	13	2	7
Arthrocnemum	1	<1	3

These estimated values were low and indicate that recycling varies with the plant and it is different from element to element. However, it should be mentioned that percentages were calculated on the basis of 1 year in a sediment column of 30 years (Caçador et al., 1996).

CONCLUSIONS

The results from this investigation show that vascular plants in salt marshes are determinant to the dynamics of the estuarine system and strongly influence the processes of accumulation and retention of heavy metals there. Basically through their subterranean components, vascular plants in salt marshes may act as temporary "sinks" for heavy metals there are possibly also an important vector to the incorporation of these metals into the salt marsh sediments. Releasing oxygen to the rhizosphere, vascular plants critically impact the biogeochemistry of the sediments, modifying dramatically the soil characteristics in areas colonised by salt marsh vegetation. This will decrease the mobility and availability of heavy metals.

REFERENCES

Adam, P. Salt marsh ecology. Cambridge University Press, Cambridge, 1990.

Alberts, J., Price, M.T. and Kania, M. Metal concentrations in tissues of *Spartina alterniflora* (Loisel) and sediments of Georgia salt marshes. Estuarine Coastal Shelf Sci 1990; 30:4–58.

Caçador, I., Vale, C. and Catarino F. Effects of plants on the accumulation of Zn, Pb, Cu and Cd in sediments of the Tagus estuary salt marshes, Portugal. In Vernet, J.-P., editor *Environmental Contamination*. Environ Sci 1993; 55:355–65.

Caçador, I., Vale, C. and Catarino F. Accumulation of Zn, Pb, Cu and Ni in sediments between roots of the Tagus estuary salt marshes, Portugal. Estuarine Coastal Shelf Sci 1996; 42:393–403.

Caçador, I., Vale, C. and Catarino F. Relationships between metal concentrations and organic matter content in the Tagus Estuary salt marsh sediments, In Lieth, H. et al. editors. Halophyte uses in different climates I. Backhuys Publishers, Leiden, The Netherlands, 1999; 105–12.

Caçador, I., Vale, C. and Catarino, F. Seasonal variation of Zn, Pb, Cu and Cd concentrations in the root-sediment system of *Spartina maritima* and *Halimione portulacoides* from Tagus estuary sallt marshes. Marine Environ Res 2000; 49:279–90.

Catarino, F.M. and Caçador, M.I. Produção de biomassa e estratégia de desenvolvimento em *Spartina maritima* e outros elementos da vegetação dos sapais do estuário do Tejo. Boletim da Sociedade Broteriana, 2ª sér. 1981; 54:384–403.

Crowder, A.A. and Macfie, S.M. Seasonal deposition of ferric plaque on roots of wetland plants. Can J Bot 1986; 64: 2120–4.

Ernst, W.H.O. Element allocation and (re)translocation in plants and its impact on representative sampling. In Lieth, H. and Markert, B., editors. Element concentration cadasters in ecosystems. VCH Verlagsgesellschaft, Weinheim, 1990; 17–40.

Gee, G.W. and Bauder, J.W. (1986). Particle-size analysis. in American Society of Agronomy-Soil Science Society of America, editors. Methods of soil analysis. Part I, Vol. 9. Physical and Mineralogical Method Agronomy Monograph, Madison, USA, 1986; 9:383–411.

Gross, M.F., Hardisky, M.A., Wolf, P.L. and Klemas, V. Relationship between aboveground and belowground biomass of *Spartina alterniflora* (smoth cordgrass). Estuaries 1991; 14:180–91.

Madureira, M.J., Vale, C. and Simões, M.L. Effect of plants on sulphur geochemistry in the Tagus salt-marsh sediments. Mar Chem 1997; 58:27–37.

Mench, M. and Martin, M. Mobilization of cadmium and other metals from two soils by root exudates of *Zea mays* L., *Nicotiana rustica* L. Plant Soil 1991; 132:187–96.

Morel, J.L., Mench, M. and Guckert, M. Measurement of Pb^{2+}, Cu^{2+} and Cd^{2+} binding with mucilage from maize (*Zea mays* L.) roots, Bio Fer Soils 1986; 2:29–34.

Otte, M.L., Rozema, J. Koster, L., Haarssma, M.S. and Broekman, R.A. Iron plaque on roots of *Aster tripolium* L.: interaction with zinc uptake. New Phytol 1989; 111:309–17.

Otte, R. *Heavy Metals and Arsenic in Vegetation of Salt Marshes and Foodplains*. PhD Thesis, Vrije Universiteit, Amsterdam, 1991.

Orson, R.A., Simpson, R.L. and Good, R.E. A mechanism for the accumulation and retention of heavy metals in tidal freshwater marshes of the Upper Delaware River. Estuarine Coastal Shelf Sci 1992; 34:171–86.

Rozema, J., Otte, M.L., Broekman, R., Kamber, G. and Punte, H. The response of *Spartina anglica* to heavy metal pollution. In Gray, A.J., editor. *Spartina anglica*. HMSO, London, 1990; 39–47.

Tinker, B. and Barraclough, P. Root-soil interactions. In Hutzinger, O., editor. Reactions and processes, 2: Part D. Springer-Verlag, Berlin, 1988; 154–71.

Vale, C. Temporal variations of particulate metals in the Tagus river estuary. Sci Total Environ 1990; 97/98:137–54.

Vale, C., Catarino, F.M., Cortesão, C. and Caçador, M.I. Presence of metal-rich rhizoconcretions on the roots of *Spartina maritima* from the salt marshes of the Tagus estuary, Portugal. Sci Total Environ 1990; 97/98:617–26.

Potential of halophytes as animal fodder in Egypt

HASSAN M. EL SHAER

INTRODUCTION

Feed resources deficiency is considered one of the basic constraints to improve animal productivity in arid and semi-arid regions in Egypt. Improving the fodder supply of desert grazing livestock (sheep, camels and goats), particularly during the prolonged dry seasons, would increase the average annual animal production by approximately 27% (FAO, 1995). Attention has been directed towards the necessity to utilize marginal resources, that is, saline soils and underground water for producing unconventional animal feed, such as halophytes.

The natural rangelands constitute the principal feed resources for livestock in the Egyptian deserts. The biomass production and quality of the natural rangelands in such areas vary considerably from season to season and from area to area, depending on several environmental factors (El Shaer, 1996). Halophytes are widely distributed throughout several regions of Egypt, because of the presence of numerous salines along the Mediterranean Sea and Red Sea shores and inlands (littoral salt marshes and inland salt marshes) as reported by Batanouny (1993).

Halophytes represent a major part of the natural vegetation in Egypt and particularly of the perennial shrubs (Zahran, 1982). The less and unpalatable plant species represent approximately 70% of the total coverage. Several attempts have been made to improve their palatability and nutritional utilization through proper processing methods. Ensiling halophytic plants with other feed ingredients appeared to be the most convenient processing method under the arid conditions of aridity in Egypt (El Shaer et al. 1991; Abou El Nasr et al., 1996). This chapter aims to show the potential of halophytes, particularly as animal fodder under the arid conditions of Egypt.

HALOPHYTE HABITATS

The halophytic vegetation of Egypt comprises 38 community types inhabiting different habitats. Although most halophytes are desert types, coastal and marine types are also prevalent (Batanouny, 1993) distinguishes for different biotops for halophytic communities.

(1) *The coastal salines*: They are prominent and form an important ecosystem at the Red Sea and Mediterranean Sea. Extensive sabkhas are present at levels slightly below or above sea level.

(2) *Great depressions and oases*: Salt marsh plants are widely distributed in oases and depressions such as Cattara Depression in Western Egypt. The common species in such habitats include *Tamarix* spp., *Juncus* spp. and *Alhagi maurorum*.

(3) *Inland saline depressions*: The saline depressions in the desert or in the wadis receive runoff water with dissolved salts. They are of limited size and occur in the Eastern desert of Egypt. These depressions support different halophytes such as *Nitraria retusa, Tamarix* spp., *Zygophylum album* and *Atriplex halimus*.

(4) *Anthropogenic salines*: Salines formed due to human activities are of common occurrence in Egypt as a result of heavy irrigation of the cultivated lands and the lack of drainage. In the oases of the Western Desert of Egypt, areas which were formerly under cultivation have become waste saline lands.

H. Lieth (ed.), Cash Crop Halophytes: Recent Studies, 111–119.

ECONOMIC BENEFITS AND USES OF HALOPHYTES

There are many usages of halophytes in Egypt, as will be listed in brief as follows:

1. *Halophytes as fuelwood and coal*: *Tamarix* trees and *Acacia* spp. trees are as woody plants as a good source for firewood and coal. The trade in coal from *Acacia* trees is very common and represents one of the main jobs of Bedouins in the South Eastern part of Egypt.
2. *As timber*: *Tamarix* spp., *Haloxylon, Prosopis* and *Avicennia* are considered the main timber in Egypt for construction of roofs and some furniture. The mechanical properties of the wood are suitable for making crates and fruit boxes.
3. *As fodders (range plants)*: Several halophytic plants represent the main source of animal feeds in Sinai and the North western coast of Egypt. Animals cannot sustain themselves on halophytes only, but they should be supplemented with other feed resources to produce milk, meat, etc.
4. *As medicine*: The ash of some halophytes is, sometimes, used in the folk medicine, especially for wounds, scabies and burns. For instance, the ash of *Zygophyllum* spp. is used for treating scabies in camels. *Avicennia* can, also, have aphrodisiac characters (Batanouny, 1993).
5. *As wind breaks*: *Tamarix aphylla* and *Prosopis* spp. have been used for a long time as wind breaks in desert areas in different Arab regions.

CHEMICAL AND MINERAL COMPOSITION

Halophytic plant species vary considerably in their chemical composition, nutritive value and palatability as indicated from data cited by many investigators (El Shaer,1981; Abd El Aziz, 1982; El Bassosy, 1983).

Data in Table 1 show that most halophytic shrubs contain moderate amounts of crude protein (CP), which it seems to cover the nitrogen requirements of grazing animals. They also contain high levels of fiber and ash contents which could limit intake and digestibility of such forages. Although, the protein content seems to be high, a large proportions of such protein occurs in the form of non-protein nitrogen. Therefore, available sources of energy should be supplemented to animals for better utilization and efficient digestion. *N. retusa, S. fruticosa* and *A. halimus* can be considered as good fodders because of their palatability for all animal species, in addition to their moderate content of protein. *A. halimus* is slightly less palatable and usually grazed during summer and autumn (El Shaer, 1981; Abd El Aziz, 1982) by sheep, goats and camels.

Data in Table 2 show that all plant species contained sufficient levels of major and trace elements. Most of the tested minerals appear to cover the requirements of livestock except for P and S contents according to the International Recommendation of Mineral Requirements for livestock (Kearl, 1982). Therefore, P and S supplementation is necessary for grazing sheep and goats in particular during the critical stages of animal productivity.

Although some halophytic shrubs show higher levels of some trace or major elements, their mineral

Table 1. The chemical composition (%, on dry matter basis) of most common halophytes in Sinai and the North Western Coast of Egypt

	*PR	DM	Ash	PR	CP	EE	NFE
Nitraria retusa	A	38.1	30.2	31.5	11.3	2.55	24.45
Atriplex halimus	A	35.0	23.1	26.2	13.2	2.30	35.20
Salsola tetrandra	A	38.0	36.1	34.2	6.77	2.41	20.52
Suaeda fruticosa	A	24.1	14.2	30.1	12.1	5.00	38.60
Tamarix aphylla	C	35.1	20.1	14.2	12.2	3.52	49.98
Halocnemun strobilaceum	C	30.1	40.1	17.0	6.92	2.15	33.83
Tamarix mannifera	G,C	42.1	25.2	13.0	8.15	3.11	50.54
Haloxylon salicornicum	Nil	45.1	18.1	29.5	17.5	5.71	34.19
Zygophyllum album	Nil	27.7	30.2	13.7	6.75	2.33	47.02

*PR: Palatability for animal species; S: sheep; G: goat; C: camel; A: All animals; DM: Dry matter; CP: Crude protein; CF: Crude fiber; EE: Ether extract; NFE: Nitrogen-free extract.

Table 2. Overall average values of element concentration in edible parts of halophytic plants (on DM basis) grown in Sinai and the North Western Coast of Egypt

	Ca%	P%	Na%	K%	Mg%	S%	Zn ppm	Cu ppm	Fe ppm	Mn ppm
Nitraria retusa	1.96	0.22	5.35	0.66	0.36	0.14	32	11	578	62
Atriplex halimus	1.69	0.32	3.91	0.57	0.32	0.17	64	10	503	51
Tamarix aphylla	3.73	0.16	2.75	0.78	0.43	0.12	38	13	274	60
Zygophyllum album	2.26	0.14	2.89	1.14	0.64	0.09	41	7.8	393	52
Suaeda fruticosa	2.11	0.41	4.06	1.29	0.30	0.20	55	13	674	88
Salsola tetrandra	3.98	0.16	5.65	1.45	0.59	0.12	44	8.9	664	79
Tamarix mannifera	3.01	0.01	2.70	0.91	0.46	0.09	45	16	291	52
H. salicornicum	4.00	0.15	5.01	1.74	0.33	0.07	73	18	603	80
H. salicovnicum	4.5	0.14	5.21	2.01	0.39	0.10	91	17	621	87
Range	1.7–4.5	0.01–0.41	2.7–5.7	0.6–2.0	0.3–0.6	0.07–0.20	32–91	8–18	274–674	51–88

concentrations are still within normal ranges and without any harmful effects on livestock production (El Shaer, 1986).

UTILIZATION OF FRESH AND DRIED HALOPHYTIC FODDER

Intake and Nutrient Utilization

Voluntary feed intake (VFI) and digestibility are considered the two major components regarding forage quality of grazing ruminants. The process of aging and maturation of the ranges was found to be associated with a decline in digestibility, CP and consequently nutritive value (El Shaer, 1981; El Bassosy, 1983).

Dry matter intake (DMI) and DM digestibility of halophytic forages were higher during the grazing season than in the drought season for both sheep and goats (Table 3). Rams consumed less DMI of forages than bucks in the drought season (34.4 vs. 44.1 g $kg^{-0.75}$) as reported by El Shaer (1981). Le Houérou (1992) reported that sheep became adapted to saltbush and increased their intake of forage over a 3–5 month period.

Mixed diets containing halophytes are acceptable to sheep and goats. Most of halophytic shrubs are high in protein content of moderate digestibility. On the other hand, DMI by sheep and goats from *T. mannifera* and *A. halimus* were markedly higher ($P < 0.05$) than that from *Z. album* and *H. strobilaceum*, since the fresh *Z. album* and *H. strobilaceum* are known to be unpalatable feeds. The poor intake of the fresh and air-dried materials of some species could be attributed to three

Table 3. Dry matter intake (g DM kg^{-1} $W^{-0.75}$) of halophytic species by sheep and goats

	Sheep (S)	Goat (G)	S/G ratio
Fresh state			
A. halimus	19.4	15.8	1.23
H. strobilaceum	18.2	14.2	1.28
T. mannifera	10.9	10.3	1.06
Z. album	2.12	2.94	0.72
Air-dried state			
A. halimus	13.2	21.0	0.63
H. strobilaceum	9.45	8.23	1.15
T. mannifera	5.12	4.75	1.08
Z. album	0.00	0.00	0.00

Source: El Shaer et al. (1990).

main factors: (1) high Na, Ca and silica, (2) higher levels of acid detergent lignin (ADL) and neutral detergent fiber (NDF) and (3) many shrubs contained higher levels of plant secondary metabolites such as tannins, alkaloids, saponins, oxalates, etc.

Neutral detergent fiber (NDF) is a good indicator for forage intake. The low NDF value (50.2%) for fresh *Potamogeton crispus* would explain their higher intake by sheep compared to fresh *Tamarix mannifera* and *Glinus lotoides* (Kandil et al., 1991). Therefore, the low CP content (approximately 6%) and higher levels of NDF, ADL and acid detergent fiber for *T. mannifera* and *G. lotoides* have been considered to limit the forage consumption and digestion, while *P. crispus* showed a contrary trend. Total digestible nutrients (TDN) and digestible crude protein (DCP) were higher for sheep fed *P. crispus* than the other two groups. Also, sheep fed *P. crispus* were only in a positive nitrogen balance while the other groups did not retain nitrogen in their bodies(Kandil et al., 1991). On the other hand, Abou El Nasr et al. (1996) added that sheep fed saltbush hay, fresh *Acacia* and *Acacia* hay were not able to sustain their maintenance requirements of DCP and showed a negative nitrogen balance which was attributed mainly to lower nitrogen intake and live body weight loss.

Growth and Feed Efficiency

In Southern Sinai, Hassan et al. (1980) indicated that sheep lost weight during allover a year (12 months), but losses were minimal in the spring season (26 g day^{-1}) as compared to 134 g day^{-1} in the summer season during a poor raining year. Also, sheep and goats were not able to maintain their live weights even when the pasture reached its best conditions low rainy seasons (Hassan et al., 1980; Warren et al., 1990; El Shaer, 1997). Similar results were obtained by El Shaer (1981) through two successive years. Rams lost weight in drought season (-47.8 g day^{-1}) but gained weight ($+24$ g day^{-1}) in grazing season, whereas, bucks gained weight in both drought and grazing seasons (22.8 and 98.1 g day^{-1}, respectively) in good rainy seasons.

UTILIZATION OF ENSILED HALOPHYTES

The silage is considered a proper conservation method for plants that could improve their palatability and nutritive value since feeding fresh materials could hardly sustain the maintenance requirements of animals. El Shaer et al. (1991) found that the ensiling process improved the quality of halophytes and enhanced their acceptability for sheep and goats.

Physical and Fermentative Characteristics of Silage

Numerous halophytic silages were tested under the arid conditions of Sinai. This chapter will focus on the most common silages that have been applied in Bedouin's farms. Eight silages were made from halophytic natural or/and cultivated shrubs in addition to some other feed ingredients as illustrated in Table 4 (Abou El Nasr et al., 1996; Anon, 1998).

Data on some physical (odour, texture and colour) and fermentation characteristics such as pH, ammonia–nitrogen (NH_3-N), and total volatile fatty acids (VFA's) of the previously mentioned halophytic silages (Table 4) are presented in Table 5 (as cited from Abou El Nasr et al., 1996; Anon, 1998). Concerning the physical traits, it seems that silages varied in their odour, colour, and texture. In general, most silages showed good normal odour, with pleasant aroma. They were moist and the colour ranged from greenish to yellowish colour. Silages containing ensiled materials of a mixture of natural and cultivated shrubs with broiler litter (BL) or anaerobic digested manure (silages 5 and 6) seemed to have better physical traits. Concerning the fermentative characteristics (Table 5), the pH was acidic (ranged from 4.00 for the silages 5 and 6 to 4.60 for the silage 3). Value of ammonia–nitrogen (NH_3–N) and total volatile fatty acids (TVFA) varied among silages. The highest

Table 4. Ensiling materials of the tested silages (%, on DM basis)*

Ingredients	S1**	S2	S3	S4	S5	S6	S7	S8
Natural shrubs								
Tamarix mannifera	30	20	20	15	10	10	–	–
Zygophyllum album	30	20	20	15	10	10	–	–
H. strobilaceum	30	20	20	15	10	10	–	–
Cultivated shrubs								
A. nummularia	–	–	–	18	12	12	80	–
A. saligna	–	–	–	27	18	18	–	80
Broiler litter	–	–	–	–	30	–	–	–
ADM*	–	30	30	–	–	30	–	–
Molasses	10	10	10	10	10	10	20	20
Total	100	100	100	100	100	100	100	100

*S: Silage, see explanation in text.

*Anaerobic digested manure (ADM) as a by-product from a biogas unit.

Table 5. Some physical and fermentative characteristics of the tested silages

Criteria	S1	S2	S3	S4	S5	S6	S7	S8
Physical traits								
Odour	Good	F. good	F. good	V. good	V. good	Good	Good	Good
Texture	F.soft	Fair	Soft	Soft	V. good	F. Soft	Soft	Soft
Colour	Y	Y	G	G	Y	Y	G	G
Fermentative traits								
pH	4.50	4.40	4.66	4.10	4.00	4.00	4.50	4.10
NH_3–N, % DM	4.70	6.50	6.00	4.70	6.90	6.10	0.26	0.18
TVFA's, meq/100ml*	0.22	0.40	0.55	0.20	0.36	0.60	4.48	5.47

F. good = Fairly good ; V. good = Very good; F. soft = fairly soft. S = Silage; G = greenish; Y = yellowish.
*meq/100 ml: milliequivalent/100 millilitre.

Table 6. Overall averages of chemical composition, sodium and potassium concentrations of ensiled ingredients and silages (as % on DM basis)

Criteria	DM	Ash	CP	CF	EE	Na	K	NDF	ADF	ADL
Ensiled ingredients										
A. nummularia	40.4	25.5	12.2	16.1	2.10	4.50	2.30	60.5	33.1	10.3
A. saligna	44.8	10.6	8.80	22.3	2.40	1.10	1.10	63.0	44.1	18.4
T. mannifera	51.5	22.3	7.60	16.0	3.50	2.75	0.79	49.0	33.3	12.2
Z.album	36.0	26.9	6.40	16.0	4.40	2.80	1.26	44.9	31.4	13.2
H.strobilaceum	60.0	29.1	6.50	14.3	3.00	5.00	1.60	41.8	22.8	11.2
Broiler litter (BL)	88.3	16.7	28.0	7.80	1.80	0.51	1.69	–	–	–
ADM	60.0	27.3	11.0	12.5	0.40	0.82	0.53	–	–	–
Silages										
S1	45.0	22.7	7.10	21.3	2.50	1.30	6.88	51.1	28.9	7.50
S2	59.2	22.7	9.10	23.4	1.30	1.67	1.07	52.0	34.1	13.8
S3	45.7	20.8	8.40	11.2	1.50	2.25	1.20	42.7	28.6	13.2
S4	47.4	22.6	7.20	17.8	1.60	1.60	0.95	50.5	33.5	13.8
S5	48.1	23.4	13.9	14.2	2.30	2.00	1.40	40.6	25.4	11.0
S6	49.2	23.3	9.50	15.8	1.50	2.10	1.73	47.4	33.8	15.4
S7	37.1	13.1	10.2	25.3	4.90	–	–	64.7	39.6	11.2
S8	33.1	22.5	11.8	20.4	4.00	–	–	60.1	38.4	8.46

ADM: Anaerobic digested manure; S7: Acacia silage; S8: Saltbush silage.

NH_3–N value (6.90% DM) was recorded for S6 followed by that of S2 (6.50% DM) which included the anaerobic digested manure (ADM). Similar findings were obtained on some ensiled halophytic shrubs by El Shaer et al. (1990 and 1991). Abou El Nasr et al. (1996) reported that total VFA's concentrations for saltbush (*A. nummularia*) and *Acacia*, as percentage of dry matter silages, were 4.48 and 5.47 which indicates good quality silages (Table 5) while NH_3–N%, DM values were lower than those of other silages as the ensiled materials of saltbush or *Acacia* did not include broiler litter or anaerobic digested manure of high non-protein nitrogen. Generally, the present results are of great importance since the available unpalatable halophytic materials could be improved and used as good quality ensiled animal feeds, particularly if halophytic shrubs are ensiled with dried broiler litter and available energy sources.

Chemical and Mineral Composition of Feed Ingredients

Average values of proximate chemical and mineral constituents of ensiled ingredients (natural and cultivated halophytic shrubs, broiler litter and anaerobic digested manure) and silages are summarized in Table 6.

Concerning the ensiled materials, all halophytes attained reasonable crude protein levels ranged from 6.59% (*H.strobilaceum*) to 12.2% (*Atriplex nummularia*) which seemed to be enough to cover the maintenance requirements of nitrogen for small ruminants (Kearl, 1982). Ash content varied greatly among halophytic species and ranged from 10.6% (*A. saligna*) to 29.1% (*H. strobilaceum*). Broiler litter was rich in nitrogen content with 28% CP. The ADM, resulted as a by-product from the biogas unit at Ras Sudr Research Station, contained moderate level of CP (11.0%) with great ash (27.3%) concentrations. All ensiled materials attained reasonable concentrations of sodium (Na) and potassium (K). The highest Na levels were showed in *H. strobilaceum* (5.0%) while the lowest concentrations were observed in the broiler litter (0.51%). *Atriplex nummularia* showed the greatest K concentrations (2.30%) whereas the minimum values were recorded in ADM (0.53%). However, the concentrations of Na and K in the ensiled materials, seemed to be above the dietary requirements of ruminants (Kearl, 1982) without harmful impact on livestock production.

Concerning the ensiled products (Table 6), it appeared that silage 5, which contained ensiled shrubs with broiler litter, contained the highest level of CP (13.9%) due to inclusion of broiler litter together with the other ensiled materials. Silages 1 and 4 showed the lowest CP content (7.10% and 7.20%, respectively) since they included ensiled shrubs only without BL or ADM. In addition the ensiling process resulted in a decrease in CP content of silages (El Shaer et al., 1990, 1991). However, the CP levels in the tested silages seemed to be reasonable to maintain the protein requirements of ruminants. Ash contents of silages were generally high and comparable among silages. Such higher ash content could be attributed to the inclusion of natural halophytic shrubs and ADM. Na and K concentrations of the tested silages appeared to be in normal ranges and enough to cover the requirements of ruminants (Kearl, 1982).

Intake of Ensiled Halophytes and Nutrient Utilization

Khamis (1988) found that ensiling of desert plants improved the palatability of livestock feed and the DMI was increased (Table 7). Maximum intakes were recorded for goats and sheep fed *H. strobilaceum* and *A. halimus* followed closely by *T. aphylla*, whereas *Z. album* silage was inferior.

The same author summarized the data of DMI, digestibility, TDN and DCP for the different silages (Table 7). It seems, that animals were able to cover their maintenance requirements. Also, voluntary intake was increased by ensiling a mixture of halophytic species, that is, *A. halimus, H. strobilaceum, T. mannifera* and *Z. album* with ground date seeds. He added, also, that ensiling of less or unpalatable halophytes with other feed ingredients would improve forage acceptability.

Sheep and goats had positive nitrogen balance except when they were fed on *Z. album* silage. Animals also consumed more free water when fed *H. strobilaceum* silage in comparison with the other three silages. On the same trend, in a study on goats and sheep fed broiler litter ensiled with some desert shrubs in Sinai, El-Shaer et al. (1990) reported that ensiling these forages with broiler litter (BL) improved their palatability and DMI as compared with

Table 7. Intake and digestion of halophytic silages

	T. mannifera		*A. halimus*		*Z. album*		*H. strobilaceum*	
Items	Goats	Sheep	Goats	Sheep	Goats	Sheep	Goats	Sheep
DMI (g Kg^{-1} $w^{-0.75}$ day^{-1})	42.15	44.17	59.69	59.87	22.53	24.54	59.41	58.21
Silage digestion (%)								
DM	51.54	50.81	60.72	58.09	49.89	47.3	66.05	61.67
CF	42.42	45.65	67.44	54.26	21.30	40.46	64.37	64.23
CP	56.60	59.00	65.67	74.29	66.14	52.01	66.99	83.50
EE	71.79	67.46	73.27	75.37	50.56	56.08	70.83	63.70
NFE	47.65	48.99	59.14	60.85	52.34	41.45	64.60	58.75
TDN (g $w^{-0.75}$ day^{-1})	17.34	18.70	32.52	32.22	8.81	9.64	29.84	28.67
DCP (g $w^{-0.75}$ day^{-1})	2.93	3.20	5.62	6.37	1.62	1.95	4.56	5.57

fresh and air-dried materials. They added also that ensiling such shrubs with BL prevents nitrogen losses and enhances the conversion of the available NPN into true protein.

Each *Acacia saligna* and *Atriplex nummularia* (saltbush) were offered to sheep in three forms: fresh, air-dried and ensiled according to the method of Abou El-Nasr et al. (1996). The authors noticed that maximum DMI were achieved by sheep fed the ensiled forages. DM, CP, EE, NFE and neutral detergent fiber nutrients were efficiently utilized and digested by sheep given the ensiled shrubs. In addition, the most nutritious diets were saltbush and acacia silage which provided sufficient digested nutrients (TDN and DCP) to meet maintenance requirements of sheep. Also, sheep fed with silage and fresh saltbush diets were in a positive nitrogen balance.

Feeding and digestibility trials were conducted by El Shaer et al. (2000) on 30 mature Baladi male goats (in 6 equal treatments) fed the above mentioned silages (the first six silages). Animals in the first three treatments T1, T2, and T3 were fed on a mixture of the fresh cultivated shrubs in addition to one of the silages: S1, S2 and S3 whereas those in T4, T5 and T6 were fed on S4, S5 and S6 as sole basal diets (El Shaer et al. 2000). Data from this study indicated that voluntary feed intake gradually increased and reached the peak in the 4th week of the feeding trial. Data from the metabolism trial indicated that the highest silage intake was attained for animals fed silage 5 (32.5 g DM Kg^{-1} BW and 5.01 g CP kg^{-1} BW) which contained the ensiled materials of natural and cultivated shrubs. Meanwhile, the lowest intakes were recorded for animals in T6. Digestibility coefficients of DM, CP and OM significantly ($P < 0.05$) varied among treatments. The maximum TDN and DCP intakes were attained by goats fed silage 5. Nitrogen retention was positive for all animal groups except for those in T2 and T6. Animals in all treatments showed a positive Na balance and retained various amount of Na. Similar trends were shown for K retention except for those in T1 and T2. The authors highly recommend silage 5 as a good quality basal diet for goats under the arid conditions of Sinai in Egypt.

Growth and Feed Efficiency

El Shaer et al. (1991) illustrated the performance of sheep and goats fed *H. strobilaceum* with broiler litter silage and the control diet (Table 8). The data show that sheep gained slightly similar weight when they were fed on berseem hay as well as on the tested silage (HS- BL silage) whereas goats fed the HS-BL silage gained more than those fed the berseem hay (72 vs. 65 g). Feeding such silage appeared to be more economic since feed costs decreased about one third in comparison with the conventional diet (berseem hay).

On the other hand, Abou El-Nasr et al. (1996) reported that low intake of the saltbush (*A. nummularia*) hay, fresh *Acacia* hay affected body weight changes of sheep as they tended to lose weight. In contrast, sheep fed saltbush silage gained weight (91.1 g day^{-1}).

Results of body weight changes of sheep fed the tested basal diets (the six silages as previously described during a feeding trial (Anon, 1999) is summarized in Table 9. Although the sheep used in this study were mature, some sheep gained noticeable and others lost weight (T4 and T6). The fluctuations in body weight changes were correlated mainly to the

Table 8. Growth of sheep and goats fed halophytic silage vs. Berseem Hay

	Sheep		Goats	
Item	Berseem hay	HS–BL[1] silage	Berseem hay	HS–BL silage
Average daily gain, g	71.6	73.3	65.0	71.7
Average daily DMI, g kg^{-1} $w^{-0.75}$	70.8	74.3	66.2	77.0
Feed conversion ratio:				
kg DM feed kg^{-1} gain	14.0	14.2	11.1	11.9
kg TDN kg^{-1} gain	6.88	7.78	5.67	5.89
Feed cost, LE[2] kg^{-1} gain	3.79	1.21	3.02	1.01

Source: El Shaer et al. (1990).

[1]HS–BL silage: *H. strobilaceum*–broiler litter silage.

[2]LE: Egyptian pounds.

Table 9. Body weight changes of native sheep fed the halophytic silages as basal diets

Criteria	T1	T2	T3	T4	T5	T6
Initial body weight, Kg	46.6	49.0	45.0	46.0	42.0	41.0
Final body weight, Kg	48.0^{a}	50.0^{a}	46.3^{a}	$44.6^{a,b}$	$44.3^{a,b}$	37.7^{b}
Body weight changes						
Kg	$1.40^{a,b}$	$1.00^{a,b}$	$1.30^{a,b}$	$-1.40^{b,c}$	2.30^{a}	-3.30^{c}
% of initial body wt.	$2.95^{a,b}$	$2.15^{a,b}$	$3.19^{a,b}$	$-2.73^{b,c}$	5.54^{a}	-7.79^{c}

^{a,b,c}values in the same row with different superscripts are different ($P < 0.01$).

DMI factor. For instance, animals in T6 showed the highest weight loss due to the least voluntary feed intake. The opposite was true for sheep fed S5 in T5 which showed higher gain associated with the highest DMI (34.8 vs. 13.2 g DM kg^{-1} BW) as shown in Table 9 (Anon, 1999).

In general, the data on processing halophytes as silages point towards the recommendation of silage mixtures of halophytic shrubs incorporated with other feed materials such as molasses, broiler litter, urea, fodder beet pulp, etc. as good animal feeds that can be used economically and efficiently. Thus, halophytes as rangeland vegetation can add another useful feed component under desert conditions.

CONCLUSION

Alternative feed resources, particularly from the less halophytic plant materials, in addition to the palatable ones could provide additional feeds for livestock on Egypt's rangelands. These halophytic plant species may play an important economic role for the bedouins, providing economic animal feeds throughout the seasons. Additionally cultivation of some halophytic species on saline non-productive soils also, should be studied in Egypt.

REFERENCES

Abd ElAziz, H.M. A study of the nutritive value of some range plants in the North-Western coastal desert. Ph.D. Thesis, Faculty of Agriculture, Ain Shams University, Egypt, 1982.

Abou El Nasr, H.M., Kandil, H.M., El Kerdawy, D.A., Khamis, H.S. and El Shaer, H.M. Value of processed saltbush and Acacia shrubs as sheep fodder under arid conditions of Egypt. Small Ruminants Res 1996; 24:15–20.

Anon. Sustainable halophyte utilization in the Mediterranean and subtropical dry region. Annual Report (June, 1998). Concerted Action Project, EC contract: IC 18CT 96-0055, Osnabrueck University, Germany, 1998.

Anon. Palatability and nutrients utilization of some halophytic silages fed to sheep. Annual Report, Concerted Action Project, EC contract: IC 18 CT 96-0055, Osnabrueck University, Germany, 1999.

Batanouny, K.H. Ecophysiology of halophytes and their traditional use in the Arab world. Advanced Course on Halophytes Utilization in Agriculture, 12–23 September 1993, Agadir, Morocco, 1993.

El Bassosy, A.A. Study of nutritive value of some range plants from Saloom to Mersa Mattroh. Ph.D. Thesis, Faculty of Agriculture, Ain Shams University, Shobra El-Khema, Egypt. 1983.

El Shaer, H.M. A comparative nutrition study on sheep and goats grazing Southern Sinai desert range with supplements. Ph.D. Thesis, Faculty of Agriculture, Ain Shams University, Egypt, 1981.

El Shaer, H.M. Seasonal variation of the mineral composition of the natural pasture of Southern Sinai grazed by sheep. In Proceedings of the 2nd Egyptian–British Conference on Animal and Poultry Production, August 1986, 26–28 , Bangor, Gwynedd, UK, 1986.

El Shaer, H.M., Salem, O.A., Khamis, H.S., Shalaby, A.S. and Farid, M.F.A. Nutritional comparison studies on goats and sheep fed broiler litter ensiled with desert shrubs in Sinai. In Proceedings of the International Goat Production Symposium, 22–26 October 1990. Tallahassee, FL, USA, 1990; 70–74.

El Shaer, H.M., Kandil, H.M. and Khamis, H.S. Salt marsh plants ensiled with dried broiler litter as a feedstuff for sheep and goats. J Agric Sci Mansoura Univ. 1991; 16:1524–34.

El Shaer, H.M. Rangelands as feed resources in the Egyptian desert: Management and Improvement. In Proceedings of the International Symposium on Desert Development in the Arab Gulf Countries, State of Kuwait, March 1996, 23–26 .

El Shaer, H.M. Practical approaches for improving utilization of feed resources under extensive production system in Sinai. In Proceedings of the International Symposium on Systems of Sheep and Goat Production, Bella, Italy, October 1997, 25–27.

El Shaer, H.M., Fahmy, A.A., Abdul-Aziz, G.M., Shalaby, A.S. and Abd El Gawad, A.M. Nutritional evaluation of some halophytic silages fed to goats in Sinai. Proceedings of the International Conference on Goats, May 2000, 15–18, Tourse, France, 2000.

FAO. Sustainable range – dependant small ruminant production system in the Near East Region. FAO Regional Office for the Near East, Cairo, Egypt, 1995.

Hassan, N.I., El-Serafy, A.M. and Farid, M.F.A. Studies on pastures indigenous to Southern Sinai. I. Chemical composition and *in vitro* digestibility. J Anim Sci 1980; 51(Suppl. 1):239.

Khamis, H.S. Nutritional studies on some agricultural-by products and some natural pasture plants in arid and semi-arid areas using sheep and goats. Ph.D. Thesis, Faculty of Agriculture, Cairo University, Egypt, 1988.

Kandil, H.M., El Shaer, H.M., Khamis, H.S. and Ahmed, A.M. Nutritional value of hyper – arid forage species for sheep in Upper Egypt. J Agric Sci Mansoura Univ 1991; 16:518–25.

Kearl, L.C. Nutrient requirements of ruminants in developing countries. Utah Agricultural Experimental Station, Logan, Utah, USA, 1982.

Le Houérou, H.N. The role of saltbushes (*Atriplex* spp.) in arid land rehabilitation in the Mediterranean Basin: A review. Agrofor Syst 1992; 18:107–48.

Zahran, M.A. Ecology of the halophytic vegetation of Egypt. In Sen, D.N. and Rajpurahit, K.S., editors. Contribution to the ecology of halophytes. W. Junk Publ., The Hague, Netherlands, 1982; 3–20.

Lipid composition of seeds of local halophytes: *Cakile maritima*, *Zygophyllum album* and *Crithmum maritimum*

M. ZARROUK, H. EL ALMI, N. BEN YOUSSEF, N. SLEIMI, A. SMAOUI, D. BEN MILED AND C. ABDELLY

INTRODUCTION

Halophytes, naturally adapted to high salinity, have been traditionally grazed by animals for a long time in arid areas of north African countries (Le Houérou, 1992; Wilson, 1994; Choukr-Allah, 1996). Several species such as *Atriplex halimus* and *A. nummularia* could be considered as potential sources of fodder in the Mediterranean basin. Outside this utilisation, most halophytes remain at present largely undomesticated. However, recent studies have allowed the consideration of a wide variety of seed-producing halophytes as new sources of grains or vegetable oils. Some salt-tolerant grains and oilseeds of species such as *Salicornia bigelovii* have already been used or examined (Shay, 1990; Glenn, 1991). In order to identify local Tunisian halophytes presenting an economical interest as oilseed-species, we have explored some salty Tunisian areas. Three interesting species have been identified as oil producers which are called *Zygophyllum album*, *Cakile maritima* and *Crithmum maritimum*. The first species is a perennial plant and has been identified in salty depressions, while the second is an annual plant localised in marine sands and the latter is found in rocky coasts in the north of Tunisia (Novikoff, 1961).

MATERIALS AND METHODS

Z. album, *C. maritima* and *C. maritimum* are characterised by a high biomass production, an important seed yield, a suitable seed size and also a homogeneous ripening of seeds. Seeds were harvested in the natural biotope of the plants, from Sebkha of Soliman (about 30 km in the south of Tunis) for *Z. album* in October and from the beach of Raoued (suburb of Tunis) for *C. maritima* in July and from rocky coasts of Kourbous (about 40 km in the east south of Tunis) for *C. maritimum* in August–July. The lipid extraction was carried out according to the procedure of Allen and Good (1971) with the following solvent mixture: chloroform/methanol/water (1:1:1, v/v/v), or using the soxhlet method (Burghart et al., 1987).

Lipid classes were separated from total lipid extract by thin layer chromatography according to the method of Mangold (1964). Fatty acid methyl esters were analysed by gas liquid chromatography in a Hewlett-Packard chromatograph equipped with flame ionisation detector (FID) and a capillary column Supelcowax 10 (30 m × 0.53 mm internal diameter, film thickness 1 μm) maintained at an isothermal temperature of 200°C during all the time of analysis. Dry matter and water contents were determined after desiccation of seeds of halophytes and rapeseed or fruits of olive, at 70°C during one week.

RESULTS

Dry Matter and Water Tissue Contents of Seeds

Considering seed size criteria, it appeared that dry weight of 1000 seeds is the highest in *C. maritima* representing 773.9 mg while it represents respectively 161.1 mg in *C. maritimum* and only 132.6 mg in *Z. album* (Figure 1).

The dry matter of *C. maritima* seeds is nearly three times higher than that of rapeseed (*Brassica napus*), a conventional oleaginous plant. In contrast, seeds of the two other halophytes contain a lot less dry matter

H. Lieth (ed.), Cash Crop Halophytes: Recent Studies, 121–124.

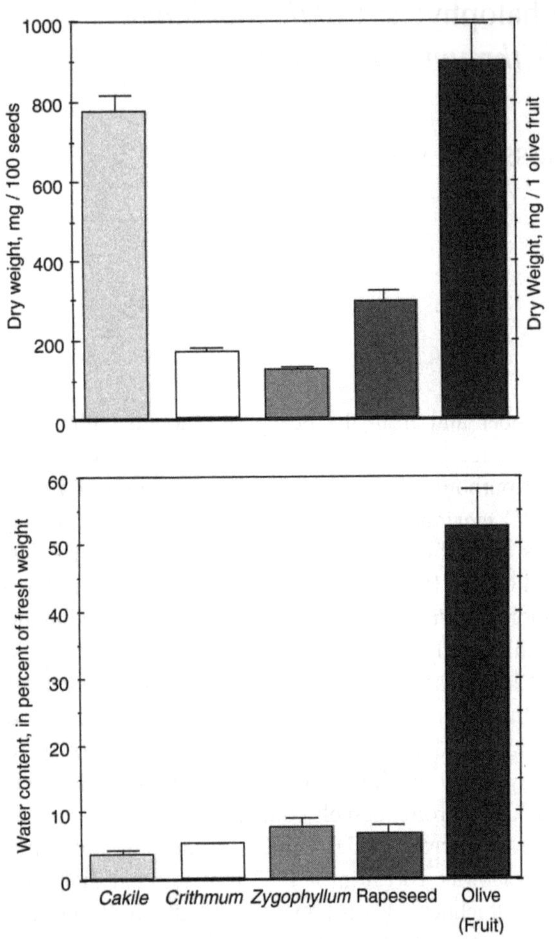

Fig. 1. Dry matter and water tissue contents in seeds of halophytes as compared to rapeseed and olive fruits.

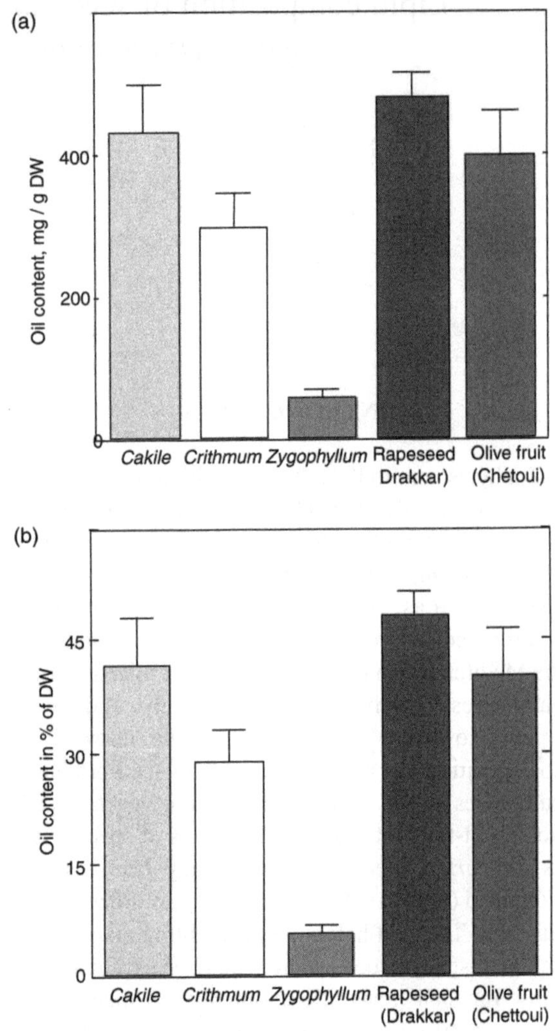

Fig. 2. Oil content of seeds of halophytes (a) in mg g^{-1} DW and (b) in percent of DW as compared to conventional oleaginous plant: olive (fruits) and rapeseed.

than *Cakile*, representing approximately 50% of that of rape ones. Dry matter of olive fruit is by far higher than that of halophytes and rapeseed. Nevertheless, the three studied halophytes produce seeds suitable for harvest size and oil extraction.

The tissue water content in ripened seeds is higher in *Zygophyllum album* than in the two other halophytic species. However, water content of seeds never exceed 10% of fresh matter and is therefore in the same range as fresh matter of the most cultivated crops (Multon, 1982), but it remained largely below value for olive fruit.

Oil Content of Seeds

The seeds of *C. maritima* and *C. maritimum* present high levels of oil content forming respectively 42% and 30% on dry weight basis. However, oil content in *Z. album* seeds is very low and constitutes no more than 6%. Therefore, such findings indicate that seeds of *C. maritima* and *C. maritimum* contain high amounts of oil as conventional oleaginous species like rapeseed or olive (*Olea europea*). Both these halophytes could be considered as oleaginous species (Figure 2).

LIPID CATEGORIES

The main lipid fractions isolated from seed oil of the studied halophytes are: triacylglycerol (TAG), free

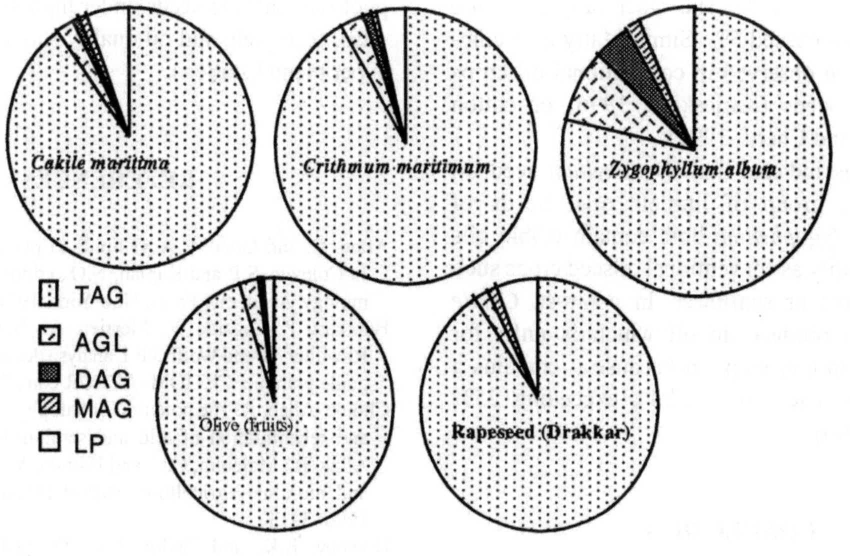

Fig. 3. Lipid composition of seeds of halophytes (in percent of total) as compared to rapeseed and olive fruits.

fatty acids (FFA), diacylglycerols (DAG), monoacylglycerol (MAG) and polar lipids (PL). The TAG, main form of lipid storage in oilseeds, constituted the major lipid class in oils extracted from the three studied halophytes (Figure 3).

Oils derived from seeds of *C. maritima* and *C. maritimum* have similar lipid composition as seeds or fruits of classic oleaginous crops like rape and olive. Oil from seeds of *Z. album* contain high level of FFA and PL.

Fatty Acid Composition

Seed oil of *C. maritimum* is high in monounsaturated fatty acids, particularly oleic acid (C18:1) which accounted for 81% of total fatty acids (Figure 4). Linoleic acid (C18:2) is less represented and forms only 12%, while saturated fatty acids [palmitic (C16:0) and stearic (C18:0)] do not exceed together 6%. Only traces of linolenic acid (C18:3) are detected, no more than 0.1%. This oil belongs to the group of oleic oils. The fatty acid composition of *Crithmum* oil is very similar as the one of olive oil (Zarrouk et al., 1996) or rapseed oil (Karleskind, 1996).

In contrast, the fatty acids of *Zygophyllum* seeds oil are characterised by a high level of diunsaturated linoleic acid, which represents 64% of total fatty acids. Oleic acid level amounts to 19% followed by

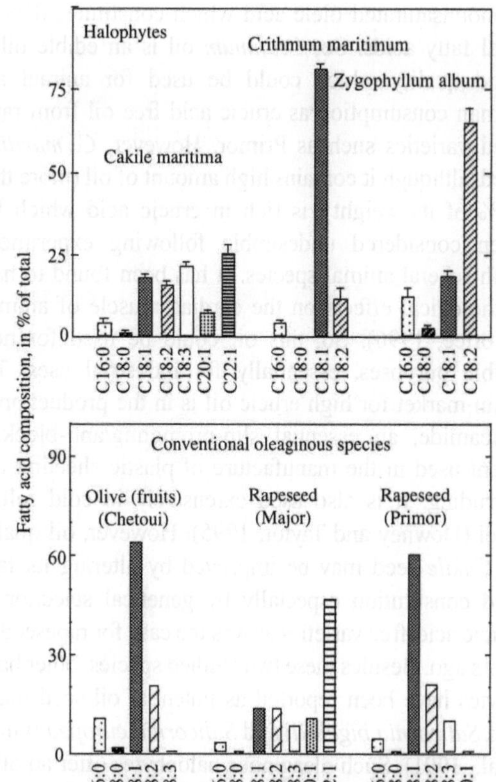

Fig. 4. Fatty acid composition of seeds of halophytes (in percent of total) and oilseed crops.

stearic acid forming 12%. The other fatty acids are minor and do not exceed 3%. Similar fatty acid composition has been observed in conventional oil crops like sunflower seeds (Karleskind, 1996) or cotton seeds (Smaoui and Chérif, 1992).

Such findings indicate that *Zygophyllum* seed oil belongs to the group of linoleic oils. Seeds of *Crithmum* and *Zygophyllum* both contain edible oils with similar quality as conventional oilseed crops such as soybean, olive or sunflower. In contrast, *Cakile maritima* seeds produce an oil which is unfit for animal or human consumption because it contains a high level of erucic acid (C22:1), exceeding 25% (Karleskind, 1996).

CONCLUSION

The two species *C. maritimum* and *C. maritima* may be considered as potential oilseed crops interesting for oil production. *Crithmum* seed contains high oil content representing 30% of its weight. The oil is rich in monounsaturated oleic acid which constitutes 81% of total fatty acids. So, *Crithmum* oil is an edible oil of good quality which could be used for animal and human consumption as erucic acid free oil from rapeseed varieties such as Primor. However, *C. maritima* seed, although it contains high amount of oil (more than 40% of its weight), is rich in erucic acid which has been considered undesirable following experiments with several animal species. It has been found to have pathological effects on the cardiac muscle of animals (Morice, 1996). So, this oil could be used for non-edible purposes, essentially for industrial uses. The main market for high erucic oil is in the production of erucamide, an essential slip-promoting/anti-blocking agent used in the manufacture of plastic sheeting and extruding. It is also used extensively in cold rolling steel (Downey and Taylor, 1995). However, oil quality of *Cakile* seed may be improved by altering its fatty acid constitution especially by genetical selection of erucic acid-free varieties as was the case for rapeseed 20 years ago. Besides these two studied species, other halophytes have been reported as potential oil seed plants like *Salicornia bigelovii* and *Salicornia europea* (Glenn et al., 1991). Such oleaginous halophytes offer an other alternative to valorise saline lands. However, domestication of these species in salty areas required other studies in order to evaluate some production parameters such as growth, nutrient uptake, photosynthesis and transport of photosynthates to seeds under high-salinity conditions. Also, seed yield and oil quality should be examined in experimental stations.

REFERENCES

Allen, C. and Good, P. Acyl lipids in photosynthetic systems. In Colowic, S.P. and Kaplan, N.O., editors. Methods in enzymology. Academic Press, New York, 1971; 23:523–47.

Burghart, P., Lagarde, F., Merrien, A., Nouat, E., Prevot, A., Ribaillier, D. and Wolff, J.P. L'analyse des graines oléagineuses, guide pratique. CETIOM–AFNOR edn., Paris, 1987; 1–226.

Choukr-Allah, R. The potential of halophyte in the development and rehabilitation of arid and semi-arid zones. In Choukr-Allah, R., Malcolm, C.C. and Hamdy, A., editors. Halophytes and biosaline agriculture. Marcel Dekker, Inc., New York, 1996; 3–13.

Downey, R.K. and Taylor, D.C. Diversification of canola/rapeseed fatty acid supply for the year 2000. In Parmentier, M. and Fanni, J., editors. Jounées Chevreul de Nancy: Oléagineux et Corps Gras du Futur. Nancy, 1995; 7–17.

Glenn, E.P., O'Leary, J.W., Watson, M.C., Thompson, T.L. and Kuehl R.O. Salicornia bigelovii Torr: An oil seed halophyte for sea water irrigation. Science 1991; 251:1065–7.

Karleskind, A., editor. Oils and fats manual, a comprehensive treatise, Vol. 1. Lavoisier Publishing, 1996; 1–806.

Le Houérou, H.N. The role of salt bushes (*Atriplex* spp.) in arid land rehabilitation in the Mediterranean basin. Rev Agrofor Syst 1992; 18:107–48.

Mangold, H.K. Thin layer chromatography of lipids. J Am Oil Chemist's Soc 1964; 41:762–73.

Morice, J. Rapeseed and mustard. In Karleskind, A., editor. Oils and fats manual, a comprehensive treatise, Vol. 1. Lavoisier Publishing, 1996; 125–33.

Multon, J.L. Conservation et stockage des grains et graines et produits derives, Vol. 1. Technique et documentation Lavoisier, Paris, 1982; 1–576.

Novikoff, G. Contribution à l'étude des relations entre le sol et la végétation halophile de Tunisie. Annales de l'Institut National de la Recherche Agronomique de Tunisie, 34:1–339.

Shay, E.G. Saline agriculture, salt tolerant plants for developing countries. In reports of a panel of the board on science and technology for international development, National Academy Press, Washington, DC, 1990; 17–39.

Smaoui, A. and Chérif, A. Glycerolipids in salt stressed cotton seeds. In Chérif, A., Ben Miled, D., Marzouk, B., Smaoui, A. and Zarrouk, M., editors. Metabolism structure and utilization of plant lipids. CNP, Tunis, 1992; 341–3

Wilson, A.D. Halophytes and shrubs in semi-arid regions of Australia: value for grazing and land stabilization. In Squires, V.R. and Ayoub, A.T., editors. Halophytes as a resource for livestock and for rehabilitation of degraded lands. Nairobi, Kenya, Kluwer Academic Publishers, Netherlands, 1994; 101–13.

Zarrouk, M., Marzouk, B., Daoud, D. and Chérif, A. Oil accumulation in olives and effect of salt on their composition. Olivae 1996; 61:41–5.

Part III

Field Scale Investigations

Introduction to Part III

H. LIETH

If we have enough information about the physiology and the chemical content of the plant biomass we need to demonstrate that the selected plants can be grown in the field according to the acceptable agronomic techniques.

For these studies different parameters and constraints are to be analysed. It is clear that a large variety of cropping techniques have to be tested, many more than those we can include here from the studies undertaken by the colleagues we had in our group. But each example included in this part offers the possibility to undertake similar studies with other species of interest or to undertake studies with other agronomically feasible techniques. The final goal for each experimental design must be the proof of a sustainable production possibility under the chosen experimental constraints. Only when this point is clarified with satisfaction we can start the analysis of an economical feasibility.

This part is not included in this volume, because it was covered in the following recent publications:

Progress in Biometeorologie, series editor H. Lieth, volume 13, Halophyte uses in different climates I, Ecological and Ecophysiological Studies, edited by H. Lieth, M. Moschenko, M. Lohman, H.-W. Koyro and A. Hamdy, Backhuys Publishers, Leiden, 1999

Progress in Biometeorologie, series editor H. Lieth, volume 14, Halophyte uses in different climates II, Halophyte Crop Development: Pilot Studies, edited by A. Hamdy, H. Lieth, M. Todorović and M. Moschenko, Backhuys Publishers, Leiden, 1999

Progress in Biometeorologie, series editor H. Lieth, volume 15, Halophyte uses in different climates III, Computer-Aided Analysis of Socio-Economic Aspects of the Sustainable Utilisation of Halophytes, Backhuys Publishers, Leiden, 2001

Halophytes in different climates IV, Cash crop halophytes for future halophyte growers, 2nd edition prepared by Helmut Lieth and Marina Mochtchenko on behalf of the International Society of Halophyte Utilization, for the QUEST Conference for the Middle East and North Africa in Doha/Qatar – January 27–30, 2002

H. Lieth (ed.), Cash Crop Halophytes: Recent Studies, 127.

The use of saltgrass (*Distichlis spicata*) as a pioneer forage crop in salty environments

AMNON BUSTAN AND DOV PASTERNAK

INTRODUCTION

Almost 1 billion hectares of land in the world is affected by salinity and over 50% of irrigated lands have salinity problems (Flowers and Yeo, 1995). Salinization has been identified as a major process leading to desertification of agricultural lands. The process of salinization of arable lands is mostly (but not solely) caused by human activities, and it is associated with malpractices in irrigation projects, mostly in arid and semiarid countries. Soil salinization can also result from natural processes, such as inland intrusion of seawater (i.e. coastal areas of West Africa and northeast China) or due to continuous evaporation from shallow aquifers in desert environments (i.e. coastal regions of north Africa and the Pampas del Tamarugal in Chile).

In irrigated lands salinization process of the soil takes place under two different conditions: (a) where evaporation rates exceed irrigation rates, or (b) where drainage systems are inefficient (or do not exist at all). Salt that is dissolved in the irrigation water is continuously accumulated in the upper soil profile due to evaporation of pure water to the atmosphere. Additional water above the amount required by plant transpiration and evaporation must be applied, at least occasionally, to leach out the salt that has accumulated during previous irrigations. The removal of the excess water from the field is essential to avoid buildup of shallow groundwater or contamination of groundwater with salt and fertilizer. This leaching procedure requires regional, efficient, carefully designed drainage systems. Of course, the quality of the irrigation water (e.g. fertilizers and salt concentrations), and soil properties have a primary influence on the rate of the salinization process. Scarcity of fresh water, particularly for agriculture, is a major concern for most semiarid and sub-humid regions throughout the world. The uses of saline water ($EC_i > 3$ dS m^{-1}) for irrigation to replace the diminishing supply of fresh water for agriculture must be conducted carefully to avoid acceleration of soil salinization in these regions.

Salt has adverse affects both on soils and plants. A soil with an EC > 4 dS m^{-1} in the saturated-soil extract is traditionally characterized as a saline soil (Richards, 1954). Salinity and sodicity affect soil structure, which must be stable for adequate air and water permeability (Shainberg and Singer, 1990). High sodium levels combined with low soil-water EC can lower soil permeability and decrease its infiltration capacity through the swelling and dispersion of clays and the slaking of aggregates. The disturbance of soil structure and the subsequent reduction in water infiltration capacity may decrease water availability to plants, thus reducing crop growth and yields. Furthermore, salinity has also direct effects on plants (Greenway and Munns, 1980; Läuchli and Epstein, 1990). Salinity depresses the external water potential, thus narrows the gap between the external and the internal water potential. The water availability to the plant is lessened, at least initially, and plants' growth and productivity might decline. High concentrations of given ions may cause disorders in mineral nutrition. For example, high sodium concentrations may cause deficiencies of other elements, such as potassium or calcium. Certain ions, such as sodium and chloride, may have toxic effects that also cause reduced growth and productivity. Most crop plants are glycophytes or non-halophytes that do not express fully their genetic potential for growth, development and productivity under salt stress, therefore they lose their economical value as salinity levels increase. Unfortunately, the salinity of the future agricultural environment is expected to increase with the spreading use of

H. Lieth (ed.), Cash Crop Halophytes: Recent Studies, 129–134.

irrigation to arid and semiarid lands and the decrease in the availability and quality of water for agriculture.

THE RECLAMATION OF SALINE SOILS

In the process of reclamation of salty soils large efforts must be put to reduce salt concentration in the soil to levels that enable revegetation. The predominant way to do so is leaching salt below plants' root zone (2 m) using large amounts of water. The efficiency of the leaching process depends on soil type and texture, and on the difference between soil and water salt concentrations. The larger the portion of large particles in soil texture the faster the leaching process would be. Saline water may be used to leach soil as long as the salt concentration of the water is significantly lower than that of the soil. The leaching process can be carried out only if the removal of the excess water away is provided. The build-up of shallow groundwater reduces the efficiency of salt leaching, and might prevent it quite soon. The danger of contamination of good-quality groundwater should be also taken into consideration. Therefore, construction of drainage systems must be included in any salt-leaching project that is carried out, if sustainability is sought.

Saline soils are often sodic. Reclamation of such soils can be done by means of chemical (mostly $CaSO_4$) amendments. The calcium replaces the sodium adsorbed to the clay particle layers, and the sodium (which is the cause of aggregate dispersion) is leached down the soil profile. In many cases gypsum (or other chemical amendment agents) are not readily available and/or their cost is not within the means of the local population.

Under these circumstances, it might be possible to reclaim sodic-saline soils through the use of special plants. This is a relatively new approach, which was stimulated by the work of Robbins (1986), who demonstrated that sorghum roots could significantly lower the pH of the rhizosphere. This lowering of pH results in increased solubility of calcium, which in turn replaces the sodium adsorbed to the clay surfaces. Furthermore, the dense sorghum roots form an effective system of canals through which water and air can infiltrate down, through the soil profile. Successful trials on the bio-reclamation of saline-sodic soils were carried out in Pakistan with the forage grass *Leptochloa fusca* and the forage legume *Sesbania aculeata*. In terms of soil amelioration, the efficiency of *S. aculeata* was the same as that of gypsum application (Qadiz et al., 1995).

USE OF SALT-TOLERANT CROPS

The USDA Salinity Laboratory has recommended the use of salt-tolerant crops for better utilization of saline soils and saline waters since the beginning of this century. The use of halophytes for solving specific problems of very high salinity was elucidated by Boyko (1966), in his book "Salinity and Aridity". More recently, Maas (1990) produced a comprehensive list of salt-tolerance limits for many agricultural crops.

It is noteworthy that the notion that halophytes can be used to "pump out" the salt from high-salinity soil is false. The amount of salt that can be removed that way is minor comparing those required for soil recovery. However, production of economically valuable halophytes taking place during the salt-leaching procedure will increase the worthiness of soil-reclamation projects.

Work on irrigation with saline water of salt-tolerant crops has been carried out successfully at the Ramat Negev Experimental Station in Israel since 1971 (Pasternak and De Malach, 1995). Since the pioneering work of Boyko (1966), many centers have conducted selection and research trials on halophytes with the idea of developing them into economically viable crops (Shay, 1990). The introduction and selection work on halophytes carried out in Israel (Pasternak and Nerd, 1995) concentrated on the identification of highly salt-tolerant fodder plants. *Atriplex* species were found to be highly productive with sea-water irrigation (Pasternak et al., 1993). However, when *in vivo* feeding trials were carried out, it was found that *Atriplex* has a very low feed value due to low overall and nitrogen digestibility but particularly due to the high salt content in the leaves. Animals fed with *Atriplex* need to spend a considerable amount of energy to metabolize tissue salts (Arieli et al., 1989).

The grasses tested in the study of Pasternak and Nerd (1995) were found less tolerant than *Atriplex* species or other chenopods. However, they did not accumulate salts in their leaves and were therefore potentially suitable as forage species.

SALT RESISTANCE IN PLANTS

Saline environment causes plant responses through several effects, among which osmotic stress and ion toxicity are of principal importance (Greenway and

Munns, 1980; Läuchli and Epstein, 1990; Munns, 1993; Gorham, 1996). Initially, salinity causes water deficit resulting in osmotic stress. When exposed to salinity for long periods or to high levels of salt, most plant species are unable to regulate ion fluxes into interior domains. The resulting salt buildup in the plant leads to ion toxicity.

Osmotic adjustment is usually mediated by the synthesis of intracellular compatible solutes (osmolytes) and by maintenance of ion homeostasis in the cytoplasm (McCue and Hanson, 1990). Ion homeostasis is controlled by selective uptake mechanisms of ions through the plasma membrane and by ion fluxes through the tonoplast (Maathuis, 1996). In plants, these processes are predominately mediated by cytoplasmic concentrations of K^+ (Wyn Jones, 1996). Under saline conditions, however, when Na^+/K^+ ratio in the environment of the roots is high, selectivity of ion uptake in root plasma membrane is disturbed and sodium leaks into the cytoplasm (Läuchli, 1996). Halophytes are expected to display efficient mechanisms of ion excursion in the plasma membrane (Serrano et al., 1999).

The complex effect of salt stress on plants requires basic, long-term solutions displayed at the whole plant level. Increased root/shoot ratio, hidden-type stomata, salt glands, or salt compartmentation to older plant tissues (Flowers et al., 1977; Cheeseman, 1988; Bohnert et al., 1995) are examples of how halophytes succeed to acquire and save water, and to avoid salt toxicity.

Two strategies of a biotic stress-resistance may be found in plants: tolerance and avoidance (Levitt, 1980). Salt-tolerant metabolic pathways were reported for some algae species but not for higher plants (Munns, 1993). Greenway and Osmond (1972) found no differences between *in vitro* salt-sensitivity of key-enzymes from halophytes and glycophytes. Salt-resistant plants, including halophytes, *avoid* salt accumulation in sensitive tissues (e.g. photosynthetic tissues). Salt avoidance mechanisms take place in membranes of various organs, mainly roots. Processes of ion exclusion, excretion, or salt compartmentation occur at the cellular and tissue levels (Cheeseman, 1988; Läuchli and Epstein, 1990; Munns, 1993). Thus, differences in the levels of salt accumulation in the aboveground organs of halophytes are due to salt exclusion and salt excretion activities in the roots. In most species of chenopodiaceae salt compartmentation in the leaves, and salt excretion through salt glands on leaf surface dominate mechanisms of salt exclusion and salt excretion in the roots. In salt-resistant grasses, on the other hand, root activity seems to be the major mechanism that limits salt accumulation in aboveground organs.

SALTGRASS (*D. SPICATA* L. GREENE)

One of the most promising grass species tested by Pasternak and Nerd (1993) was the American saltgrass (*Distichlis spicata*). *D. spicata* is a perennial rhizomatous grass, belonging to the Poaceae family. It is a species with an exceptional range of resistance to abiotic stress. Saltgrass originates in the American continent. It is found in cold coastal (Hansen et al., 1976) areas of Canada and in Oregon and New England (Bertness et al., 1992), in warm Louisiana (Sigura and Hudnall, 1991), in marshlands bordering the Atlantic Ocean, and in the tidal area of the San Francisco Bay (St. Omer, 1994). *D. spicata* also grows in cold inland regions such as Montana (Prodgers and Inskeep, 1991), Saskatchewan in Canada (Dodd and Coupland, 1966; Ungar, 1974) and in the hot and arid Death Valley of California (Hansen et al., 1976), in Mexico (Gonzales Vincente, 1982), and in Argentina (Ginzo et al., 1986). The authors have collected *D. spicata* in the Pampas del Tamarugal in Chile and in Patagonia in Argentina (D.P.), and in Utah, Arizona, and California (inland ecotypes), and in California, Delaware, Georgia and Alabama (coastal ecotypes) (A.B.). In spite of this impressive distribution over a wide range of climate conditions, *D. spicata* is a typical warm-season, C_4 grass. In regions of prolonged cold winter period, *D. spicata* initiates later than C_3 grasses, but it grows faster and blooms earlier (Dodd and Coupland, 1966). The aboveground plants' organs die at the first winter frost and only the underground parts remain alive but dormant until the next growing season.

D. spicata can grow with sea-water salinity levels (3.5% salt in the water, EC_i of about 50 dS m^{-1}), but at such high salt levels yields were reduced to 30% of the yield at 1% salt in water (Gallagher, 1985). Among seven members of the grass subfamily Chloridoideae examined by Marcum (1999), *D. spicata* was the most resistant to salinity (up to 500 mM NaCl). Screening of saltgrass ecotypes for growth response to salinity level revealed significant variation among plants according to their origin in various eco-geographic regions in America (Bustan et al., unpublished).

However, all ecotypes displayed resistance to salinity up to 10 dS m^{-1}).

In saline marshlands, *D. spicata* forms a classic behavior of a pioneer plant. Due to its rapidly growing rhizomes it invades into disturbed patches of the marsh where salinity is too high for other plant species. The sharp rhizomes can penetrate the most compact soils providing infiltration of water and air into deeper soil layers, thus promoting changes in the local microenvironment. When these changes take place and salinity levels decrease *D. spicata* is rejected by less salt-resistant but stronger plant competitors (Warren and Niering, 1993; Shumway, 1995). *D. spicata* is, therefore, a good candidate as a pioneer plant for the reclamation of saline soils, and it can be easily replaced by superior crops when soil salinity decreases.

D. spicata tolerates prolonged periods of flooding (Bertness et al., 1992), and its root will grow in saturated soils without any detectable oxygen concentration (Cooke et al., 1993). This is due to continuous aerenchimatous ducts connecting the root tips and rhizomes to the stems and leaves (Hansen et al., 1976). On the other hand, *D. spicata* is also known to withstand long periods of drought in relatively arid environments (Hansen et al., 1976). *D. spicata* survives prolonged dry seasons by the form of dormant below-surface rhizomes, and emerge new, rapidly growing shoots during the wet season. The inland type (sometimes named *D. stricta*) also thrive under inundation, as well as the coastal type.

D. spicata tolerates both acid and alkaline soils (Prodgers and Inskeep, 1991). In a hydroponic culture trial conducted by Fuller and Richardson (1986) there was no reduction in dry matter production at a solution pH of 10.5.

D. spicata is resistant to high concentrations of heavy metals and is therefore used for reclamation of disturbed mined lands. Observations in western Montana showed that *D. spicata* grew well on soils containing high levels of Cu, Zn and Mn, all these metals being excluded from its tissues. It also withstands relatively high levels of Al (Fuller and Richardson, 1986).

Although saltgrass is a clonal plant species in general, sexual reproduction also takes place. Seeds are then stored in the soil forming seed-banks (Smith and Kadlec, 1983) or transferred by birds (Vivian-Smith and Stiles, 1994), animals, wind and water (Smith and Kadlec, 1985) over considerable distances. Seed production and dispersal might cause problems of uncontrolled invasiveness that must be taken care of anywhere saltgrass is introduced.

ECONOMIC UTILIZATION OF SALTGRASS

Attempts to select ecotypes of *D. spicata* for forage production have been carried out by Gallagher (1985) and by Yensen and Bedell (1993). Gallagher demonstrated large variability in the productivity among selected ecotypes. For example, in a trial at Sadat City an accession from Seabrook gave a dry matter yield of 560 g m^{-2} while an accession from Lewes yielded only 179 g m^{-2}. Yensen et al. (1995) claimed that they selected superior forage cultivars of *D. spicata* with an "alfalfa-like growth with similar nutritional values". This statement has, however, not been validated by scientific evidence. Gallagher (1985) also demonstrated that there is a large apparent genetic variability in salt content in leaves. Ash content in the leaves from a selection from Delaware was 7% (on dry weight basis), whereas that of Peruvian selection was 17.4%.

Some attempts to utilize *D. spicata* as a forage species have been reported from South America. In Argentina, cattle graze during the summer months in saline habitates, where saltgrass may comprise from 8% to 24% of the dry weight diet of the cattle. In a study of the forage value of *D. spicata*, Briznela et al. (1990) concluded that this species may serve as a medium- to high-feed quality forage source during its fast growing stage. In another study, *D. spicata* was planted in the dried bed of the former Lake Texcoco, near Mexico City, over a 20,000 ha area and irrigated with reclaimed sewage water (Shay, 1990). The soil there is both highly alkaline and saline. *D. spicata* then served as the sole grazing species for a herd of cows.

Pasternak et al. (1993) conducted, over a three-year-period, a comparative field study of five salt-resistant grasses (including *D. spicata*) and alfalfa. Yields (amount and quality) were evaluated under optimum water and fertilizer regimes at five levels of soil EC_e (3.0–12.0 dS m^{-1}). In the same trial, the effect of the amount of irrigation on the yield at two levels of water salinity (EC_i 1.2 and 10.0 dS m^{-1}) was also studied. At a soil EC_e of 12.0 dS m^{-1} *D. spicata* (Seabrook accession) exhibited the same productivity (4 kg m^{-2} DW) as the Suwannee cultivar of bermudagrass

(*Cynodon dactylon*) but was significantly more salt- and drought-resistant than the other species. *D. spicata* had a relatively low ash content (11%) as compared with alfalfa, and medium crude protein concentration (~15%), but very high (~42%) fiber content compared with alfalfa (~26%). As a result of this trial it was concluded that *D. spicata* has the potential to become an important forage plant resistant to salt, alkalinity and flooding under diverse climatic conditions, provided that the fiber content could be markedly reduced. It is thought that the reduction in fiber content could be achieved through selection and breeding for "leafiness", that is, for a high ratio of leaf blade biomass to total plant biomass. Leafiness should thus be the major trait for selection.

CONCLUDING REMARKS

The above-described outstanding ecological plasticity of *D. spicata* together with its economic potential make this species highly suitable for reclamation of saline soils in regions of diverse climates, for revegetation of saline marshlands and for use as a potential crop for irrigation with water of high salinity. In many countries throughout the globe, where soil salinization has taken large areas out of production but a certain level of income must be provided to the farmer during soil reclamation, *D. spicata* appears to be a plant of first choice for the initiation of the process.

Yet, saltgrass is still a wild plant species and efforts should be made toward domestication. Selection objectives must include: (a) higher yield and quality as forage, (b) suitability to specific climatic and edaphic conditions, including certain range of salinity, and (c) sexual sterility to avoid invasiveness.

In Israel, we have started a selection program as such, and projects of reclamation of salinized lands are embarking in various regions such as Central Asian Republics, Middle East and West Africa. It should be kept in mind, however, that the use of halophytes for soil reclamation must be a part of a comprehensive process, in which the establishment of salt leaching and efficient drainage is pivotal.

REFERENCES

Arieli, A., Naim, E., Benjamin, R.W. and Pasternak, D. The effect of feeding salt bush and sodium chloride on energy metabolism in sheep. Animal Product 1989; 49:451–7.

Bertness, M.D., Wikler, K. and Chat, A. Flood tolerance and the distribution of *Iva frutescens* across New England salt marshes. Oecologia 1992; 91:171–8.

Bohnert, H.J., Nelson, D.E. and Jensen, R.G. Adaptations to environmental stresses. The Plant Cell 1995; 7:1099–111.

Boyko, H. editor. Salinity and aridity – new approaches to old problems. Dr, Junk, The Hague, 1966; 408 pp.

Briznela, M.A., Cid, M.S., Minon, D.P. and Fernandez-Grecco, R. Seasonal utilization of saltgrass (*Distichlis* spp.) by cattle. Animal Feed Sci Technol 1990; 30:321–5.

Cheeseman, J.M. Mechanisms of salinity tolerance in plants. Plant Physiol 1988; 87:547–50.

Cooke, J.C., Batler, R.H. and Madole, G. Some observation on the vertical distribution of vesicular arbucular mycorrhizae in roots of salt marsh grasses growing in saturated soils. Mycologia 1993; 85:547–50.

Dodd, J.D. and Coupland, R.T. Vegetation of saline areas in Saskatchewan. Ecology 1966; 47:958–67.

Flowers, T.J., Troke, P.F. and Yeo, A.R. The mechanisms of salt tolerance in halophytes. Annu Rev Plant Physiol 1977; 28:89–121.

Flowers, T.J. and Yeo, A.R. Breeding for salinity tolerance in crop plants: where next? Aust J Plant Physiol 1995; 22:875–84.

Fuller, R.D. and Richardson, C.J. Aluminate toxicity as a factor controlling plant growth in bauxite residue. Environ Toxical Chem 1986; 5:905–15.

Gallagher, J.L. Halophytic crops for cultivation at seawater salinity. In Pasternak, D. and San Pietro, A., editors. Biosalinity in action. Martinus Nijhoff, Dordrecht, The Netherlands, 1985; 323–36.

Ginzo, H.D., Collantes, M.B. and Capo, O.H. Fertilization of halophytic natural grassland in Argentina. Turriabla 1986; 36:453–9.

Gonzales Vincente, C.E. Alternative plant covering in the water of the former lake Texcoco. Ciencia Forestales 1982; 7:3–24 (in Spanish).

Gorham, J. Mechanisms of salt tolerance in halophytes. In Choukr-Allah, R. editor. Halophytes and biosaline agriculture, Marcel Dekker Inc., New York, 1996; 31–53.

Greenway, H. and Munns, R. Mechanisms of salt tolerance in non-halophytes. Annu Rev Plant Physiol 1980; 31:149–90.

Greenway, H. and Osmond, C.B. Salt responses of enzymes from species differing in salt tolerance. Plant Physiol 1972; 49:256–9.

Hansen, D.J., Dayanadan, P., Kaufman, P.B. and Brotherson, J.D. Ecological adaptation of salt marsh grass, *Distichlis spicata* (Gramineae) and environmental factors affecting its growth and distribution. Am J Bot 1976; 63:635–50.

Läuchli, A. Salinity – potassium interaction in crop plants. In Oosterhuis, D.M. and Berkowitz, G.A., editors. Frontiers in potassium nutrition: new perspectives on the effects of potassium on physiology of plants. Proceeding of a Symposium, Potash & Phosphate Institute of Canada, 1996; 71–6.

Läuchli, A. and Epstein, E. Plant responses to saline and sodic conditions. In Tanji, K.K. editor. Agricultural salinity assessment and management. ASCE manual No. 71, New York, 1990; 113–37.

Levitt, J. Response of plants to environmental stresses, Vol. 1, 2nd edn. Academic Press, New York.

Maas, E.V. Crop salt tolerance. In Tanji, K.K., editors. Agricultural salinity assessment and management. ASCE manual No. 71, New York, 1990; 262–304.

Maathuis, F.J.M. and Sanders, D. Plasma membrane transport in context – making sense out of complexity. Curr Opin Plant Biol 1999; 2:236–43.

Marcum, K.B. Salinity tolerance mechanisms of grasses in the subfamily Chloridoideae. Crop Sci 1999; 39:1153–60.

McCue, K.F. and Hanson, A.D. Drought and salt tolerance: towards understanding and application. Biotechnology 1990; 8:358–62.

Munns, R. Physiological processes limiting plant growth in saline soils: some dogmas and hypotheses. Plant Cell Environ 1993; 16:15–24.

Pasternak, D. and De Malach, Y. Crop irrigation with saline water. In Pessarakli, M., editor. Handbook of plant and crop stress. Marcel Decker Inc. 995; 599–622.

Pasternak, D., Nerd, A. and DeMalach, Y. Irrigation with brackish water under desert conditions IX. The salt tolerance of six forage crops. Agric Water Manage 1993; 29:321–34.

Pasternak, D. and Nerd, A. Research and utilization of halophytes in Israel. In Choukr-Allah, R. Hamdy, M. and Malcolm, C.M. editors. Halophytes and biosaline agriculture. Marcel Decker Inc., New York, 1995; 325–48.

Prodgers, R.A. and Inskeep, W.P. Heavy metal tolerance of inland salt grass (*Distichlis spicata*). Great Basin Naturalist 1991; 51:271–8.

Qadiz, M., Ahmad, N. Quereshi, R.H. and Ilyas, M. Role of salt tolerant plants to reduce salinization and sodification. In Kahn, A. and Ungar, I.A. editors. Biology of salt tolerant plants. University of Karachi Publishers, Karachi, 1995; 109–34.

Richards, L.A. Diagnosis and improvement of saline and alkali soils. USDA handbook No. 60. Government Printing Office, Washington DC, 1954; 160pp.

Robbins, C.W. Sodic calcareous soil reclamation as affected by different amendments and crops. Agron J 1986; 78:916–20.

Serrano, R., Culiaz-Maci, F.A. and Moreno, V. Genetic engineering of salt and drought resistance with yeast regulatory genes. Scientia Horticulturae 1999; 78:261–70.

Shainberg, I. and Singer, M.J. Soil response to sodium and sodic conditions. In Tanji, K.K. editor. Agricultural salinity assessment and management. ASCE, New York, 1990; 91–112.

Shumway, S.W. Physiological integration among clonal ramets during invasion to disturbance patches in a New England salt marsh. Ann Bot 1995; 76:225–33.

Shay, G. Saline agriculture, salt tolerant plants for developing countries. National Academy Press, Washington DC, 1990.

Sigura, G.C. and Hudnall, Q.H. Gypsum and water management interactive for revegetation and productivity improvement of brackish marsh in Louisiana. Soil Sci Plant Anal 1991; 1721–39.

Smith, L.M. and Kadlec, J.A. The effect of disturbance on marsh seed banks. Can J Bot 1995; 63:2133–7.

Smith, L.M. and Kadlec, J.A. Seed banks and their role during drawdown of a North American marsh. J Appl Ecol 1993; 20:673–84.

St. Omer, L. Soil and plant characteristics in a dyked and tidal marsh in San Francisco Bay. Am Midland Naturalist 1994; 1:32–43.

Ungar I. In Queen, W.H. and Reimold R.J., editors. Ecology of Halophytes. Academic Press Inc., New York, 1974.

Vivian-Smith, G. and Stiles, E.W. Dispersal of salt marsh seeds on the feet and feathers of waterfowl. Wetland 1994; 14:316–19.

Warren, S.W. and Niering, W.A. Vegetation changes on a northern tidal marsh: interaction of sea-level rise and marsh accretion. Ecology 1993; 74(1): 96–103.

Wyn Jones, R.G. Cytoplasmic potassium homeostasis: review of the evidence and its implications. In Oosterhuis, D.M. and Berkowitz, G.A., editors. Frontiers in potassium nutrition: new perspectives on the effects of potassium on physiology of plants. Proceeding of a Symposium, Potash & Phosphate Institute of Canada, 1996; 13–22.

Yensen, N.P. and Bedell, Y.L. Consideration of the selection, adaptation, and application of halophyte crops of highly saline desert environments as exemplified by the long term development of cereal and forage cultivation of *Distichlis* spp. (Poaceae). In Lieth, H. and Al Massoum, A. editors. Towards the rational use of high salinity tolerant plants. vol. 2. Kluwer Academic Publishers, Dordecht, 1993; 305–13.

Yensen, N.P., Bedell, Y.L. and Yensen, S.B. Domestication of *Distichlis* as a grain and forage. In Khan, M.A. and Ungar, I.A., editors. Biology of salt tolerant plants. Dept. of Botany, University of Karachi Publishers, Karachi, 1995; 388–93.

Primary analysis of four salt tolerant plants growing in Hai-He Plain, China

BAI YAJUN, LIU XIAOJING AND LI WEIQIANG

INTRODUCTION

In China, the saline soil area, as an important soil resource, is about 99,130,000 ha, in which the modern saline soil area is 36,930,000 ha, and the potential saline soil area is about 17,330,000 ha (Wangzunqin, 1993). The saline soil is mainly distributed in low and coastal regions, such as Hebei, Shandong, Liaoning, Jiangsu provinces. On the other hand, in inland, Inner Mongolia and Xinjiang, there is a larger area of saline soil too in arid regions.

As an important part of the North China Plain, Hai-He Plain (mainly in Hebei province) is a well-known low-yield area with frequent disasters in the history. Soil salinization used to be a major problem in this area. While Global Environmental Change is widely concerned by all over the world, soil salinization becomes to an essential research field.

Though the saline soil environment is odious, some plants are still growing very well that are halophytes or salt tolerant plants. In China there are 423 halophytic species. These plants can tolerate stress environment and may have high-economic value (Zhao Kefu, 1999). In some rural areas, farmers used to produce vegetable oil with halophytes such as *Suaeda sala* (L.) *Pall* and *Descurainia ophia* (L.) *Webb. ex Prantl*, etc. Some halophytes are important Chinese herbs such as *Lycium chinese* Mill, *Glycyrrhiza* Fisce, *Apocymum* L., etc. Some are used as food, vegetable, fodder, fiber, greening trees, etc. With the development of biotechnology, halophyte sustainable utilization will be the future for reclamation of saline soil and seawater or saline water irrigation agriculture in 21st century.

In China the studies on halophytes started in 1950. These researches were mainly focused on the physiology and biology of halophytes. Recently some studies were conducted on halophyte utilization. Qingdao Ocean Institute has studied the utilization of *Suaeda salsa* (L.) *Pa* and manufactured the product-conjugated linoleic acid (CLA). However, due to few studies on halophyte utilization, no cultivated variety was selected and about 20 species of halophytes at the vanishing edge. Since 1995 we have conducted research in the coastal areas, Bohai sea and finished the investigation of halophytes distribution. Soil samples and water quality was analyzed. One halophyte garden has been established in Huanghua City, Hebei province. *Descurainia sophia* was seeded in Nanpi station.

MATERIALS AND METHODS

Natural Situation of the Study Sites

The samples of halophytes were gotten in Nanpi station and Haixing station of Shijiazhuang Agricultural Modernization Institute of CAS. The Nanpi station and HaiXing station belong to Cangzhou City of Hebei province. The area was part of Hai-He Plain with semi-arid and semi-humidity and monsoon climate. Nanpi station is located at N38°06′ latitude, E116°40′ longitude. Haixing station is located at N38°17′ latitude, E117°18′ longitude, the climate and geographic characters are similar to Nanpi station.

Analysis Methods

Four kinds of wild salt-tolerant plants were collected in August 1998, they are *Nitraria sibiria Pall, Suaeda sala* (L.) *Pall, Chenopodium glaucum* and *Descurainaia sophia*. Table 1 shows the natural conditions of sampling site and their characteristics for four varieties of plants.

H. Lieth (ed.), Cash Crop Halophytes: Recent Studies, 135–138.

Table 1. The natural conditions of sampling site and the characteristics of four kinds of plants

Name	Place of sample collection	Salt content of soil (salt/100 g dry soil)	Possible utilization
Nitraria sibiria Pall	Haixing station	2%	The fruit can be eaten directly.
Suaeda salsa (L.) *Pa*	Haixing station	1.5%	Young shoot as vegetable. Seed oil edible or used in industry. The remains of seed can be fed to animals.
Chenopodium glaucum	Road side in Haixing	1.2%	Young shoot is vegetable. Oil content in seed is very high.
Descurainia sophia	Farm land in Nanpi station	0.4%	Potential oil crop rich in polyunsaturated fatty acids.

Table 2. The methods of analyzing different ions

The type of ions	Analyzing method
CO_3^{2-} and HCO_3^-	Neutralization titration in two kinds of indicators
Cl^-	$AgNO_3$ titration
Ca^{2+} and Mg^{2+}	EDTA titration
K^+ and Na^+	Calculated by difference of cation and anion

The total salt contents of soil were analyzed with the method of adding all the ions (cations and anions) together. The dry soil sample was mixed with distilled water (dry soil : distilled water = 1 : 5) and then the salt content of filtrate was measured. The methods of analyzing different salt content were as following (Table 2). The content of soil water was gotten by oven drying at the temperature of 108°C for 8 hours.

Rough lipids content of the seeds are analyzed according to GB 14772-93 method. Fatty acids were analyzed through GB/T 17377-1998 lipid acid formal ester method with gas chromatography analysis. The machine type is ShimaThuma GC-9A capillary gas chromatography. The seeds structures of the plants were observed through scan electron microscope.

RESULTS AND DISCUSSION

The rough lipids content and fatty acid composition of four kinds of wild plants were measured (Table 3). They are *Nitraria sibiria Pall, Suaeda salsa* (L.) *Pa, Chenopodium glaucum and Descurainaia sophia.* Figure 1 shows the seeds' transverse microstructure of the plants through SEM. The content of rough lipids in *Nitraria sibirica Pall* seed is 7.84%, *Suaeda salsa* (L.) *Pa* is 20.5%, *Chenopodium glaucum* is 16.4%, *Descurainia sophia* is 35.1%. Table 3 shows the fatty acid composition in the rough lipids of four plant seed.

Oil content in *Descurainia sophia* is high with 76.65% polyunsaturated fatty acid and the linolenic acid content is 53.7% with α-linolenic acid, γ-linolenic acid and arachidonic acid content is 5.35%. γ-linolenic acid and arachidonic acid is a necessary pre-substance for producing prostaglandin (PG) in human body. PG is useful for people's health with lots of forms, such as: PGE_1, PGE_2, $PGE_{1\alpha}$, $PGF_{1\alpha}$, $PGF_{2\alpha}$, TXA_2, TXB_2. PGE_2 may extend blood vessel and increase blood vessel of brain and prevent oxygen-deficiency. To the breath system, PGE may expand smooth muscle and alleviate asthma. To the stomach and intestines, PGE and PGI may restrain hypochlorhdria and cure gastric ulcer. To the blood platelet, TXA_2 may promote platelet assemble into thrombus while PGI_2 may prevent the form of thrombus. Therefore, the balance of TXA_2–PGI_2 is important to artery thrombus, coronary heart artery and hypertension. PG may affect some kinds of hormones inflammation and immunity system (The Cyclopedia of Chinese Medicine – Basic Medicine, 1995).

There is another very interesting thing. The rough lipids content and fatty acid composition of *Descurainia sophia* grows in the common soil of Chuan-Xi altiplano in Sichuan province is very different with that of *Descurainia sophia* grows in Nanpi station. Table 4 shows that the linolenic acid content of *Descurainia sophia* which grows in Nanpi is higher than that in Hongyuan(Sichuan) and Guangyuan(SiChuan). In fact, lots of halophytes are traditional Chinese medicine. Therefore, there are lots of work to be done to promote the research and use of halophytes.

Table 3. The fatty acid composition in the rough lipids of four plants seed(%)

Name of halophyte	*Nitraria sibiria Pall*	*Suaeda salsa* (L.) *Pa*	*Chenopodium glaucum*	*Descurainia sophia*
Palm acid	7.03	14.6	11.4	8.68
Stearic acid	2.62	2.84	2.17	1.72
Oleic acid	22.4	19.4	23.6	11.7
Linoleic acid	65.3	54.1	53.78	16.6
Linolenic acid	1.21	3.77	4.70	53.7
Arachidonic acid	—	—	—	5.35
Total lipid content of dry matter	7.84	20.5	16.4	35.1

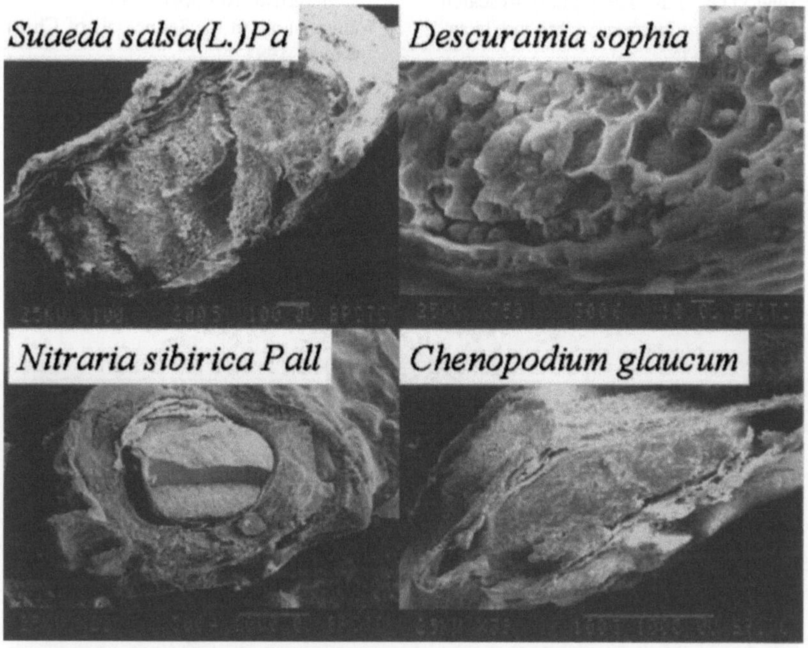

Fig. 1. The transverse structure of the plants seeds through SEM.

Table 4. The rough lipids content and fatty acid composition of *Descurainia sophia* growing in three different places

Place of sample	Soil salinity	Total lipids content	Palm acid	Oleic acid	Linoleic acid	Linolenic acid	Habitation
Nanpi, farm field	0.4%	35.10	6.6	11.7	16.6	53.7	In saline soil
Hongyuan, SiChuan	<0.1%	44.20	5.84	13.06	16.53	37.1	In table land
Guangyuan, Chendu	<0.1%	32.23	6.03	14.49	15.52	35.96	In farm land

CONCLUSIONS

Based on the analysis above, we found that the four kinds of salt-tolerant plants are rich in rough lipids, especially the polyunsaturated fatty acid content is very high in the seed. The seed of *Nitraria Sibiria Pall* has the highest linoleic acid of the four (65.3% of total rough lipid), the output of which is greatly varied with the growing condition; *Descurainia Sophia* has the highest linolenic acid of the four (53.7% of total rough lipid), the output of which, we had measured it, can be over 3000 kg/ha.

These four salt-tolerant plants could be the potential cash crops in Hai-He plain which can be planted in salt and drought regions. Some of them can be irrigated with sea water. This is only a primary analysis, the further studies should be carried out on the effect of soil salt content on the changes of seed fatty content and composition.

REFERENCES

The editor committee of Basic Medicine. The Cyclopedia of Chinese Medicine, basic medicine. Science and Technology Publishing Company, ShangHai, 1995; 393–5.

Mou Zhilan Luo Peng Yang Yi Gao Fuli. The Primary Research of High Linolenic acid Plant *Descurainia Sophia* in Chuan-Xi Meadow. Prataculture Sci 14(1).

Wang Jiayun Wang Yan Li Zhen. Using usable resource to produce health caring edible oil science and technology of agric and forest in Ningxia 1998; 3.

Zhangmin Wang Changhong Chen Shanquan. The Clinic Comparison Observing of the Reduce Function of *Perilla Frutescens* seeds oil to Plasma Fatty. Chin Med J Liaoning 1999; 26(3).

Zhao Kefu, Li Fazeng. The halophytes of China. Science Publishing House, 1999.

Tian Kuixiang, Toshiaki Tadano. The improvement and evolution of saline soil in Hai-He Plain in recent 20 years. Proceedings of China–Japan Joint Symposium. 12–13 August, Xi'an China.

Li Huiyin et al. Investigation of aquatic environmental quality and analysis of the dynamics of pollutants in coastal zone. Eco-agricult Res 5(4): 50–3.

Wang Zunqin et al. Saline soil of China. Science Publishing. House, 1993.

Field scale experiments on plant tolerance to saline irrigation water

VITO SARDO

INTRODUCTION

In fulfilment of an activity within the CEC-funded Concerted Action "Sustainable Halophyte Utilisation in the Mediterranean and Subtropical Dry Regions," a field was installed in 1997 with experimental and demonstration purposes near the shore of the Ionian sea, in Eastern Sicily.

Researches were jointly conducted by three institutes of the local Faculty of Agriculture (Istituto di Idraulica Agraria, Istituto di Coltivazioni Arboree, Istituto di Chimica Agraria) throughout the years 1998 and 1999; although the contract with the CEC has now expired, further research continues with additional support of a fourth institute (Istituto di Orticoltura e Floricoltura).

DESCRIPTION OF THE STRUCTURES

The field, with a surface of 5000 m^2, is on a coarse quartzitic sandy soil underlain by a saline (about 15 dS m^{-1}) aquifer at the depth of about 3 m. The distance from the seashore is of about 100 m; since a full description has been given elsewhere (Sardo, 1999), here only a brief outline will be reported. The whole field was subdivided into an *experimental* and a *demonstration* area.

The experimental area includes three benches (Bench A, B and C), where 270 20-L microlysimeters were installed, which can be automatically drip-irrigated with various seawater/freshwater ratios and various leaching fractions (Figures 1 and 2).

Additionally, a "quick-test" set up was installed, as reported in Cutore et al. (1999), as well as a facility for the assessment of ureaformaldehyde foam impact on vetivergrass tolerance to salinity (Belligno et al., unpublished data). The latter included eight treatments as resulting from Table 1. The demonstration area included a subsurface irrigated and a sprinkler irrigated section, where, as explained elsewhere (Sardo, 1999), a number of plant species were tested; presently,

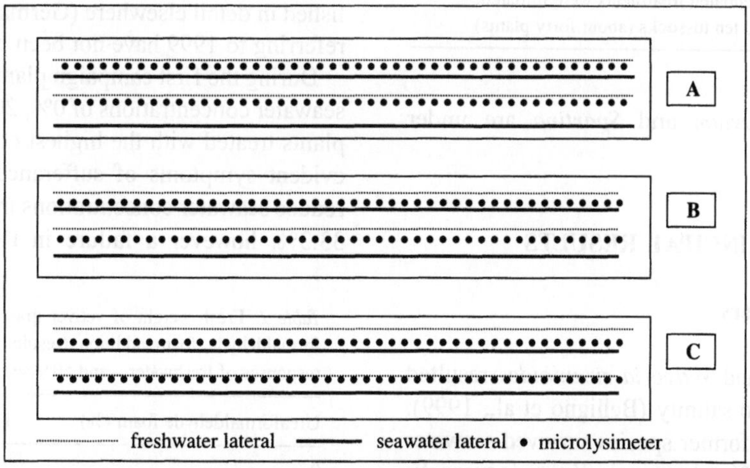

Fig. 1. Schematic representation of the three benches.

H. Lieth (ed.), Cash Crop Halophytes: Recent Studies, 139–142.

Fig. 2. One bench with plants in microlysimeters and collecting tanks.

Table 1. Lay out of the experiments with ureaformaldehyde foam

foam 50 %	*foam 75 %*	***50 % seawater***
control	*foam 25 %*	
foam 50 %	*foam 75 %*	*freshwater*
control	*foam 25 %*	

- Plants planted in 27 L pits;
- percentages sand/foam in volume;
- planted 6–7 November 1988/Harvested 3 February 2000-05-25;
- bare-rooted vetiver plants cv *Monto*;
- plants aged 1 year;
- all plants were irrigated initially with freshwater (till 21 June 1999); thereafter differentiated treatments were initiated;
- each treatment included ten tussocks (about forty plants)

however, only *Sesuvium* and *Spartina* are under observation.

THE PRINCIPAL RESULTS

Grasses in Lysimeters

Elytrigia elongata and *Vetiveria zizanioides* resulted surprising tolerant to salinity (Belligno et al., 1999): all the plants of the former species survived seawater concentrations of 60%, whereas six *Vetiveria* tussocks out of nine survived such treatment.

Fresh and dry matter production in leaves decreased with increasing salinity; opposite to that, fresh and dry weight in roots did not significantly decrease. Chemical analysis evidenced an ash percentage in leaves and roots increasing with salinity and decreasing with leaching fraction (Belligno et al., unpublished data).

Na and K exhibited an opposite behaviour, since the former was increased by a higher salinity while the latter was decreased. A difference between leaves and root response was also evidenced with phosporus uptake: roots in fact showed a marked decrease in P uptaking with salinity, whereas leaves were not significantly affected (Belligno et al., ibid).

Vetivergrass and Ureaformaldehyde Foam

A strong action of salinity was evidenced in 50% seawater-irrigated plants vs those irrigated with freshwater, the average total fresh weight (leaves + roots) being 1929 and 1014 g/tussock, respectively. Also plant height was influenced by salinity since it decreased from an average 95.75 cm in freshwater-irrigated plants to 72.30 cm in those irrigated with 50% seawater. 25% ureaformaldehyde addition positively impacted total fresh weight, but the action was not significant at 50% and negative at 75% addition (Table 2).

Woody Plants

Experiences were conducted through the summer months in 1998 and 1999 with plants of the genera *Olea, Citrus, Pistacia, Punica* and *Ficus* (the latter only in 1999). While results referring to 1998 are published in detail elsewhere (Germanà et al., 1999) those referring to 1999 have not been published.

During the first campaign plants were irrigated with seawater concentrations of 0%; 20%; 40%; 60%; since plants treated with the highest concentrations showed evident symptoms of sufferance, it was planned to reduce seawater concentrations in 1999 to 0%; 16.5%; 33.3%, however a failure in the freshwater supply

Table 2. Fresh weight of vetiver grass as influenced by ureaformaldehyde foam addition (results referring to cumulated responses of freshwater – and 50% seawater – irrigated plants)

Ureaformaldehyde foam (%)	Fresh weight (g/tussock)
0	1307
25	1717
50	1431
75	1056

system raised such percentages to 0%; 25%; 50%. The electric conductivity in drained water was in the range of 2.6 to 4.7; 12 to 20 and 19 to 34 dS m^{-1}, respectively: due to the very large leaching fraction (up to 97%) these EC values closely reflected those in irrigation water. In both years in fact water was daily applied in volumes which largely exceeded the potential evapotranspiration rate, to secure an efficient salt leaching.

Various biometric and physiological parameters were monitored in the course of the two campaigns, including trunk section increase, shoot length, net photosynthetic activity, stomatal conductance, xylem potential: all of them resulted affected by water salinity, yet plant tolerance on the whole resulted higher than expected.

In synthesis, seawater concentration negatively impacted net phothosynthesis, trunk growth, shoot length and xylem pressure while enhanced stomatal conductance.

The Quick Test

A set of experiments was conducted on plants of the genera *Chamaerops, Typha* and *Phragmites* (sixteen plants for each genus) in the twin goal of assessing plant tolerance to salinity and salt balance in the soil.

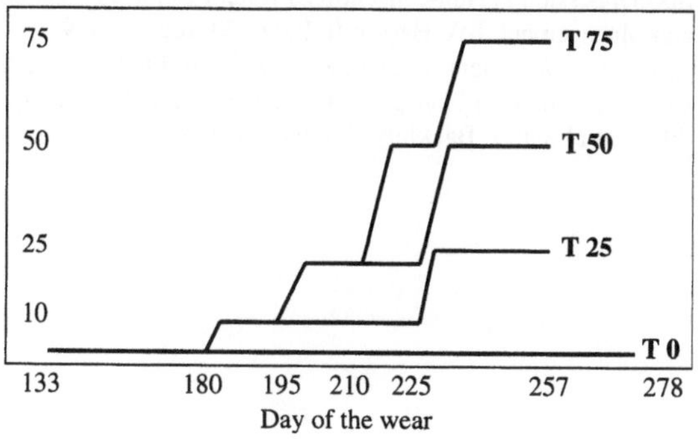

Fig. 3. Seawater concentrations in the various treatments.

Since the experimental set up and results have been described in detail elsewhere (Cutore et al., 1999) here only a synthesis will be reported.

Experiments were conducted in 25 L, freely draining plastic containers irrigated during almost 3 months with increasing seawater percentages; applied and leached water and salt amounts were recorded and a balance was elaborated. The underlying idea was to work out a system able to supply a first fast screening of plant tolerance. Seawater concentrations in the various treatments (from T0, corresponding to zero seawater concentration, up to T75, corresponding to 75% seawater concentration) and their variation in time are reported in Figure 3.

Plant tolerance to salinity resulted much lower in *Typha* and *Phragmites* than in *Chamaerops*: the latter in fact could survive 50% seawater concentration for over one month whereas *Typha* and *Phragmites* hardly survived 25%.

One interesting aspect of the research was related to salt balance, since results showed that freshwater addition permitted to recover quickly enough the initial concentration even after two months' irrigation with 50–75% seawater addition.

The Demonstration Activity

This activity included planting, propagating, establishing and tending a number of halophytes, observing their subsequent response to various environmental conditions and to full-strength seawater irrigation.

Plants which demonstrated unable to thrive under the typical Mediterranean climate included *Avicennia, Batis, Conocarpus, Thespesia*. Plants of the genera *Atriplex, Kosteletzkya, Leucaena, Lippia* showed a good development but were not seawater-irrigated due to the limited number of plants available and the consequent need to avoid the risk of severely damaging them. Plants of *Sesuvium* and *Spartina* were seawater-irrigated by means of a subsurface and a sprinkling

Table 3. Water and salt balance in 1999

	Water volumes (L)			Salt amounts (kg)	
	Applied	Evapotranspired	Leached	Applied	Leached (6 oct)
Bench A	4817	230	4587	53.12	36.83
Bench B	4410	265	4145	49.52	38.17
Bench C	2945	247	2698	32.07	25.52

irrigation system; *Spartina* had a stunted development with the subsurface seawater application and soon perished when sprinkled; on the opposite, *Sesuvium* survived sprinkling irrigation and developed well with the subsurface irrigation. *Sesuvium* development and tolerance to adverse environmental conditions (e.g. from −2°C to 47°C) gives promise for its use in the fixation of sandy coastal dunes.

The Sustainability

Some aspects of the impact of saline water irrigation on the soil were explored through the analysis of the water and salt balance in the microlysimeters and in the quick-test containers.

The significance of the results achieved is limited by the particular texture (unstructured coarse sand) of the soil used in the experiments. Both groups of experiments, in microlysimeters and in quick test containers, showed that a salt buildup can be avoided as long as a high-leaching fraction coupled to sufficient winter precipitations are secured.

A different aspect referring to soil permeability was also studied: measurements with a double ring infiltrometer showed that after 2 years of daily seawater irrigation, the permeability values decreased to 22 and 954 mm h^{-1} in sprinkler and subsurface irrigated plots, respectively, as opposed to 1620 mm h^{-1} in the control. The remarkable reduction in the sprinkler-irrigated plots is likely due both to the formation of a saline crust in the surface (as revealed by the chemical analysis) and the sealing action of the kinetic energy-rich water drops from the sprinklers.

CONCLUSIONS

The 3-year activity has supplied an abundance of information – some of them rather surprising – on technical aspects for the application of seawater in irrigation and plant response to salinity.

Although the data so far collected are certainly far from complete, the need for a radical revision of currently accepted guidelines is evident: for instance, the claim to state percentages of yield reduction exclusively as a function of water salinity is misleading and clearly lacks a serious experimental basis.

Results collected in the course of this activity concur to show that salt tolerance in some glycophytes is higher than expected while *Sesuvium* unsuspected ability to expand its runners under adverse conditions makes it a choice plant for fixing coastal sandy dunes: all such considerations concur to encourage in insisting in this most promising field of research.

REFERENCES

Belligno, A., La Loggia, F., Sambuco, F. and Sardo, V. Experiences on Vetivergrass Salinity Tolerance of Plants. In Proceedings of the Special Session of Non-conventional Water Resources Practices and Management, IAV Hassan II, Rabat, Morocco, 1999.

Cutore, L., La Loggia, F. and Sardo, V. A quick test to assess salinity tolerance of plants. In Proceedings of the Special Session of Non-conventional Water Resources Practices and Management, IAV Hassan II, Rabat, Morocco, 1999.

Germanà, C., Cutore, L. and Sardo, V. Assessing tolerance to irrigation water salinity in five woody plants. In Proceedings of the Special Session of Non-conventional Water Resources Practices and Management, IAV Hassan II, Rabat, Morocco, 1999.

Sardo, V. Establishment of an experimental field for saline irrigation in Sicily. In Hamdy et al., editors. Halophyte uses in different climates. Backhuys Publishers, 1999.

The influence of environmental factors and zonal distribution of *Phragmites australis* and *Phacelurus latifolius* in salt marsh, Central Japan

ICHIRO YOKOYAMA, KEIICHI OHNO AND YUKIRA MOCHIDA

INTRODUCTION

For many years physiologists, ecologists and plant sociologists have studied the halophytes and their habitats (Ellenberg, 1988). And we can see many papers studied the relationship between environmental factors and vegetation for halophilous plants, and topographical study in salt marsh or tidal flat (e.g. Linke, 1939; Gillner, 1960; Ellenberg, 1992). The common reed, *Phragmites australis* (Cav. Ex Roem. Schult.) Trin ex Steud., is a cosmopolitan, large perennial grass, with a presence on all continents (Holm et al., 1977), and well studied comparatively (e.g. Miroalav, 1999). On the other hand, *Phacelurus latifolius* (Steud) Ohwi is a halophyte whose habitat is in salt marsh and whose distribution is in East Asia, hardly we can see a paper for a relationship between environmental factors and vegetation or topographical study with it.

Halophytic vegetations in Japan were mainly studied as phytosociologically (e.g. Miyawaki and Ohba, 1969, 1972; Miyawaki et al., 1975; Okuda, 1978), and the zonal distribution of them was well known. On the other hand, according to Yamada et al. (1998) who made phytosociological study in Koajiro area, salt marsh vegetations, such as *Caricetum scabrifoliae*, *Phaceluretum latifolii* and *Cladietum chinensis*, are very rare in Kanto district, Japan.

P. australis is a hygrophyte whose habitat is marshlands, and we can see its habitats in many wetlands in the plains of central Japan. In the study site, its habitats are mostly by the seashore and in the back marsh. *P. latifolius* is a halophyte whose habitat is salt marshes occurring on tidal flats (Miyawaki and Okuda, 1972), and there are few places where we can see its habitat.

STUDY SITE

The tidal flat of Koajiro, on the Miura Peninsula, Kanagawa Prefecture, central Japan, is a small, flat spread in the estuary of a small stream that flows through a warm temperate forest. There are various salt marsh vegetation dominated by tall grasses on the tidal flat, influenced by brackish water and tidal changes. Two Gramineae plants, *P. australis* and *P. latifolius*, are characteristically distributed in a banded zonation pattern.

In the study site, the *P. latifolius* community is at the backside of the stunted *P. australis* community by the seashore and in the front of the back marsh. This suggests that the zonal distribution of plants is made along the environment gradient. We researched what kinds of environmental factors and physical conditions influenced the habitats of these plants.

The study site is shown in Figure 1. The tidal flat of Koajiro surrounded by warm-temperate evergreen forests, occupied 0.4 ha, and halophytic vegetation accompanied the zonation pattern (Figure 2).

As regards of climate, Miura Peninsula is in warm temperate zone. Mean annual temperature is 16.6°C, and the Warmth Index (Kira, 1949) is 139.7°C month, and mean annual precipitation is 1,451 mm (1990–1999).

METHODS

Four line transects were established across the zonal distribution of the plants. We studied the following factors.

1. *Micro-topography on the study site.* We measured the ground height on the tidal flat by Automatic Level

H. Lieth (ed.), Cash Crop Halophytes: Recent Studies, 143–149.

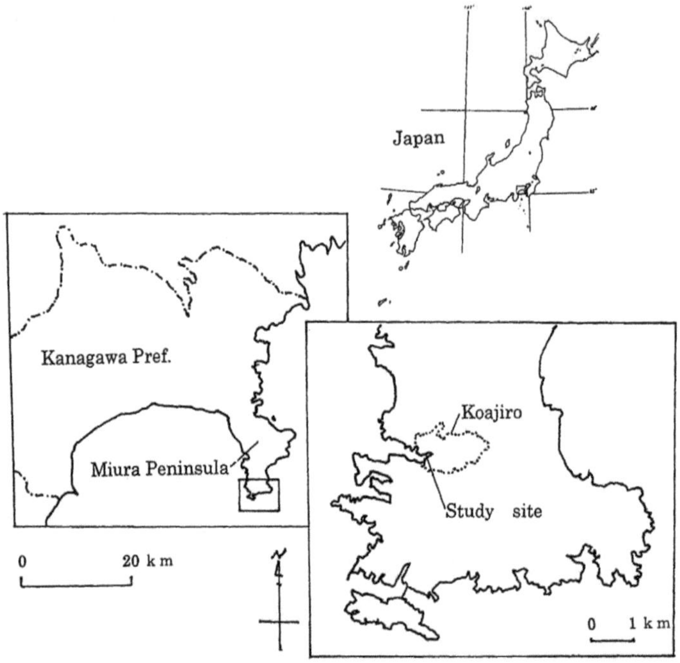

Fig. 1. Index and location maps of the study site, the tidal flat of Koajiro, the Miura Peninsula, Kanagawa Prefecture, central Japan.

Fig. 2. The zonal distribution of the halophytic vegetation composed of *P. australis* and *P. latifolius*.

(Nikon AE-5C), and made a site map with a 10-cm contours.

2. *Zonation pattern of the vegetation*. According to phytosociological method (Braun-Blanquet, 1964) and the measurement of the vegetation height, the zonation pattern of the plant distribution was drawn on the site map.

3. *Environmental factors*. Salinity, seawater level, and underground water level were measured along four line transects. As the water table gauge (Toma and Ogino, 1995), we inserted some pipes, 20 mm diameter and with many holes, into the ground vertically for measuring salinity and water level. These were established along four line transects with suitable spaces. Using the pipes, a thin pipe like a straw, and a stethoscope, we measured water level and extracted water for measuring salinity. Our research was carried out for each stand with an hour interval from ebb tide to high tide for half a day.

4. *Biomass and below-ground biomass*. Along the line transect I, we established three stands, st.1 was in stunted *P. australis* community, st.2 was in *P. australis* – *P. latifolius* community, and st.3 was in *P. latifolius* community. We cut all plants on the ground 50 cm × 50 cm at each stands, and measured their dry weight for each species to estimate above the ground biomass. And we trenched the ground 50 cm × 50 cm at each stands, and extracted soil blocks by 30 cm depth to estimate below the ground biomass. Rhizomes and roots were washed with a lot of water by 2 mm mesh basket from the soil block, and were measured their dry weight for each species and organs.

5. *Sediment type*. Soil samples were extracted for each soil blocks, and were decided their sediment type and color with Standard Soil Color Charts (Koyama and Takehara, 1990).

RESULTS AND DISCUSSION

Four line transect were similar about their environment gradient. But two line transects of right side had no back marsh because there were shore protection and road behind the vegetation. On the other hand, two line transects of left side had back marsh which were supplied fresh water from backside cliff. As we thought that line transect I was typical, it was shown in Figure 3. And in the back marsh, all tall plants fell down and were not in any ear.

Based on the results of our studies, we found three communities of plants, namely:

- Stunted *P. australis* community,
- *P. australis* – *P. latifolius* community,
- *P. latifolius* community.

Stunted *P. australis* community was composed of *P. australis* only. These plants were short, and their growth rate was low. They were in front of the seashore where there was a high frequency of submergence. *P. australis* – *P. latifolius* community was a mixed community composed of *P. australis*, *P. latifolius*, and *Carex scabrifolia*. *C. scabrifolia* is a short, halophyte whose habitat was in front of the seashore, and its coverage value was not so high in the study site. In this community, *P. australis* was growing taller toward the backside than in the front, and the proportion of *P. latifolius* was gradually increasing from the front to the back. *P. latifolius* community was on the highest ground height, where the upper mean highest water level (MHWL) was found (Figure 3).

We clarified the relationship between the plants and the environmental factors using the line transect method. The zonation pattern of these communities was affected by the ground height, the level of underground water, and salinity. But the most important environmental factor was the frequency of submergence decided by the water level (Figure 4). The physical conditions made by tidal changes correlated with the frequency of submergence. There were two distinguishable lines: a borderline (A) between Stunted *P. australis* community and *P. australis* – *P. latifolius* community, and the other one was a borderline (B) between *P. australis* – *P. latifolius* community and *P. latifolius* community. The former line was caused by the frequency of submergence and the higher salinity. The borderline (B) was also confirmed by our observation immediately at time of full tide on the flood tide day. In front of (A), there was a high frequency of submergence and the salinity was in the upper 25‰ after high tide (1–2 hours after). The latter, the borderline (B), was decided by MHWL. The zone between (A) and (B) was where the plant height escalated, the salinity and frequency of submergence decreased.

Salinity showed its interesting movement at high tide and after high tide. Because tidal change, fresh

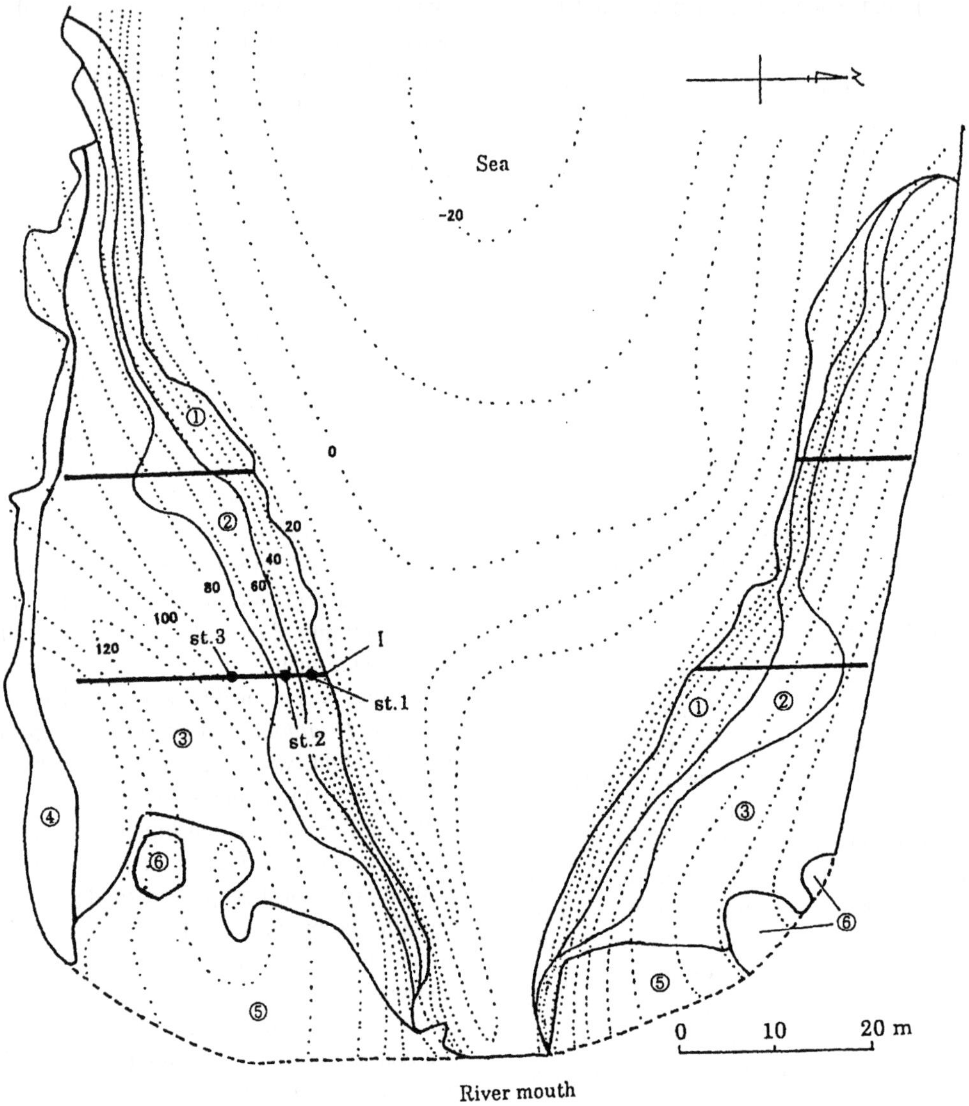

Fig. 3. Zonal distribution of the plant communities and micro-topography. Zone ①: Stunted *Phragmites australis* community: Zone ②: *Phragmites australis – Phacelurus latifolius* community: Zone ③: *Phacelurus latifolius* community: Zone ④: Back marsh: Zone ⑤: *Pleioblastus chino* community: Zone ⑥: Shrub: I: Line Transect I. (Dotted lines are 10-cm contours.)

water supply from river mouth of a small stream, and water evaporation from ground surface exerted irregular changes on salinity. In addition, water movement under the ground was complicated about between brackish and fresh water (Petersen, 1999). It was necessary for us to make more research about salinity and water movement in the site.

According to Miroalav (1999), there were a lot of papers that studied in salt concentrations with *P. australis*. Halsam (1973) said that the lethal concentration of salt solution was in the range of 10–15‰. Mook and van der Toorn (1982) mentioned that seedlings were particularly sensitive to increases in salt concentrations and the germination potential decreased as low as 6‰. Roman et al. (1984) and Sinicrope et al. (1990) pointed out that it was possible to endure for high salt concentrations as seawater for *P. australis*, but its colonies showed poor growth and reduction. Matoh (1998) made a study *in vitro*, the growth of *P. australis* was affected by NaCl solutions

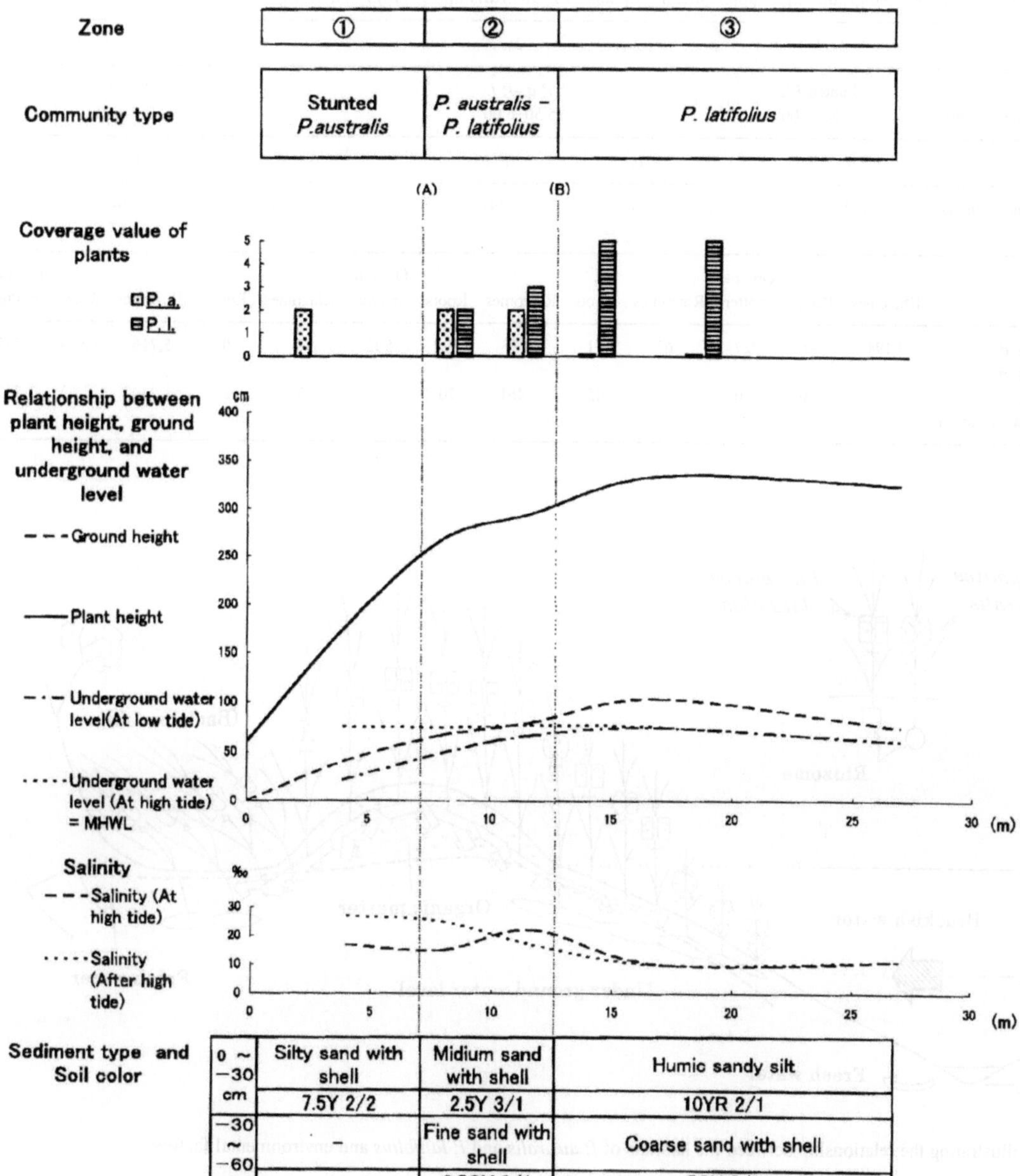

Fig. 4. Relationship between plant communities and environmental factors, and the sediment types, along line transect I.

of 16.6‰. As a result of our studies, *P. australis* was affected by in the range of 15–20‰ for salinity, and habitat of *P. latifolius* was in nearby 10‰ for salinity.

The aboveground biomass and subterranean biomass of either species were estimated (Table 1). The rhizomes and roots of *P. australis* existed deeper in the ground than those of *P. latifolius*, and the rhizomes of *P. australis* were soft and contained a lot of water. On the other hand, Miyawaki and Okuda (1972) pointed out that *P. latifolius* had very strong and thick rhizomes. Our study showed immediately that the rhizomes and roots of *P. latifolius* were shallow in the ground, there were a great deal more of them than with *P. australis*, and there was a lot of organic matter formed by itself.

Table 1. The biomass above the ground and subterranean biomass of *P. australis* and *P. latifolius*, along line transect I

Stand no.	1			2					3				
Community type	Stunted *P. a.*			*P. a.–P. l.*					*P. l.*				
Distance from the sea (m)	1.50~2.00			5.50~6.00					9.50~10.00				
Individuals	*P. a.*			*P. a.*		*P. l.*			*P. a.*		*P. l.*		
Above the ground biomass (g m^{-2})	352			156		344			0		736		
	Rhizomes	Roots	Organic matter	Rhizomes	Roots	Rhizomes	Roots	Organic matter	Rhizomes	Roots	Rhizomes	Roots	Organic matter
Bellow the ground 0~−30 cm (g m^{-3})	3,194	346	927	63	11	4,623	809	502	0	0	5,716	1,849	3,547
Bellow the ground −30~−60 cm (g m^{-3})	0	0	0	297	45	461	70	242	7	1	314	62	1,732

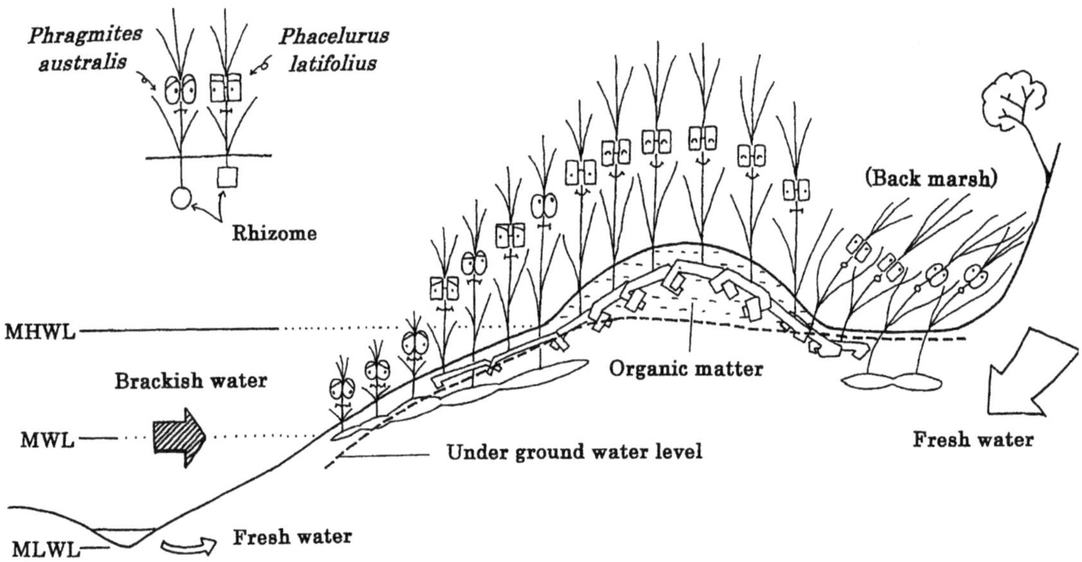

Fig. 5. Model illustrating the relationship between the habitats of *P. australis* and *P. latifolius* and environmental factors.

Not only were the *P. latifolius* rhizomes hard and tight, but also their sociability was very strong. As a result, *P. latifolius* community extended gregariously, with a high coverage value and sociability throughout the *P. australis* habitat (Figure 5).

Sediment type and soil color were shown in Figure 3. Because shell fragments were recognized in all soil blocks without upper one of under *P. latifolius* community (std. 3), these habitats were made on marine deposit in early stage. Therefore, there was humus deposit under *P. latifolius* community. And *P. latifolius* community had formed its habitat by itself while made the deposit.

The above matter suggested that the ecological and physiological habitat of *P. australis* community had a wide range and surrounded *P. latifolius* community in the study site. Then, *P. latifolius* community extended especially in its optimum habitat limited by salinity and ground height by the seashore. So the physiological range for *P. latifolius* was narrow, and its habitat was apt to be under the influence of human impacts. However, it was expected that it had many utilities as a buffer zone for the conservation of the coastal line, because *P. latifolius* had very gregarious and strong rhizomes and roots. After more physiological studies with *P. latifolius*, its community was able to put practical use for tidal conservation or natural shore protection.

REFERENCES

Braun-Blanquet, J. Pflanzensozilogie. Grundzuge der Vegetationskunde, Vol. 3. Aufl. 865. Springer-Verlag, Wien, 1964.

Ellenberg, H. Vegetation ecology of Central Europe, 4th edn. 1988; 348–67.

Ellenberg, H. Zeigerwarte von Pflanzen in Mitteleuropa. Scripta Geobotanica 1992; 18:258 s.

Gillner, V. Vegetation und Standortsunterssuchungen in den Strandwiesen der schwedischen Westküste. Acta Phytogeogr Suecica 1960; 43:198s.

Halsam, S.M. Some aspects of the life history and autecology of *Phragmites communis* Trin., a review. Pol Arch Hydrobiol 1973; 20(1):79–100.

Holm, L.G., Plucknett, D.L., Pancho, J.V. and Herberger, J.P. World's worst weeds. Univ. Press Hawaii, Honolulu, 1977; 609.

Kira, T. Ringyo Kaisetsu. Tokyo: Nippon Ringyo Gijutsu Kyokai. 1949; 17:105–41 (in Japanese).

Koyama, M. and Takehara, H. Revised standard soil color charts. Nihon Iroken Jigyo Co. Ltd., 1990.

Linke, O. Die Biota des Jadebusenwattes. Helgoländ Wiss Meeresunters 1939; 1:201–348.

Matoh, T., Matushita, N. and Takehashi, E. Salt tolerance of the reed plant *Phragmites communis*. Physiol Plant 1998; 72:8–14.

Miroalav, M. Ecology and use of *Phragmites communis* in Eastern Canada. Bull Kansai Organ Nature Conservat 1999; 21(2):289–99.

Miyawaki, A.u. Ohba, T. Studien über Strand-Salizwiezengesellschaften auf Honsyu, Shikoku, und Kyusyu (Japan). Sci Rep Yokohama–Nat Univ Sec II 1969; 15:1–23.

Miyawaki, A.u. Okuda, S. Pflanzensoziologische Untersuchungen über die Auen-Vegetation des Flusses Tama bei Tokyo, mit vergleichenden Betrachtung über die Vegetation des Flusses Tone. Vegetatio 1972; 24:229–311.

Miyawaki, A., Okuda, S., u. Suzuki, K. Vegetation in der Umgebung der Bucht von Tokyo. Sci Rep Yokohama–Nat Univ 1975; 23–29.

Mook, J.H. and van der Toorn, J. The influence of environmental factors and management on stands of *Phragmites australis*. 2. Effects on yield and its relationships with shoot density. J Appl Ecol 1982; 19(2):501–18.

Okuda, S. Pflanzensoziologische Untersuchungen über die Auenvegetation der Kanto-Ebene. Bull Inst Environ Sci Tec Yokohama–Nat Univ 1978; 4(1):43–112 (in Japanese).

Peterson, J. Die Dünentalvegetation der Watternmeer-Inseln in der südlichen Nordsee. Eine pflanzensoziologische und ökologische Vergleichsuntersuchung unter Berücksichtigung von Nutzung und Naturschutz. Diss. Univ. Hannover, 203 s., Hannover, 1999.

Roman, C.T., Niering, W.A. and Warren, R.S. Salt marsh vegetation change in response to tidal restriction. Environ Manage 1984; 8(2):141–50.

Sinicrope, T.L., Hine, P.G., Warren, R.S. and Niering, W.A. Restration of an impounded salt marsh in New England. Estuaries 1990; 13(1):25–30.

Toma, T. and Ogino, K. Soil Water Movement of a Mangrove Forest in Halmahera Island, East Indnesia. TROPICS 1995; 4:187–200.

Yamada, A., Ohno, K. and Okuda, S. The vegetation of Koajiro area and its vicinity, the Southern Miura Peninsula, Kanagawa Prefecture. ECO-HABITAT: JISE Res 1998; 5(1):29–52 (in Japanese).

Part IV

Relevant Accompanying Contributions

Introduction to Part IV

H. LIETH

Developing production systems with halophytes requires a solid knowledge of the local flora and fauna in the adjacent vegetation and ecosystem.

In many cases we need to import new species, a process which is known to cause ecological problems with the native vegetation. For halophytes this problem is mostly restricted to the area which is used for saline irrigation since most halophytes have little competitive power in low-saline environments. The floristic investigations will concentrate, therefore, on saline wetland and coastal areas.

Such investigation may be undertaken for the identification of local species with the potential for utilisation in saline production systems or for checking the potential of introduced species to become obnoxious in the existing ecosystems around newly created saline production systems. Such considerations are especially important if genetically selected or modified species are included in pilot plantations.

For the cases mentioned, this part contains some examples. There is much more to be analysed before a saline production system can be established. Careful planning is needed prior to the establishment of a pilot plantation. This will require also construction and operational planning as well as marketing and economic analyses as we have outlined in our home page http://www.usf.uni-osnabrueck.de/projects/expo2000.

H. Lieth (ed.), Cash Crop Halophytes: Recent Studies, 153.

Genetically modified halophytes in agro-ecosystems: ecological and environmental considerations

DETLEF BARTSCH

PROGRESS IN PLANT TRANSFORMATION WITH SPECIAL REGARD TO HALOPHYTES

Plants with exciting new properties of great human values are under development, and new methods in plant breeding are used now on the molecular level. Benefits of genetically engineered improvements include prevention of billions of dollars of losses by introduction of resistance to plant diseases, increased benefits for the environment due to reduced chemical pesticide applications, and improved human health because of reduced exposure to chemical pesticide residues (Kahl and Winter, 1995). Such new genetically modified plants (GMPs) must be safe in regard to health and the environment. Therefore it is of importance to assess GMPs carefully for their effects on the environment and to understand the genetical and biochemical basis of their effectiveness for sustainable crop production.

Salinity in soil has an important effect on plant growth. About 7% of the world's land surface and about 5% of cultivated land are effected by salinity, and about 20% of irrigated land has suffered from secondary salinization and 50% of irrigation schemes are affected by salts (Flowers et al., 1997). Consequently, the development of salt-resistant crops is seen as an important area of research. On the other hand existing halophytes could be used in saline areas for agricultural and landscape improvement (Glenn et al., 1999). Modern biotechnology provides tools for the genetic improvement of halophyte plants (Yeo, 1998; Serrano et al., 1999). Several genetically modified (transgenic) crops have been released into the environment. In the US alone, dozens of transgenic crop varieties have been deregulated [1]. Among these plants, only sugar beet (*Beta vulgaris* L.) can be regarded as a halophyte. There is yet no sign of commercial approach to add salt-tolerance traits to conventional cultivars, but several concepts are currently being developed (Bohnert and Jensen, 1996; Winicov, 1998).

RISK ASSESSMENT: ECOLOGICAL CONSEQUENCES OF THE MODIFIED PHENOTYPE

Until now it is not clear if the products of genetic engineering will have negative ecological impacts that differ from those of traditional breeding. Therefore, environmental research is essential to identify the risks related to the release of GMPs (Tiedje et al., 1989). The ecological advantage of a transgenic plant trait can lead to a spread of this plant. For the concern that a gene transfer would possibly alter the pest status or effect the food quality of human nutrition etc. are overviews available (Metcalfe et al., 1996). A general overview of early biosafety research is given by Harding and Harris (1997). Scientific based risk assessment for GMPs is based on both hazard identification and the likelihood of hazard occurring (Sharples, 1991).

Establishment and Spread of Transgenes

It would be interesting to know why many people in the Western industrial countries are critical about genetically engineered plants, when they are introduced into the environment. In the past thousands of non-indigenous species have been spread actively from continent to continent and region to region. There are numbers of reports available for cultivars escaped into natural ecosystems (Williamson, 1993; Bartsch et al., 1993; Bartsch and Ellstrand, 1999). In their outstanding review, Ellstrand et al. (1999)

H. Lieth (ed.), Cash Crop Halophytes: Recent Studies, 155–159.

demonstrated that 12 of the 13 world-wide most important crops hybridize with wild relatives somewhere in their cultivation area (Table 1). Transgenic plants will be no exception from this phenomenon, but there is no evidence that transgenic plants are a risk *per se*. Since risk is a product both of probability and hazard, it is clear that biosafety research on environmental effects should not only target the probability of gene flow, but also must focus on the consequences (and potential hazard) of successful transgene flow to relatives of transgenic crops. That is, biosafety research should address the phenotype (especially, the fitness phenotype) of the transgenic hybrid versus that of non-transgenic controls.

The influence of transgenic attributes on the establishment and spread of halophytes stands in the center of consequence assessment: Tolerance against salinity is one of the determining factors controlling the establishment in salt marsh and salt desert environments. Competition is hypothesized to play a key role in determining both the upper and lower limits of species distribution along a salinity gradient (Ungar, 1998). The relationship between the level of salt tolerance of species and their ability to compete with glycophytes in less saline habitats seems to be reciprocal. Halophytes – and sugar beet is a good example – are not competitive in non-saline habitats. Their competitiveness increases in saline habitats like seashores.

Allelopathic effects have been reported in salt desert habitats, but have not been reported along salinity gradients in salt marshes (Ungar, 1998). Some species of halophytes that are salt accumulators have the ability to change soil chemistry. Chemical inhibition of intolerant species occurs when high concentrations of sodium are concentrated in the surface soils of salt desert plant communities that are dominated by salt-accumulating species. Establishment of less salt-tolerant species is inhibited in the vicinity of these salt-accumulating species. Herbivory is reported to cause both an increase and a decrease in plant diversity in saline habitats. Thus, genes mediating insect protection might change the ecological behavior of genetically modified plants.

Unwanted Effects: The Weed Concept

There are various concepts defining weeds and there is no classical approach generally accepted (Amman et al., 2000). Lambelet-Haueter (1990) divides weed definitions into popular, economical and ecological concepts whereas Holzner (1982) groups them similarly into subjective and ecological ones. *Popular* as well as *subjective concepts* define weeds as plants of *any* kind growing in the wrong place, causing damage, being of no benefit and suppressing cultivated plant species. *Economical concepts* reflect the view of agronomists who concentrate on the reduction of yield, thereby stressing the damage aspect. A weed problem is solved as soon as the plant no longer creates considerable damage in the fields, a state which is reached by means of adjusted weed control (crop rotation, tillage, herbicide application). In contrast to the previous concept, *ecological definitions* include habitats outside agrosystems colonized by weeds. The usual preference of weeds for habitats disturbed by man is stressed, like for cultivated fields and gardens as well as disturbed areas on road sides, recently built artificial slopes and others. An aggressive weed can cause damage not only in agrosystems but also in (semi-)natural plant communities by outcompeting weak species.

Table 1. 12 of the 13 world-wide most important crops that hybridize with wild relatives somewhere in their cultivation area. Sugar beet is added to the list of Ellstrand et al. (1999) due to their European importance

Crop	Relative(s)	Crop	Relative(s)
1. Wheat	Wild *T. turgidum* subspecies, some *Aegilops* species	8. Millets	*Eleusine coracana* ssp. *africana*, Wild *Pennisetum* species
2. Rice	Wild *Oryza* species	9. Beans	Wild *Phaseolus* species
3. Maize	Wild *Zea mays* subspecies	10. Oilseed rape	Some wild *Brassicaea* species
4. Soybean	*Glycine gracilis, G. soya*	11. Peanut	No report
5. Barley	*Hordeum spontaneum*	12. Sunflower	Wild *Helianthus annuus*
6. Cotton	Wild *Gossypium* species	13. Sugarcane	Wild *Saccharum* species
7. Sorghum	Wild *Sorghum* species	27. Sugar beet	Some wild *Beta* species

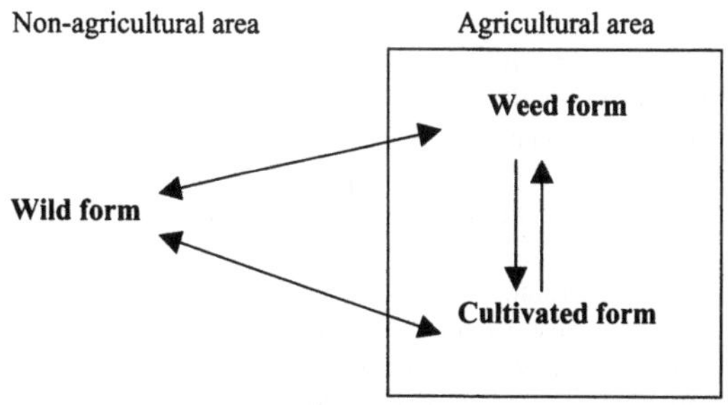

Fig. 1. Gene flow in cultivar-weed complexes (Harlan, 1965, modified).

Following Holzner (1982), it is sometimes difficult to call a plant a weed because one and the same species may be considered in some parts of its area as a harmless component of natural vegetation, in others as a weed and again in others, even as a useful plant species. Traditional breeding has so far focussed on biomass improvement efforts. In comparison to wild relatives, cultivars are in general genetically less diverse and therefore less adapted to natural environments. Reports of fitness advancement for hybrids in natural ecosystems are rare. In opposite, hybridization and escape of cultivar genes has led to disadvantages for wild plant populations in some documented cases. Crop-to-weed gene flow have had important practical and economic consequences since it promotes the evolution of more aggressive weeds (e.g. Anderson, 1949; Barrett, 1983). Hybridization with domesticated species has also been implicated in the extinction of certain wild crop relatives (e.g. Ellstrand and Elam, 1993; Small, 1984). One of the most important goals is the conservation of genetic resources, in this case the genetic diversity of wild relatives of crop plants.

CONSERVATION MONITORING OF PLANT GENETIC RESOURCES: THE SUGAR BEET EXAMPLE

The genus *Beta* is endemic to the Old World. Beets have been cultivated for more than 2000 years in the eastern Mediterranean region. *Beta vulgaris* comprises to an extraordinary variable group, in which cultivated and wild forms are often difficult to distinguish. This is mainly due to the extensive use of sea beet (*Beta vulgaris* ssp. *maritima* ARCANG) gene resources in conventional breeding programs. Sea beet is largely a coastal taxon, with a wide distribution from the Cape Verde and Canary Islands in the west, northward along Europe's Atlantic coast to the North and Baltic Seas. It also extends eastward through the Mediterranean region into Asia where it occurs in Asia Minor, in the central and outer Asiatic steppes, and desert areas as far as western India (Letschert, 1993). Sea beet varies from self-compatible, annuals to self-incompatible, iteroparous perennials with a life span between 1 and 5 years. Cultivated *B. vulgaris*, including Swiss chard, red garden beet and sugar beet, are biennial. The latter is partially self-incompatible due to the extensive use of male sterility genes in sugar beet breeding. All cultivated and wild subspecies of *B. vulgaris* are mostly wind-pollinated, although some insect pollination has been noted. Besides wild and cultivated forms, a weed beet form is also known to grow in sugar beet fields, where, due to its genetically based annuality, it bolts late in the season. Producing lignescent inflorescence and large amounts of seed instead of commercially desirable fleshy beet roots, weed beets can be a serious problem for sugar beet farmers. Other weed forms of *B. vulgaris* are reported in several winter crops of the Nile Delta region inside cultivated fields as well as irrigation canals and drains (Shaltout et al., 1992, 1995). Other weed types of an additional *Beta* species, *B. macrocarpa*, are known from irrigated desert areas in Morocco and California (Bartsch and Ellstrand, 1999).

Due to the similarity of their seedling morphology and physiology to sugar beet, conventional methods do not control annual weed beets. Relatively little is known about the persistence and dynamics of weed beet populations, but the species *Beta vulgaris* is a good example for a wild/cultivar complex (Mücher et al., 2000), in which gene flow and hybridization can be observed according to the model of Harlan. Hybridization and gene flow between cultivated and wild species is an important source for the origin of weeds (Boudry et al., 1993; Desplanque et al., 1999). F1 hybrids formed by spontaneous crosses between cultivated beets and south European wild types are typically annual, because this life cycle trait is genetically dominant and common in southern wild populations. To this end, we carried out a monitoring program with transgenic sugar beet (*Beta vulgaris* ssp. *vulgaris* var. *altissima* DÖLL) with beet necrotic yellow vein virus (BNYVV) coat protein (*cp*), phosphinothricin-acetyl-transferase (*bar*) and neomycin-phospho-transferase (*nptII*) genes. We found that the virus-tolerant phenotype of hybrids with wild sea beet were not significantly different to naturally tolerant genotypes. In addition, there is no evidence that the virus has any ecological role, since the virus cannot be found in sea beet populations. We conclude that ecological implications due to the introduction and spread of virus resistant transgenic hybrids will be minimal in this special case. However, for the assessment of any future effect the collecting of basic data on geographic distribution and genetic diversity of wild plant populations is absolutely necessary (Bartsch and Pohl-Orf, 1996; Bartsch et al., 1996).

We also studied the ecological impact of a century of gene flow from traditionally bred cultivated beets into the wild sea beet populations of north-eastern Italy (Bartsch et al., 1999). We demonstrated that gene flow from a crop to a wild relative does not necessarily result in a decrease in the genetic diversity of the wild plant, although the cultivated beets are less diverse and outnumbered the wild relatives by the factor 10,000 to 1.

OUTLOOK: ENVIRONMENTAL RISKS AND BENEFITS

The Biological Diversity Convention [2] targets in some chapters potentially adverse effects of transgenic plants. Every introduction of transgenic halophytes should be weighted carefully against the benefits case-by-case and step-by-step. Productivity increase by crop improvement in combination with sustainable agriculture is the task of the future.

NOTES

1. For details see http://www.aphis.usda.gov/bbep/bp/petday.html.
2. For details see http://www.biodiv.org/.

REFERENCES

Ammann, K., Jacot, Y. and Rufener Al-Mazyad, P. Safety of genetically engineered plants. An ecological risk assessment of vertical gene flow. VIB, 2000, in press.

Anderson, E. Introgressive hybridisation. John Wiley & Sons, Inc., New York; Chapman & Hall, Limited, London, 1949.

Barrett, S.C.H. Crop mimicry in weeds. Econom Bot 37: 255–82.

Bartsch, D. and Ellstrand, N.C. Genetic evidence for the origin of Californian wild beets (genus *Beta*). Theoret Appl Genet 1999; 99:1120–30.

Bartsch, D., Lehnen, M., Clegg, J., Pohl-Orf, M., Schuphan, I. and Ellstrand, N.C. Impact of gene flow from cultivated beet on genetic diversity of wild sea beet populations. Molecular Ecol 1999; 8:1733–41.

Bartsch, D., Schmidt, M., Pohl-Orf, M. Haag, C. and Schuphan, I. Competitiveness of transgenic sugar beet resistant to beet necrotic yellow vein virus and potential impact on wild beet populations. Molecular Ecol 1996; 5:199–205.

Bartsch, D., Sukopp, H. and Sukopp, U. Introduction of plants with special regard to cultigens running wild. In Wöhrmann, K. and Tomiuk, J., editors. Transgenic organisms. Birkhäuser Verlag, Basel, Switzerland, 1993; 135–51.

Bartsch, D. and Pohl-Orf, M. Ecological aspects of transgenic sugar beet: Transfer and expression of herbicide resistance in hybrids with wild beets. Euphytica 1996; 91:155–58.

Bohnert, H.J. and Jensen, R.G. Strategies for engineering water-stress tolerance in plants. Trends in Biotechnol (TIBTECH) 1996; 14:89–97.

Boudry, P., Mörchen, M., Saumitou-Laprade, P., Vernet, P. and Van Dijk, H. The origin and evolution of weed beets: consequences for the breeding and release of herbicide-resistant transgenic sugar-beets. Theoret Appl Genet 1993; 87:471–8.

Desplanque, B., Boudry, P., Broomberg, K., Saumitou-Laprade, P., Cuguen, J. and Van Dijk, H. Genetic diversity and gene flow between wild, cultivated and weedy forms of *Beta vulgaris* L. (*Chenopodiaceae*), assessed by RFLP and microsatellite markers. Theoret Appl Genet 1999; 98: 1194–201.

Ellstrand, N.C. and Elam, D.R. Population genetic consequences of small population size: implications for plant conservation. Ann Rev Ecol Syst 1993; 24:217–42.

Ellstrand, N.C., Prentice, H.C., and Hancock, J.F. Gene flow and introgression from domesticated plants into their wild relatives. Ann Rev Ecol Syst 1999; 30:539–63.

Flowers, T.J., Garcia, A., Koyama, M. and Yeo, A. Breeding for salt tolerance in crop plants: the role of molecular biology. Acta Physiologiae Plantarum 1997; 19:427–33.

Glenn, E.P., Brown, J.J. and Blumwald, E. Salt tolerance and crop potential of halophytes. Crit Rev Plant Sci 1999; 18:227–55.

Harding, K. and Harris, P.S. Risk assessment of the release of genetically modified plants, A review. Agro-Food-Industry Hi-Tech 1997; 8:8–13.

Holzner, W. Concepts, categories and characteristics of weeds. In Holzner, W. and Numata, N., editors. Biology and ecology of weeds. Dr. W. Junk Publishers, The Hague, 1982; 3–20.

Kahl, G. and Winter, P. Plant genetic engineering for crop improvement. World J Microbiol Biotechnol 1995; 11: 449–60.

Lambelet-Haueter, C. Mauvaises herbes et flore anthropogène. II. Classifications et catégories. Saussurea 1991; 22:49–81.

Letschert, J.P.W. *Beta* section *Beta*: biogeographical patterns of variation and taxonomy. Ph.D. thesis. Wageningen Agricultural University Papers 93-1, 154 pp.

Metcalfe, D.D., Astwood, J.D., Townsend, R., Sampson, H.A., Taylor, S.L. and Fuchs, R. Assessment of the allergenic potential of foods from genetically engineered crop plants. Crit Rev Food Sci Nutrit 1996; 36:35–56.

Mücher, T., Hesse, P., Pohl-Orf, M., Ellstrand, N.C. and Bartsch, D. Characterization of weed beets in Germany and Italy. J Sugar Beet Res 2000.

Serrano, R., Culianz-Macia, F.A. and Moreno, V. Genetic engineering of salt and drought tolerance with yeast regulatory genes. Scientia Horticulturae 1999; 78:261–9.

Shaltout, K.H., Sharaf El-Din, A. and El-Fahar, R.A. Weed communities of the common crops in the Nile Delta region. Flora 1992; 187:329–39.

Shaltout, K.H., El-Kady, H.F. and Al-Sodany, Y.M. Vegetation analysis of the Mediterranean region of Nile Delta. Vegetatio 1995; 116:73–83.

Sharples, F.E. Ecological aspects of hazard identification for environmental uses of genetically engineered organisms. In Levin, M.A. and Strauss, H.S., editors. Risk assessment in genetic engineering. McGraw-Hill, New York, 1991; 18–31.

Small, E. Hybridization in the domesticated weed-wild complex. In Grant, W.F., editor. Plant biosystematics. Academic Press, Toronto, 1984; 195–210.

Tiedje, J.M.R., Colwell, R.L., Grossman, Y.I., Hodson, R.E., Lenski, R.E., Mack, R.N. and Regal, P.J. The planned introduction of genetically engineered organisms, ecological considerations and recommendations. Ecology 1989; 70:298–315.

Ungar, I.A. Are biotic factors significant in influencing the distribution of halophytes in saline habitats? Bot Rev 1998; 64:176–99.

Williamson, M.H. Invaders, weeds and the risk from genetically manipulated organisms. Experientia 1993; 49:219–24.

Winicov, I. New molecular approaches to improving salt tolerance in crop *plants*. Ann Bot 1998; 82:703–10.

Yeo, A. Molecular biology of salt tolerance in the context of whole-plant physiology. J Exp Bot 1998; 49:915–29.

Halophytic grasses as vital components of cash crop halophytes: *Sporobolus madraspatanus* Bor.

A.J. JOSHI AND MANOJ MISRA

INTRODUCTION

In order to assess its usefulness in increasing forage production on saline wastelands, seed germination and seedling growth of highly salt-tolerant grass *Sporobolus madraspatanus* growing in Gujarat were studied. Seed germinability up to 16 dS m^{-1}, its ability to grow in 32 dS m^{-1} seawater and greater accumulation of organic solutes such as alanine, asparagine, aspartic acid, proline, sugars and of inorganic ions Na^+ and Cl^- in leaves, stems and roots of 61-day-old plants collectively showed its reasonably good growth in saline condition. Thus, *Sporobolus madraspatanus* could be a vital resource component of the cash crop halophytes.

Halophytic grasses constitute a vital component in the warm continents and they also luxuriantly grow in upper parts of the intertidal zone all along the coast of India (Chapman, 1964). Understanding their salt-tolerance mechanism would help in utilising such grasses for increasing forage production of saline wastelands, an important part of the integrated coastal zone management (Clark, 1964).

Sporobolus madraspatanus Bor. is extensively used as forage for cattle by the natives in Gujarat and elsewhere and it is possible to introduce it in barren saline areas (Joshi and Misra, 1992). The present investigation describes the effects of seawater concentrations on seed germination and growth of young plants of *Sporobolus madraspatanus*. It further examines the impact of seawater stress on accumulation of amino acid, sugars and inorganic ions in leaves, stems and roots of young plants.

MATERIAL AND METHODS

Seeds of *Sporobolus madraspatanus* were collected from Bhal area (22° 25′ N; 72° 22′ E) forming a part of the coastal belt in Gujarat. The seeds were germinated in distilled water and 4–56 dS m^{-1} seawater concentrations on Whatman filter paper No. 1 in petridishes at room temperature (30–35°C) and six replications of each treatment were maintained.

For raising young plants, the seeds after soaking were transferred to perforated plastic pots filled with sand. The pots were dipped in distilled water every alternate day to keep the sand bed uniformly wet. After 15 days, half strength Hoagland's nutrient solution was added to the medium and from day 31 salinity was imposed with a daily increment of 4 dS m^{-1}. The plants were harvested on day 61 for growth data and biochemical analysis.

Proteins were estimated by folin–phenol reagent (Lowry et al., 1951); sugars by Anthrone reagent (Umbreit et al., 1959); reducing sugars by Folins and Malmros method (1929); other sugars (Mishra and Rao, 1960) and amino acid by paper chromatography (Joshi, 1986). Na^+ and K^+ by flame photometry; Ca^{2+} and Mg^{2+} by EDTA titration (Vogel, 1978) and chlorides on chloride meter (Elico EE-34 Model).

RESULTS

Although the seeds germinated up to 16 dS m^{-1} seawater maximum germination was observed in distilled water (Figure 1a) and the seeds, which had been subjected to high concentration of seawater showed a greater degree of recovery in distilled water reflecting the osmotic effects of concentrations manifested ionic and other effects on the process.

The shoot and root length of 61-day-old control plants was 9.5 and 14.0 cm, respectively. However, a noticeable decrease in these parameters was observed for those plants which were subjected to 16 and 32 dS m^{-1} seawater concentration (Figure 1b and c). Salt stress

H. Lieth (ed.), Cash Crop Halophytes: Recent Studies, 161–166.

remarkably reduced these values for plants raised in 16 and 32 dS m^{-1} seawater concentrations though dry matter percentage of such plants increased considerably (Figure 1c). Amounts of alkali soluble proteins in vegetative organs ranged between 82 and 191 mg g^{-1}, while that of ethanol and water soluble ones varied between 23 and 103 mg g^{-1} and their accumulation was not related to increased levels of salinity in the growth medium (Figure 2).

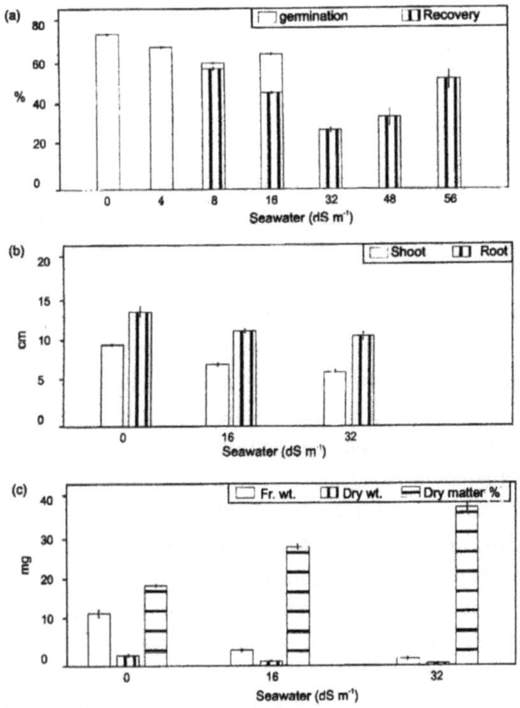

Fig. 1. Effects of seawater salinity on seed germination (a) and on growth (b, c) on 61-day-old plants of *S. madraspatanus*. Each value represents mean ± SEM of ten observations.

Findings of amino acid showed that accumulation of alanine (69–2277 μg g^{-1}), asparagine (144–939 μg g^{-1}), glutamic acid (75–364 μg g^{-1}), phenylalanine (128–1013 μg g^{-1}), proline (74–3179 μg g^{-1}) and serine (47–311 μg g^{-1}) was greater in leaves and stems of plants subjected to 16 and 32 dS/m seawater than those of the control plants (Table 1). Nevertheless, a noticeable decrease in their accumulation was observed in roots. On the other hand were concentration of methionine (17–108 μg g^{-1}), serine (47–311 μg g^{-1}) and valine (813–57 μg g^{-1}) in the vegetative organs of seawater treated plants less than those of control plants.

Concentration of total (14.16–27.46 mg g^{-1}) and reducing sugars (3.92–11.62 mg g^{-1}) in vegetative organs of plants raised in 16 and 32 dSm^{-1} seawater concentration was greater than amounts recorded for control plants (Figure 3B). Likewise, accumulation of arabinose (0.18–0.57 mg g^{-1}); galactose (0.18–0.64 mg g^{-1}); glucose (0.32–1.43 mg g^{-1}) and rhamnose (0.19–0.83 mg g^{-1}) in vegetative organs of plants raised in saline conditions was obviously greater than those of the control plants (Figure 3a–c).

Mineral composition of the plants indicated an increase in the ash, sodium and chloride content in the tissue of plants raised in seawater concentrations; whereas accumulation of potassium, calcium and magnesium was not related to the concentration of seawater in the growth medium (Figure 4).

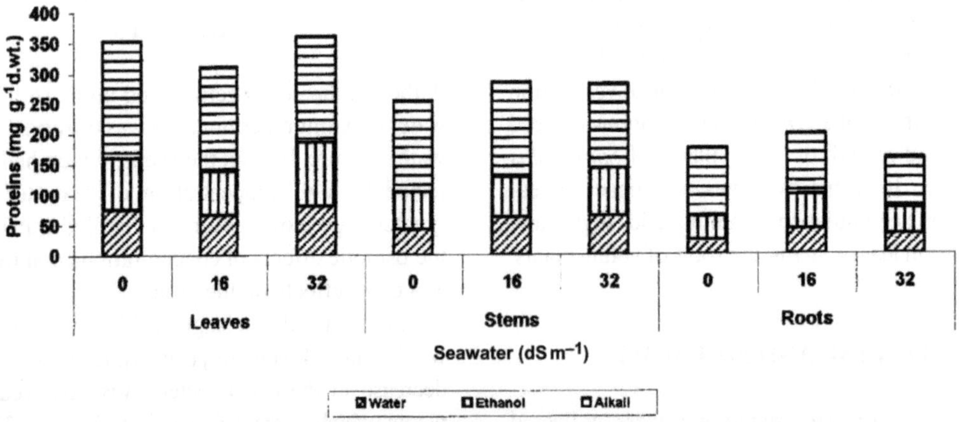

Fig. 2. Effects of seawater salinity on protein content (mg g^{-1} d.wt.) in 61-day-old plants of *S. madraspatanus*. Each value represents mean ± SEM of four observations.

Table 1. Effects of seawater salinity on free amino acids ($\mu g\ g^{-1}$ d wt.) in vegetative organs of 61-day-old-plants of *S. madraspatanus*. Each value represents mean ± SEM of four observations

		ds m^{-1}		
Amino Acids		0	16	32
Alanine	Lvs	213 ± 10	287 ± 06	2277 ± 18
	St	255 ± 10	334 ± 19	295 ± 17
	Rt	153 ± 07	138 ± 04	69 ± 05
Asparagine	Lvs	282 ± 79	297 ± 18	616 ± 18
	St	609 ± 107	517 ± 40	939 ± 42
	Rt	274 ± 23	188 ± 04	144 ± 08
Aspartic Acid	Lvs	319 ± 57	761 ± 14	1404 ± 180
	St	782 ± 150	1647 ± 84	1820 ± 154
	Rt	289 ± 49	605 ± 32	370 ± 16
Glutamic Acid	Lvs	167 ± 15	49 ± 33	249 ± 15
	St	307 ± 30	250 ± 9	364 ± 13
	Rt	164 ± 28	136 ± 6	75 ± 5
Glycine	Lvs	75 ± 11	129 ± 34	90 ± 7
	St	136 ± 36	104 ± 12	254 ± 11
	Rt	132 ± 24	58 ± 4	42 ± 4
Iso-leucine	Lvs	9 ± 2	27 ± 4	19 ± 1
	St	44 ± 8	45 ± 6	37 ± 3
	Rt	24 ± 7	11 ± 1	6 ± 1
Leucine	Lvs	10 ± 2	35 ± 6	23 ± 1
	St	47 ± 12	54 ± 3	57 ± 3
	Rt	24 ± 6	18 ± 2	11 ± 2
Methionine	Lvs	241 ± 53	108 ± 15	84 ± 3
	St	185 ± 20	88 ± 6	62 ± 4
	Rt	55 ± 8	38 ± 2	17 ± 1
Phenylalanine	Lvs	212 ± 49	678 ± 59	697 ± 17
	St	751 ± 47	1013 ± 77	822 ± 43
	Rt	204 ± 33	207 ± 9	128 ± 12
Proline	Lvs	15 ± 0	74 ± 7	1995 ± 61
	St	20 ± 4	3179 ± 73	3004 ± 178
	Rt	25 ± 5	1183 ± 94	1794 ± 12
Serine	Lvs	126 ± 7	99 ± 28	150 ± 7
	St	82 ± 22	124 ± 10	311 ± 28
	Rt	158 ± 23	130 ± 8	47 ± 2
Threonine	Lvs	73 ± 14	54 ± 9	52 ± 4
	St	156 ± 12	89 ± 8	177 ± 5
	Rt	75 ± 9	59 ± 4	14 ± 2
Valine	Lvs	122 ± 27	57 ± 8	32 ± 7
	St	94 ± 13	54 ± 4	48 ± 3
	Rt	35 ± 2	21 ± 3	13 ± 6

Lvs-Leaves; St-Stem; Rt-Roots.

DISCUSSION

A failure of *Sporobolus madraspatanus* seed germination beyond 16 dS m^{-1} seawater reflects a general characteristic owned by other salt tolerant grasses (Pollack and Waisel, 1971). However, 60% seed germination in *Sporobolus wrightii* at 20 dS m^{-1} salinity manifested still higher degree of salt tolerance (Everitt, 1983). The present study evidently shows that half the number of seeds of *Sporobolus madraspatanus* do not loose their viability, even after exposure to high seawater concentrations.

Survival and adequate growth of 61-day-old plants of *Sporobolus madraspatanus* in 32 dS m^{-1} seawater

(a)

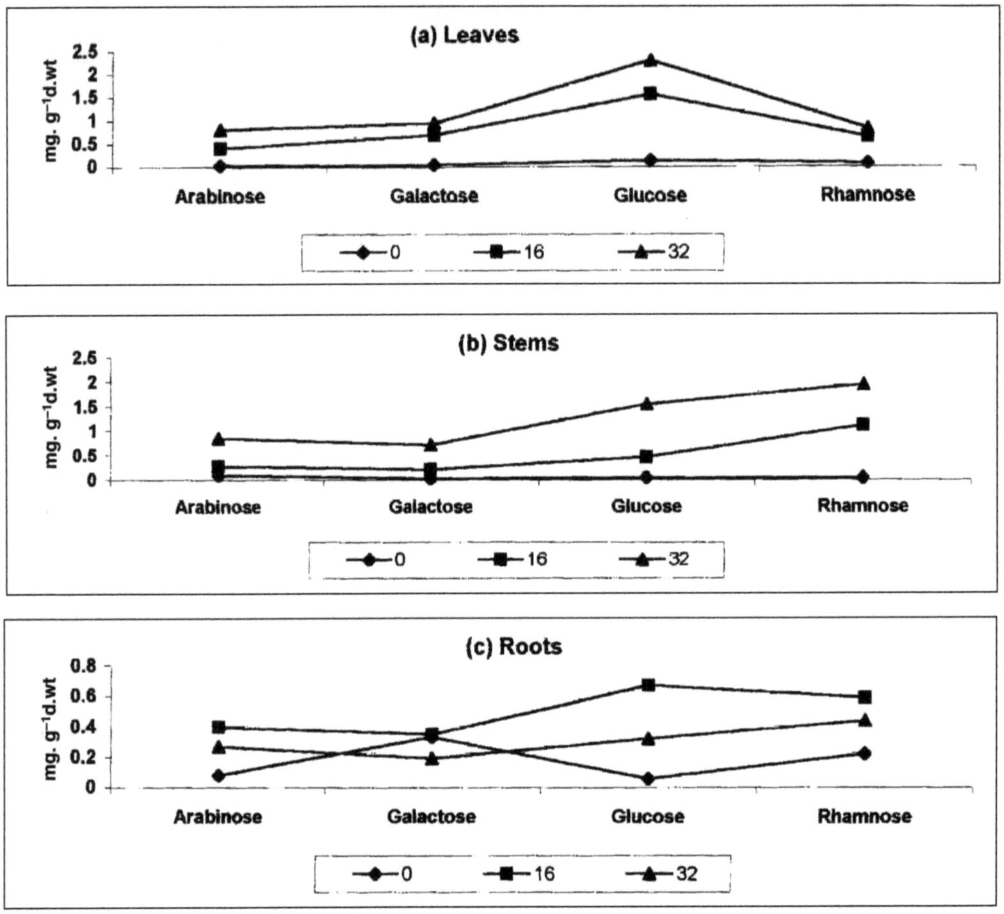

(b)

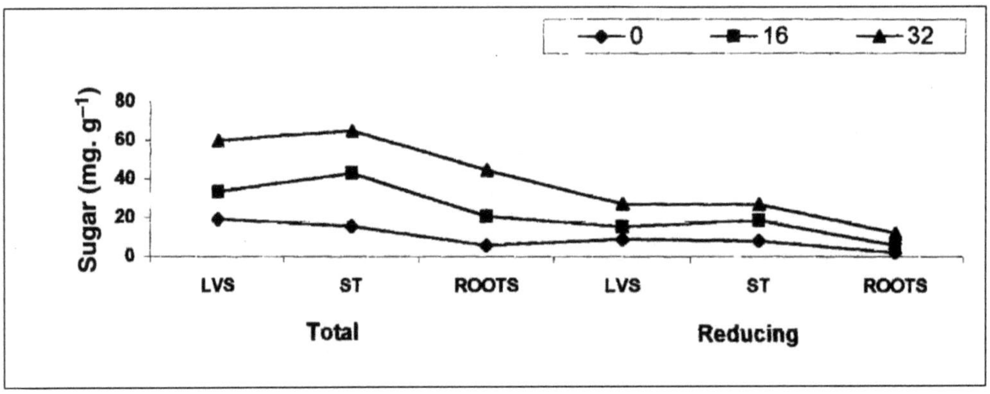

Fig. 3. (a) Effects of seawater salinity on sugar content (mg g^{-1} d wt.) in leaves (a), stems (b) and roots (c) of 61-day-old plants of *S. madraspatanus*. Each value represents mean ± SEM of four observations. (b) Effects of seawater salinity on accumulation of total and reducing sugars (mg g^{-1} d wt.) in 61-day-old plants of *S. madraspatanus*. Each value represents mean ± SEM of four observations.

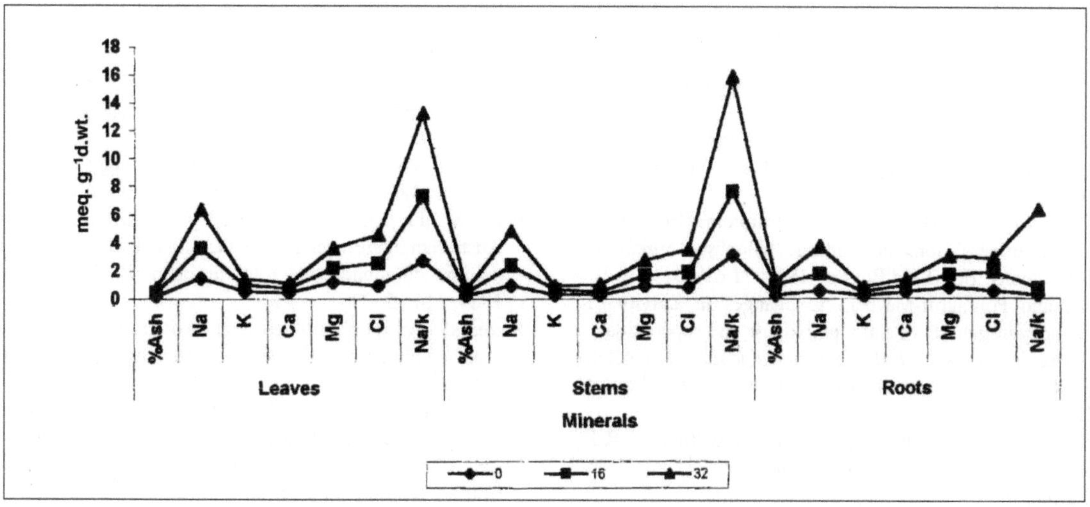

Fig. 4. Effects of seawater salinity on accumulation of mineral ions (meq g^{-1} d wt.) in 61-day-old plants of *S. madraspatanus*. Each value represents mean ± SEM of four observations.

obviously shows a remarkable degree of salt tolerance, perhaps known for the first time in Indian halophytic grasses, because optimum growth of salt tolerant plants is normally observed in growth media containing 100–300 mol m^{-3} NaCl. Thus, the experimental evidence shows the possibility to sow this grass direct on soils having salinity equivalent to 20 dS m^{-1}.

Kalir and Poljakoff-Mayber (1983) had reported an increase in leaf proteins of *Halimione portulacoides* grown in 2% NaCl solution suporting an earlier view of enhanced protein content in halophytes under saline conditions and which in turn, reflect an adaptive characteristic of salt-tolerant plants (EL-Shorbagy and Kishk, 1975. See also Costa and Cacador of this volume). The present study indicates that the protein content in 61-day-old plants of *Sporobolus madraspatanus* was not much affected by seawater concentrations.

As noted earlier (Table 1), salinity stress increased the concentration of alanine, asparagine, aspartic acid, glutamic acid, phenylalanine, serine and proline in young plants of *Sporobolus madraspatanus*. Although the osmoregulatory role of proline, glycinebetaine, sorbitol, piperolique acid etc. has been elucidated (Popp et al., 1984). The physiological role of other amino acids occuring in quite high concentration in salt-tolerant plants including *Sporobolus madraspatanus* yet remains to be explained.

Our findings showing an increase in amounts of sugar with increasing salinity in *Sporobolus madraspatanus* are similar to other halophytic species (Rozema, 1976) but no consensus prevails regarding a definite role of sugars in salt-tolerance mechanism of plants (Misra, 1989).

A positive relationship between concentration of salts and mineral ions in vegetative organs of young plants and seawater concentrations in growth media reflects the ability of the grass to maintain the internal osmotic pressure. However, 20% salt content found in the tissues must be taken as a precautionary parameter in deciding the forage value of *Sporobolus madraspatanus*.

The present study collectively shows that the *Sporobolus madraspatanus* can be introduced successfully in coastal saline areas for increasing biomass and forage production and thus, it can be considered as an important cash crop halophytes in the developing countries facing acute shortage of the fodder.

REFERENCES

Costa, A.L. and Cacador, I. Chemical composition variation during root decomposition in Tagus estuary salt marshes. Kluwer Academic Publishers, Dordrecht, The Netherlands, 2002.

Chapman, V.J. Coastal Vegetation. Robert Maxwell Publishers. Vol. 111, Pergamon Press, New York, 1964.

Clark, J.R. Coastal zone management. Lewis Publishers. CRC Press, New York, 1996.

EL-Shobagy, M.N. and Kishk, H.T. Sodium chloride effects on sugar metabolism of several plants. Phyton Australia 1975; 17:101–8.

Everitt, J.H. Effects of single salt solution and PH on seed germination of two native range grasses. J Rio Grande Val Hortic Sic 1983; 36:23–8.

Folin, O. and Malmros, H. Micro methods for glucose estimation in blood. J Biol Chem 1929; 83:115.

Joshi, A.J. Effect of seawater on amino acid and mineral ions composition in S. *brachiata*. Roxb. J Plant Physiol 1986; 123:497–502.

Joshi, A.J. and Misra, M. Fluctuations of proteins, free amino acids and mineral ions in a forage halophyte *Sprobolus madraspatanus*. Bor. J Ind Bot Soc 1992; 71:61–4.

Kalir, A. and Polljakoff-Mayber, A. Changes in activity of malate dehydrogenase, catalase, peroxidase and superoxidase dismutase in leaves of *Halimione portulacoides* (L) Allen exposed to high sodium chloride concentrations. Ann Bot 47:75–85.

Lowry, O.H., Rosenbrough, N.J., Farr, A.L. and Randall, R.J. Protein measurements with folin phenol reagent. J Biol Chem 1951; 193:265–75.

Mishra, S.D. and Rao, M.U.K. Quantitative estimation of carbohydrates by paper partition chromaatography. J. Sci Indust Res 1960; 19C.

Misra, M. Studies on salt tolerance in *Sporobolus madraspatanus* Bor. A forage halophytic grass. Ph.D. Thesis, Bhavnagar University, Bhavnagar, 1989.

Pollak, G. and Waisel, Y. Reproductive biology of *Aeluropus litoralis* (Willd) Parl. Ann Arid Zone 1971; 10:165.

Popp, M., Larher, F. and Weigel, P. Chemical composition of Australian mangroves III. Free amino acids, total methylated onium compounds and total nitrogen. Z. Pflanzenphysiol 1984; 114:15–25.

Rozema, J. An ecophysiological study on the response to salt of four halophytic and glycophytic *Juncus* species. Flora 1976; 165:197–209.

Umbreit, W.W., Burris, R.H. and Stauffer, J.E. Manometric techniques. Burgees Publishing Co., Minnesota, 1959.

Vogel, A.I. A text book of quantitative inorganic analysis. ELBS Edition, London, 1978; 506.

An ecological overview of halophytes from Pakistan

M. AJMAL KHAN

INTRODUCTION

In recent years there has been increasing recognition that the scarcity of fresh water and the salinification of agricultural areas are becoming much more global problems. Both trends represent a threat to the world's food supply. Currently the domestic, industrial and agricultural use of fresh and ground water is increasing so quickly that fresh water shortage can be expected all over the world. This is already the case in several different places. Surface and ground water in agricultural areas in many places in the world are rapidly becoming increasingly brackish and saline, particularly in arid tropical and subtropical areas. Furthermore, salt deserts (caused by a lack of fresh water) and saline inland basins (caused by the level of saline ground water rising as a result of leakage of drainage water) are being created. FAO data show that at least 40% of the world is affected by salinification in some form [10]. The actual impact of this estimate in not entirely clear. However, it is known that large areas in Australia, India, Pakistan, Egypt, Central Asia, South America, Mexico and the United States [34] are faced with salinification to greater or lesser extent.

At the same time there are growing indications that cultivation of crops with a high-salt tolerance can be seen as interesting option for utilizing saline soils and conserving fresh water (59, 60, 97). Saline agriculture is a type of agriculture on the saline soil in which crops (halophytes) that can withstand a higher salt content than normal agricultural crops are grown. Potential halophytic crops could be broadly grouped into three categories: (1) plants with a high salt tolerance: they grow in water with salt contents equal to or even higher than sea water; (2) crops with average salt tolerance: they grow in brackish water; and (3) crops with moderate salt tolerance: they grow in slightly brackish water that is not suitable for conventional agriculture.

Some halophytes stand out because of their spectacular growth and production in saline conditions. Nevertheless, the number of examples where halophytes are actually employed for a particular practical purpose is extremely low, although it is demonstrated that there are many potential halophytic crops. There could be several reasons for this limited utilization: (1) So far, there was virtually no urgency yet anywhere because agricultural crops met the needs. (2) Essentially all halophytes are species that occur naturally. There has been no selection or improvements of halophytes with a view of practical applications. (3) It is not unthinkable that as soon as real research efforts are started, halophytes with unique properties can be found. (4) Little information is available about the diversity of halophytes in the most areas of the world. (5) Saline agriculture is a new technology and it requires effort, and special skills to make it successful.

There are several efforts made to compile the halophytic flora of the world [4, 5, 107] as well list of regional halophytes. However, still the information regarding halophyte is far from completion. Although flora of Pakistan is near completion and it has information about the halophytes buried in somewhere. Current effort is made to compile a list of halophytes distributed in Pakistan with some relevant details.

PHYSIOGRAPHY, SALINITY AND CLIMATE

Pakistan has varied physiography and climate. It stretches about 1,600 km from the subtropical Arabian Sea to temperate northern mountains covering an area of 800,000 km^2. The country could be divided into seven major landscape units: (1) The northern

H. Lieth (ed.), Cash Crop Halophytes: Recent Studies, 167–187.

mountains, (2) the Hindu Kush and the western mountains, (3) the potwar plateau and salt range, (4) the Balochistan plateau, (5) the Indus plain, (6) Cholistan and Thal desert and (7) coastal areas. Indus is a major river, which passes through Pakistan with an approximate annual flow of 115 billion cubic meters [3, 120]. It originates on the Tibetan plateau at an altitude of about 5,500 m and flows south to the Arabian Sea. In addition other major tributaries, the Chenab, the Jehlum, the Sutlej, the Beas and the Ravi join the Indus at upper Indus plain.

Pakistan is primarily arid and semiarid, except for a narrow belt in the north, with low and variable rainfall. Annual precipitation ranges from 1,500 mm on the southern step of Himalayas to less than 100 mm in the Western Balochistan coast. About 69% of the country receives rainfall less than 250 mm of the rain per year. The rain primarily falls during the monsoons (June–September), however, Southwestern Balochistan receives winter rain under mediteranean trend and some northwestern areas have both winter and summer rains.

TECHNICAL INFORMATION

There are numerous ways of defining halophytes. For our purpose best definition seems to be "plants that complete their life cycle in saline habitats" [178], where salt concentration of soil solution is about 5 g/l total dissolved solids (85 mM NaCl or 7–8 dS m^{-1}, 11). This list is organized alphabetically by botanical family, and within each family, by genus and species. Generic and familial taxonomy follows "Flora of Pakistan" [4]. However, the flora being continuously updated and revised, these names could be changed later.

Distribution

We have followed the distribution of halophytic species based on general landscape of Pakistan but has added one more category of plants which are cosmopolitan in distribution. They are as follows: (1) The northern mountains (NM), (2) the Hindu Kush and the western mountains (HK), (3) the potwar plateau and salt range (PP), (4) the Balochistan plateau (BP), (5) the Indus plain (IP), (6) Cholistan and Thal desert (CT), (7) coastal areas (CO) and Cosmopolitan (CM).

Life Form

Only one life form is given per species, even though many species show a certain amount of plasticity in this regard. A = annual, AQ = aquatic, CH = chamaephyte (small shrub, under 0.5 m average height), CHN = nano-chamaephyte (dwarf shrub, under 0.25 m average height), H = hemicryptophyte, HP = herbaceous perennial, PG = perennial grass, SH = shrub, T2 = tree, average height 2 m, T4 = tree, average height 4 m, and $T(x)$ = tree, average height (x) m.

Plant Type

This category bases on the habitats in which the taxon is distributed. Hyphal = hydrohalophytes (present in salt marshes), Xeorh = Xerohalophyte = salt desert species, Psamm = Psammophytes (sand loving plants found on littoral or inland sand dunes), Xero = xerohalophytes (desert species suspected as halophytes), Chasm = Chasmophytes (cliff-dwelling species), Weedy = fugitive species, Phrea = phreatophytes.

Max dS m^{-1}: This heading gives the maximum reported salinity tolerance of a taxon as cited in the references.

GENERAL OVERVIEW

The list showed that halophytic vegetation of Pakistan is quite diverse (Table 1) with 380 species with varied level of salt tolerance. Menzel and Lieth [107] reported that about 2,200 halophytic species were found in the literature worldwide. The halophytes of Pakistan constitute about 18% of this halophytic flora. Most of the halophytic species in Pakistan are present in the playas of northern mountains [95], whereas, others are in Indus plains [78], coastal areas [77], Hindu kush and the western mountains [34], the Potwar plateau [33], Cholistan and Thal desert [20] and 37 halophytes are cosmopolitan in distribution (Figure 1). The data showed that mountainous areas of Pakistan are more diverse in comparison of flat plains. The halophytes of Pakistan represent 58 families. The highest number of halophytes is present in the family Chenopodiaceae [87], followed by Poaceae [72], Cyperaceae [27], Papilionaceae [23], Tamaricaceae [16], Asteraceae [11] and Verbenaceae [10] while other families are represented by less than 10 halophytes (Tables 1 and 2).

Table 1. Alphabetical listing of halophytes from Pakistan

Genus, species and author	Distribution	Plant type	Life form	Max dS/m	Ref.
Aizoaceae					
Aizoon canariense L.	CO	Xero	A		18
Mesembryanthemum crystallinum L.	IP	Xero	A	56	40
Sesuvium sesuvioides (Fenzl.) Verdc.	CO, IP	Psamm.	HP		122
Trianthema portulacastrum L.	NM, IP	Xero	A		37, 55
Trianthema triquetra Rottl.ex Willd.	CO, IP	Xero	A		37, 55
Amaranthaceae					
Aerva javanica (Brum. f.) Juss. ex J.A. Schultes	CO, IP	Xero	SH		135
Asclepediaceae					
Calotropis gigantea (L.) Aiton	Cosm.	Xero	T2	8	117, 21
Calotropis procera (Ait.) Ait.	HK, NM, BP, IP	Xero	T2	8	117, 7
Cynanchum acutum L.	NM	Xero	V		135
Leptadenia pyrotechnica (Forssk.) Dene.	CO, CT	Xero	SH		91
Pentatropis spiralis (Forssk.) Dene.	CT, IP	Xero	HP		50, 100
Pergularia tomentosa L.	CO, CT	Xero	V		100
Asteraceae					
Achillea millefolium L.	NM	Psamm.	CH		134
Artemisia brevifolia Wall. ex DC	NM	Psamm.	CH		4
Artemisia maritima L.	NM	Psamm.	CH		97
Artemisia scoparia Waldst. & Kit.	NM, IP	Psamm.	CH		66
Inula brittanica L.	NM	Psamm.	HP		66
Lactuca tatarica (L.) C. A. May	NM	Psamm.	HP		66
Launea sarmantosa (Willd.) Alsoton	CO	Psamm.	HP		66
Pulicaria hookeri Jafri	CO	Psamm.	HP		66
Sonchus maritimus L.	BP, IP	Xero	HP		121
Sonchus tenerrimus L.	CO	Xero	HP		121
Xanthium sibiricum Patrin.	BP	Xero	HP		66
Avicenniaceae					
Avicennia marina (Forssk.) Vierh	CO	Hyphal	T8	58	13, 73
Boraginaceae					
Heliotropium aucheri DC	BP	Weedy	HP		110
Heliotropium bacciferum Forssk.	CO	Xero	HP		110
Heliotropium currassavicum L.	CO, IP	Weedy	HP		15, 49
Brassicaceae					
Conringia persica Boiss.	BP	Psamm.	A		4
Dilophia salsa Thompson	NM	Hyphal	CH		4
Lepidium cartilagineum (J. May) Thell.	BP	Hyphal	A		110, 137
Lepidium pinnatifidum Ledeb.	PP, NM	Hyphal	AP		137
Lepidium latifolium L.	NM	Hyphal	A		38, 97
Lobularia maritima (L.) Desv.	IP	Psamm.	A		137
Raphanus raphanistrum L.	HK, PP, BP	Psamm.	A		137
Caryophyllaceae					
Cerastium glomeratum Thuill.	NM	Weedy	HP		47
Spergularia diandra (Guss.) Heldr & Sart.	BP, NM	Weedy	HP		121, 142
Spergularia marina (L.) Griesb.	IP	Weedy	HP	50	97, 110
Spergularia media (L.) Presl.	BP	Weedy	A	50	97, 110
Caesalpiniaceae					
Caesalpinea bonduc (L.) Roxburgh.	IP	Hyphal	SH	107	39, 108
Chenopodiaceae					
Aellenia acutifolia (Moq.) Jafri	BP	Xerohal.	SH		4
Aellenia auricula (Moq.) Ulbr.	BP	Xerohal.	SH		110
Agathophora alopecuroides (Dellie.) Fenzl ex Bunge	BP	Xerohal.	SH		110

Table 1. (Cont.)

Genus, species and author	Distribution	Plant type	Life form	Max dS/m	Ref.
Anabasis haussknechtii Bunge ex Boiss.	BP, PP	Xerohal.	SH		110
Anabasis setifera Moq.	BP	Xerohal.	CHN		94, 110
Arthrocnemum indicum (Willd.) Moq.	CO	Hyphal	CHN	90	52, 108
Arthrocnemum macrostachyum (Moric.) C. Koch	CO	Hyphal	CHN	90	75, 85
Atriplex aucheri Moq.	NM	Xero	A		4
Atriplex canescens James	BP	Xerohal.	SH	59	115, 117
Atriplex dimorphostegia Kar. & Kir.	BP	Xero	AP	56	5, 117
Atriplex griffithii Moq.	NM	Xerohal.	SH		5, 117
Atriplex halimus L.	IP	Xerohal.	SH	56	115, 117
Atriplex hortensis L.	NM	Xero	A	10	116, 117
Atriplex lasiantha Boiss.	NM, IP, BP	Xero	A		4
Atriplex leucoclada Boiss.	CO, IP	Xero	SH	79	5, 117
Atriplex pamirica Iljin	NM, IP	Xerohal.	A		4
Atriplex schugnanica Iljin	NM	Xero	A		4
Atriplex stocksii Boiss.	CO	Xerohal.	SH	36	74, 89
Atriplex tatarica L.	NM	Xero	A		66, 137
Bassia hyssopifolia (Pall.) O. Kuntze	NM	Weedy	A	58	57, 113
Beta vulgaris ssp *maritma* (L.) Arcangeli	BP, IP	Weedy	HP	20	97, 117
Bienertia cycloptera (Bunge ex Trautv.) Bunge ex Boiss.	CO	Hyphal	A		18, 110
Camphorosma monspelicata L.	BP	Xero	SH		66, 110
Ceratocarpus arenarius L.	NM, BP	Xero	A	15	66
Chenopodium album L.	NM, IP	Weedy	A		65, 66
Chenopodium ambrosoides L.	NM, IP	Weedy	A		104, 113
Chenopodium botrys L.	NM, HK	Weedy	A		4
Chenopodium ficifolium ssp *blomianum* (Aellen) Aellen	Cosm.	Weedy	A		4
Chenopodium foliosum Ascher	NM	Hyphal	A		4
Chenopodium glaucum L.	NM, BP	Weedy	CH		66, 108
Chenopodium murale L.	IP	Weedy	A	20	57, 58
Coriospermum korovinii Iljin	NM	Xero	A		66
Coriospermum tibeticum Iljin	NM	Xero	A		66
Cornulaca monocantha Del.	CO, BP	Xerohal.	A		92, 93
Gamanthus gamocarpus (Moq.) Bunge	BP	Xerohal.	A		121
Girgensohnia oppositiflora (Pall.) Fenzl	BP	Xero	A		92, 137
Halimocnemis pilifera Moq.	BP	Xerohal.	A		135
Halocharis clavata Bunge	BP	Xerohal.	A		130
Halocharis hispida (Schrenk ex C. A. Mey) Bunge	BP	Xerohal.	A		130
Halocharis lachnantha E. Korov.	BP	Xerohal.	A		130
Halocharis sulpheurea (Moq.) Moq.	BP	Xerohal.	A		110
Halocharis violacea Bunge	BP	Xerohal.	A		110
Halocnemum strobilaceum (Pall.) M. Bieb.	CO	Xero	CH		18
Halogeton glomeratus (M. B.) C. A. Mey	NM	Xerohal.	SH		66, 94
Halogeton tibeticus Bunge	NM	Xerohal.	A		66
Halopeplis perfoliata (Forssk.) Bunge ex Schweinf.	CO	Hyphal	HP	10	18, 93
Halostachys belangerana (Moq.) Botsch.	CO	Hyphal	SH		131, 137
Haloxylon griffithii (Moq.) Boiss.	NM	Xero	SH		117
Haloxylon persicum Bunge ex Boiss.	BP	Psamm.	SH		21, 28
Haloxylon salicornicum (Moq.) Bunge ex Boiss.	IP	Xero	CHN		110
Haloxylon scoparium Pomel.	IP	Psamm.	SH		25
Haloxylon stocksii (Boiss.) Benth. & Hook.	Cosm.	Psamm.	SH	35	78, 87
Kochia indica Wight	NM, BP, IP	Xero	A		93, 94
Kochia iranica Litw. ex Bornm	BP	Xero	A		46, 177
Kochia prostrata (L.) Schrad.	NM, BP	Xero	A		66, 110
Kochia scoparia (L.) Schrad.	Cosm.	Xero	A		92, 130
Kochia stellaris Moq.	NM	Xero	A		93

Table 1. (Cont.)

Genus, species and author	Distribution	Plant type	Life form	Max dS/m	Ref.
Krascheninnikovia ceratoides (L.) Gueldena	NM, BP	Xerohal.	SH		4
Salicornia brachiata	CO	hyphal	A	40	75, 76
Salsola arbuscula Pall.	BP	Xerohal.	CH		112, 135
Salsola canescens (Moq.) Boiss.	BP	Xerohal.	CHN		130
Salsola chorassanica Botsch	BP	Xerohal.	A		110
Salsola collina Pall.	NM	Hyphal	A		66
Salsola crassa ssp *turcomanica* Pall.	BP	Xerohal.	A		134
Salsola cyclophylla Baker	BP	Hyphal	SH		134
Salsola drummondii Ulbr.	BP, IP	Hyphal	SH		66
Salsola griffithii (Bunge) Freitag & Akhani	BP	Psamm.	CH		4
Salsola imbricata Forssk.	IP, PP, BP	Xerohal.	SH		4
Salsola incanescens C.A. Mey	BP	Xerohal.	A		110
Salsola jacquemontii Moq.	NM	Xerohal.	A		4
Salsola makranica Freitag	BP	Xerohal.	CHN		4
Salsola nitraria Pall.	BP	Xerohal.	A		110
Salsola orientalis S.G. Gmeln	BP	Xerohal.	AP		27, 110
Salsola paulsenii ssp *praecox* (Litw.) Rilke	BP	Psamm.	A		66
Salsola richteri (Moq.) Karel.	BP	Psamm.	CH		27, 114
Salsola rubescens Franch	BP	Xerohal.	CH		4
Salsola sclerantha C.A. Mey	BP	Psamm.	A		27, 110
Salsola tomentosa (Moq.) Spach	BP	Xerohal.	CHN		110
Salsola tragus L.	Cosm.	Xerohal.	A		4
Seidlitzia florida (M. Bieb.) Boiss.	BP	Xerohal.	A		110
Suaeda acuminata (C. A. Mey) Moq.	NM	Xerohal.	A		66, 110
Suaeda arcuata Bunge	BP	Weedy	A		4
Suaeda aegyptiaca (Hasselq.) Zohary	BP	Hyphal	AP		18, 48
Suaeda olufsenii Paulsen	NM	Hyphal	A		18, 48
Suaeda fruticosa (L.) Forssk.	Cosm.	Xerohal.	HP	97	82, 88
Suaeda heterophylla (Kar. & Kir.) Bunge	NM	Hyphal	A		4
Suaeda monoica Forssk.	BP, IP	Hyphal	SH	56	93, 115
Cistaceae					
Helianthemum lippii (L.) Pers.	CO, BP	Xero	SH		4
Convolvulaceae					
Cressa cretica L.	IP, BP, CT, CO	Hyphal	HP	97	71, 84
Evolvulus alsinoides (L.) L.	IP, BP, CT, CO	Hyphal	HP		113
Ipomoea alba L.	Cosm.	Psamm.	HP		15
Ipomoea pes-caprae (L.) R. Br.	CO	Psamm.	HP		133
Cyperaceae					
Blysmus rufus (Huds.) Link.	NM	Hyphal	HP	16	97, 103
Bolboschoenus affinis (Roth.) Drobov	Cosm.	Hyphal	HP		130
Bolboschoenus maritimus (Roth.) Drobov	NM, PP, HK, BP	Hyphal	HP		134
Carex divisa Hudson	BP, HK, NM	Hyphal	HP		38, 50
Carex orbicularis Boott	BP, HK, NM	Hyphal	HP		66
Carex songorica Kar. & Kir	HK, NM	Hyphal	HP		118
Carex stenophylla Wahl.	BP, HK, NM	Hyphal	HP		135
Carex stenophylla ssp. *interrupta* (V. Krecz.) Egor	Cosm.	Hyphal	HP		135
Cyperus arenarius Retz.	CO, IP	Hyphal	HP		4
Cyperus conglomeratus ssp. *conglomeratus* Rottb.	CO	Hyphal	HP		18, 110
Cyperus conglomeratus ssp. *cuvulus* Rottb.	CO	Hyphal	HP		18, 110
Cyperus corymbosis Rottb.	CO	Hyphal	HP		16, 123
Cyperus laevigatus L.	IP, PP, CO, BP	Hyphal	HP		16, 51
Cyperus malaccensis Lam.	CO	Hyphal	HP		70, 111
Cyperus pachyrhizhus Nees ex Boeck	CO	Hyphal	HP		4
Cyperus rotundus L.	Cosm.	Hyphal	HP		66

Table 1. (Cont.)

Genus, species and author	Distribution	Plant type	Life form	Max dS/m	Ref.
Cyperus serotinus Rottb.	NM, PP, HK	Hyphal	HP		66
Cyperus stoloniferous Retz.	Co	Hyphal	HP		52
Eleocharis palustris (L.) Roem & Schultes	NM. IP,	Hyphal	HP		51
Eleocharis uniglumis (Link.) Schultes	NM, PP, HK, BP	Hyphal	HP	16	97
Fimbristylis cymosa R. Br.	CO	Hyphal	HP		66
Pycerus polystachyos (Rottb.) P. Beauv	CO	Hyphal	HP		66
Schoenoplectus lacustris (L.) Palla	NM	Hyphal	HP		66
Schoenoplectus littoralis (Schrad.) Palla	NM, IP	Hyphal	HP		66
Schoenoplectus triqueter (L.) Palla	NM	Hyphal	HP		66
Schoenus nigricans L.	Cosm.	Hyphal	HP		133
Scirpoides holoschoenus (L.) Sojak	NM, BP	Hyphal	HP		4
Elaegnaceae					
Elaeagnus angustifolia L.	NM, BP, IP	Hyphal	TX		66
Hippophae rhamnoides L.	NM	Hyphal	SH		66
Euphorbiaceae					
Andrachne telephioides L.	NM, PP, BP	Weedy	HP		43
Euphorbia granulata Forssk.	Cosm.	Weedy	A		117
Euphorbia serpens Kunth.	IP	Xero	A		123
Euphorbia thymifolia L.	HK, PP, BP, CT	Psamm.	AP		122
Flueggea leucopyrus Willd.	PP, IP	Xero	SH		4
Frankeniaceae					
Frankenia pulverulenta L.	BP, IP, PP	Psamm.	AP		135
Gentianaceae					
Centaurum spicatum (L.) Fritsch	IP	Xerohal.	HP		29
Goodeniaceae					
Scaevola plumieri (L.) Vahl.	CO	Psamm.	SH	60	39, 54
Scaevola taccada (gaertn.) Roxb.	CO	Psamm.	SH	10	117
Hydrocharitaceae					
Halophila ovalis R. Br.	CO	Hyphal	AQ	58	44
Iridaceae					
Iris ensata Thunb.	Cosm.	Hyphal	AQ		110
Juncaceae					
Juncus bufonius L.	IP	Hyphal	A		113
Juncus gerardii Lois.	BP	Hyphal	HP	50	97
Juncus maritimus Lam.	CO,IP	Hyphal	HP	87	97, 117
Juncus punctorius L.	BP	Hyphal	HP		4
Triglochin maritima L.	NM	Hyphal	HP	48	36
Triglochin palustris K.	NM	Hyphal	HP		113, 137
Lauraceae					
Litsea monopetala (Roxb.) Pers.	PP	Hyphal	V		135
Liliaceae					
Asparagus officinalis L.	NM	Psamm.	HP		137
Urginea indica Kunth.	NM	Psamm.	HP		117, 135
Malvaceae					
Hibiscus tiliaceous L.	CO	Psamm.	T12	58	39, 108
Thespesia populnea (L.) Sol.	CO	Hyphal	T4	58	39, 117
Thespesia populneoides (Roxb.) Kostel.	CO	Hyphal	T6	58	39, 108
Mimosaceae					
Acacia cornigera (L.) Willd.	IP	Xero	T4		15
Acacia jacquamonti Benth.	BP, IP	Xero	T6		8
Acacia leucophloea (Roxb.) Willd.	CT	Xero	T6		8

Table 1. (Cont.)

Genus, species and author	Distribution	Plant type	Life form	Max dS/m	Ref.
Acacia nilotica (L.) Delile	CO	Xero	T6		4
Acacia sphaerocephala Schl. & Chem.	IP	Xero	T2		135
Prosopis cineraria (L.) Druce	PP, IP, BP, CT	Xero	T8		18, 42
Prosopis farcta (Banks & Sol.) Macbride	HK	Weedy	CH		18, 41
Prosopis juliflora (Swartz)	DC, IP	Xero	T8		30, 53
Molluginaceae					
Glinus lotoides L.	CO, CT, IP	Psamm.	A		4
Moraceae					
Ficus microcarpa L.	CO	Hyphal	T8		4
Myrsinaceae					
Aegiceras corniculata (L.) Blanco	CO	Hyphal	T6	58	39, 49
Ardisia solanacea Roxb.	CO	Hyphal	SH		135
Najadaceae					
Najas graminea Delile	PP	Hyphal	SAQ		113
Najas marina L.	IP, NM	Hyphal	SAQ	55	2, 108
Najas minor All.	IP	Hyphal	SAQ		135
Nyctaginaceae					
Bougainvillea spectabilis Willd.	IP	Xero	V	8	20
Pisonia grandis R. Br.	CO	Psamm.	T4		1
Orobanchaceae					
Cistanche tubulosa (Schrenk) Hook	CO, IP, BP, PP	Hyphal	A		110
Palmae					
Cocos nucifera L.	CO	Hyphal	T8	58	35, 106
Hyphaene thebaica (L.) Mart.	CO	Hyphal	T8		51, 121
Livistonia chinenesis (N. J. Jacquin) R. Brown	CO	Hyphal	T8		117
Pheonix canarienses L.	CO	Hyphal	T8		5, 26
Pheonix dactylifera Linn.	CO, IP	Hyphal	T8		5, 136
Papilionaceae					
Alhaji maurorum Medic.	Cosm.	Hyphal	SH		18, 93
Amorpha fruticosa L.	IP	Xero	SH		66
Astragalus kahiricus DC.	HK,IP	Xero	CH		110
Dalbergia sissoo Roxb.	IP, BP,	Xero	T8		118
Erythrina herbacea Linn.	IP	Hyphal	SH		33
Glycyrrhiza glabra L.	NM, BP	Hyphal	CH		66
Glycyrrhiza uralensis Fisch ex DC.	HK	Hyphal	CH		135
Lespedeza juncea var serica (Thunb.) Lace & Hemsley	HK, PP, BP	Xero	HP		4
Macroptilium lathyroides (L.) Urb.	IP	Hyphal	A		4
Medicago falcate L.	NM	Chasm	PH		66
Medicago minima (L.) Grufb.	NM	Chasm	A		22, 56
Melilotus alba Desr.	NM	Chasm	A		110
Melilotus indica (L.) All.	Cosm.	Chasm	A		110, 134
Melilotus officinalis (Li.) Pall.	NM	Chasm	A		110
Oxytropis glabra DC.	NM, HK	Chasm	CHN		66
Oxytropis microphylla (Pall.) D.C.	NM	Psamm.	CH		66
Pongamia pinnata (L.) Merrill	IP	Hyphal	T20	58	39
Robinia pseudoacacia L.	PP	Hyphal	T10		66
Sophora alopecuroides L.	NM	Psamm.	SH		66
Sophora japonica L.	PP	Phre	T20		66
Sophora occidentalis L.	IP	Hyphal	SH		66
Tephrosia purpurea (L.) Pers.	IP	Hyphal	A		91, 125
Trifolium fragiferum L.	NM	Psamm.	HP	16	30, 97

Table 1. (Cont.)

Genus, species and author	Distribution	Plant type	Life form	Max dS/m	Ref.
Plantaginaceae					
Plantago coronopus L.	BP	Xero	A	50	38
Plantago depressa Willd.	NM	Hyphal	A		86
Plantago lanceolata L.	NP, PP	Xero	HP		108
Plantago major L.	Cosm.	Psamm.	HP		137, 143
Plumbaginaceae					
Limonium gilsei (Hemsl.) Rech.	NM	Hyphal	HP		4
Limonium sinuatum (L.) Miller	Cultivated	Hyphal	HP	56	117
Limonium stocksii (Boiss.) O. Ktze	CO, BP	Hyphal	HP	50	18, 117
Psylliostachys spicata (Willd.) Nevski	BP	Hyphal	A		52, 110
Poaceae					
Aeluropus lagopoides (L.) Trin. ex Thw.	CO, RD, IP, BP	Hyphal	PG	56	18, 62
Aeluropus littoralis (Gouan) Parl.	BP	Hyphal	PG	56	18, 51
Aeluropus macrostachys Hack.	BP	Phyphal	PG		4
Agropyron cristatum (L.) Gaertn.	NM	Psamm.	PG		12
Agropyron junceum (L.) P. Beauv.	NM	Psamm.	PG		135
Agrostis stolonifera L.	NP, HK, PP,	Psamm.	HP	60	64, 124
Arundo donax L.	NP, BP, IP, PP	Weedy	PG		52, 117
Calamagrostis holciformis Jaub. & Spach.	NM	Psamm.	PG		66
Calamagrostis pseudophragmites (Hall. F.) Koeler	BP, HK, NM	Psamm.	PG		66
Cenchrus ciliaris Rich.	Cosm.	Psamm.	PG		55, 125
Chloris gayana Kunth.	NM, IP	Psamm.	PG	16	31, 55
Chloris virgata Sw.	BP	Psamm.	A		110, 128
Coelachyrum piercei (Benth.) Bor.	BP	Psamm.	PG		110
Crypsis aculeate (L.) Ait.	Cosm.	Weedy	A		66
Crypsis schoenoides (L.) Lam.	Cosm.	Weedy	A		66
Cynodon dactylon (L.) Pers.	Cosm.	Weedy	PG	17	55, 117
Dactyloctenium aegyptium (L.) P. Beauv.	Cosm.	Weedy	PG		21, 55
Dactyloctenium scindicum Boiss.	Cosm.	Xerohal.	PG		27, 73
Desmostachya bipinnata (L.) Stapf.	Cosm.	Xerohal.	PG	5.6	21, 55
Dichantheum annulatum (Forssk.) Stapf.	Cosm.	Xero	PG		128
Digitaria bicornis (Lam.) Roem & Schulte	Co	Psamm.	A		66
Digitaria longifolia (Retz.) Pers.	PP	Hyphal	PG		99
Digitaria sanguinalis (L.) Scop.	BP, HK, NM, PP	Psamm.	A		135
Diplachne fusca (L.) P. Beauv.	IP, CT	H	PG	15	18, 102
Eleusine indica (L.) Gaertn.	NM, IP, HK	Hyphal	PG		24, 51
Elymus dasytachys Trin.	Northern Area	Psamm.	PG		135
Eragrostis curvula (Schrad.) Nees.	CT, IP, BP, PP	Psamm.	PG		4
Eragrostis superba Peyr.	CO, CT	Psamm.	PG		4
Festuca pamirica Tzvelev	NM, HK	Psamm.	PG		4
Festuca rubra L.	NM, HK	Psamm.	PG	27	97, 117
Halopyrum mucronatum (L.) Stapf.	CO	Psamm.	PG	35	18, 86
Hordeum bogdanii Wilensky	BO, NM	Hyphal	PG		4
Hordeum brevisubulatum (Trin.) Link	NM, HK	Hyphal	PG		66
Hordeum marinum Huds.	NM	Hyphal	A	20	97, 101
Hordeum maritimum ssp *glaucum* (steud.) Tzvelev	BP, NM, HK, PP	Hyphal	A		4
Imperata cylindrica (L.) Raeuschel.	Cosm.	Psamm.	PG		135
Lasirus sciendicus Forssk.	Cosm.	Psamm	PG		128
Leymus secalinus (Georgi) Tzvelev	HK, NM	Psamm	PG		4
Lolium multiflorum Lam.	HK, BP	Psamm.	PG		22, 113
Lygeum spartum Loefl. ex Linn.	NM	Psamm.	PG		18, 121
Orinus thoroldii (Staps. ex Hemsl.) Bor.	NM	Hyphal	PG		66
Orthochloa compressa (Forssk.) Hilu	Cosm.	Hyphal	PG		4
Oryza coarctata Roxb.	CO	Hyphal	HP		135

Table 1. (Cont.)

Genus, species and author	Distribution	Plant type	Life form	Max dS/m	Ref.
Parapholis incurva (L.) C.E. Hubb.	BP, PP, HK	Hyphal	PG		97
Paspalidium geminatum (Forssk.) Stapf	CT	Psamm.	PG		18
Paspalum paspioides (Michex) Scribner	Cosm.	Hyphal	HP		32, 127
Pennisetum hohenackeri Hochst. ex Steud.	CT	Psamm.	PG		4
Phalaris arundinaceae L.	NM, PP	Hyphal	PG		66
Phragmites australis (Cav.) Trin. ex Steud.	NM, IP	Hyphal	PG	40.6	18, 47
Phragmites karka (Retz.) Trin. ex. Steud.	Cosm.	Hyphal	PG		108
Poa bulbosa L.	NM, BP, HK	Psamm.	PG		134
Poa pratensis L.	NM, BP, HK	Psamm.	PG		134
Polypogon monspeliensis (L.) Desf.	Cosm.	Psamm.	PG		119, 121
Puccinellia distans (Wahlb.) Parl.	NM	Hyphal	PG	27	97, 133
Puccinellia gigantia (Grossh.) Grossh.	BP	Hyphal	PG		134
Puccinellia tenuiflora (Griesb.) Scribn.	NM	Hyphal	PG		66
Sacchraum bengalense Retz.	HK, PP, IP, CT	Hyphal	PG		4
Sacchraum spontaneum L.	NM, PP, HK, IP	Hyphal	PG		66
Setaria viridis (L.) P. Beauv.	NP, BP, HK, PP	Psamm.	A		66
Spinifex littoreaus (Brum. F.) Merr.	Cosm.	Psamm.	PG		66
Sporobolus arabicus Boiss.	Cosm.	Psamm.	PG		18, 110
Sporobolus helvolus (Trin.) Dur. & Schinz.	PP, IP	Psamm.	PG		8, 52
Sporobolus piliferus (Trin.) Kunth.	IP	Psamm.	PG		4
Sporobolus tourneuxii Coss.	CT	Psamm.	PG		4
Sporobolus tremulus (Willd.) Kunth.	PP	Hyphal	A		21, 55
Sporobolus virginicus (L.) Kunth.	CO, CT	Psamm.	PG	73	18, 117
Stipa splendens Trin.	BP, HK, NM	Psamm.	PG		135
Tetrachne dregei Nees ex Boeck	BM	Psamm.	PG		110
Trikeraia hookeri (Staps.) Bor.	NM	Psamm.	PG		4
Urochloa panicoides P. Beau.	NM, HK, PP	Psamm.	PG		4
Urochondra setulosa (Trin.) C.E. Hubb	CO	Xerohal.	PG	30	62
Vulpia unilateralis (L.) Stace	NM	Psamm.	A		4
Polygonaceae					
Polygonum aviculare L.	NM	Xero	A		47, 83
Polygonum bellardii All.	NM	Xero	A		135
Polygonum delicatulum Meissn.	NM	Xero	A		4
Polygonum patulum M. Bieb.	NM	Xero	A		44
Polygonum sibiricum Laxm.	NM	Xero	A		66
Polygonum viviparum L.	NM	Xero	A		4
Rumex crispus L.	NM	Xero	A		17, 47
Portulacaceae					
Portulaca oleracea L.	Cosm.	Xero	AP		125
Portulaca pilosa L.	NM, PP	Hyphal	A		108
Primulaceae					
Anagallis arvensis L.	NM	Xero	A		47, 117
Glaux maritime L.	NM	Xero	HP	52	97, 133
Samolus valerandi L.	NM, PP, BP	Hyphal	HP		97
Punicaceae					
Punica granatum L.	NM, HK, BP	Xero	T2		4
Residaceae					
Ochradenus baccatus Del.	CO, BP	Xero	SH	58	142
Oligomeris linifolia (Vahl) Macbride	CO, IP, BP, PP	Xero	A	58	52
Rhamnaceae					
Zizyphus nummularia (Burm. f.) Wight and Arn.	NM, CO	Xero	T4		7
Rhizophoraceae					
Brugiera gymnorhiza (L.) Savigny	CO	Hyphal	T8	58	18, 108

Table 1. (Cont.)

Genus, species and author	Distribution	Plant type	Life form	Max dS/m	Ref.
Ceriops decandra (Griff.) Ding.	CO	Hyphal	T9	58	95
Ceriops tagal (Perr.) C.B. Robinson	CO	Hyphal	T8	58	73, 108
Rhizophora apiculata Blume, Bijdr.	CO	Hyphal	T8	58	66, 108
Rhizophora mucronata Poir.	CO	Hyphal	T8	59	18, 73
Rosaceae					
Potentilla anserina L.	NM	Hyphal	V		66
Potentilla bifurca L.	NM	Hyphal	V		66
Potentilla supine L.	NM	Hyphal	V		66
Rubiaceae					
Galium verum L.	NM	Hyphal	A		66
Ruppiaceae					
Ruppia maritima L.	Cosm.	Hyphal	AQ	58	109, 133
Salicaceae					
Populus euphratica Olivier	IP, BP, PP	Xero	T8		4
Salvadoraceae					
Salvadora oleoides Dene	CO, IP	Xero	T8		21, 63
Salvadora persica L.	CT, IP	Xero	T9	8	18, 63
Scrophulariaceae					
Euphrasia officinalis L.	NM	Xero	HP		135
Limosella aquatica L.	NM	Hyphal	HP		113
Simaroubaceae					
Suriana maritima L.	CO	Hyphal	SH	58	140
Solanaceae					
Lycium edgeworthii Dunal	IP, PP, BP	Xero	SH		4
Lycium shawii R. & S.	BP	Xero	SH		51
Solanum incanum L.	IP, PP, BP	Xero	SH	9	117
Sonneratiaceae					
Sonneratia caseolaris (L.) Engl.	CO	Hyphal	TX	58	109, 133
Tamaricaceae					
Reaumuria alternifolia (Labill) Britten	BP	Hyphal	CHN		110
Reaumuria palaestina Boiss.	BP	Hyphal	CHN		110
Reaumuria stocksii Boiss.	BP, CO	Hyphal	CHN		110
Tamarix androssowii Litw.	BP	Xero	T3		110
Tamarix aphylla (L.) Karst.	Cosm.	Phrea	T8	56	18, 19
Tamarix arceuthoides Bunge	NM	Xero	T2		110
Tamarix indica Willd.	Cosm.	Hyphal	T8		19
Tamarix karelini Bunge	BP	Xero	T2		110
Tamarix kotschyii Bumge	BP	Xero	T2		66, 110
Tamarix leptostachya Bunge	NM	Xero	T2		66, 110
Tamarix mascatensis Bunge	BP	Xero	SH		19, 110
Tamarix pakistanica Qaiser	IP, BP, CO	Phrea	T2		4
Tamarix passernioides Del.ex Desv.	BP, IP, CT	Xero	T2		18, 19
Tamarix ramosissima Ledeb.	CT	Xero	SH		18, 66
Tamarix smyrensis Bunge.	BP	Xero	T2		4
Tamarix tetragyna (Boiss.) Boiss.	BP	Xero	T2	26	18, 19
Tiliaceae					
Grewia tenax (Forssk.) Fiori	Cosm.	Xero	T2		108, 112
Typhaceae					
Typha domingensis Pers.	Cosm.	Hyphal	HP	26	17, 18
Typha latifolia L.	NM	Hyphal	HP	21	45, 133

Table 1. (Cont.)

Genus, species and author	Distribution	Plant type	Life form	Max dS/m	Ref.
Umbelliferae					
Ammi visnaga (L.) Lamk.	HK	Xero	T2		133, 137
Apium graveolens L.	Cosm.	Hyphal	AQ	16	97
Centella asiatica (L.) Urban	NM	Hyphal	HP		4
Verbenaceae					
Clerodendrum inerme (L.) Gaertn.	Cultivated	Hyphal	SH	58	66, 108
Phyla nodiflora (L.) Greene	IP	Hyphal	HP	15	117, 131
Verbena officinalis L.	NM, PP, HK	Hyphal	HP		135
Vitex trifolia L.	Cultivated	Hyphal	SH		66
Zygophyllaceae					
Fagonia bruguieri DC. Prodr.	BP	Xero	CHN		43
Fagonia indica ssp. *schweinfurthia* Hadidi	IP, CT,	Xero	CHN		135
Nitraria retusa (Forssk.) Aschers	CO	Xero	SH	90	18, 93
Nitraria schoberi L.	BP	Xerohal.	SH	66	18, 117
Seetzenia lanata (Willd.) Bullock	IP, BP	Psamm.	HP		23, 91
Tribulus terrestris L.	Cosm.	Xero	HP		23, 91
Zygophyllum fabago L.	BP	Xerohal.	SH		66
Zygophyllum propinquum Decne	CT, IP	Xero	SH		110
Zygophyllum simplex L.	CT, IP, CO	Xerohal.	A	29	81, 142

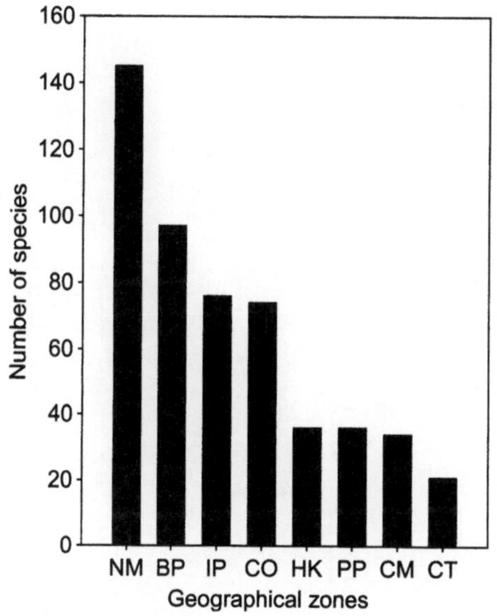

Fig. 1. Distribution of halophytic species in the major geographical zones of Pakistan.

In addition among the total 380 halophytes about 137 of them could be classified as hydrohalophytes, followed by xerophytes [91], Psammohalophytes [74], xerohalophytes [44], Chasmophytes [6], Weedy [23] and Phreatophytes [3]. Fifty percent of halophytes from Chenopodiaceae are found in arid environment while all halophytic members of Cyperaceae are found in aquatic conditions (Table 2).

Life forms of these plants also showed a high degree of variation. Annuals [98] are the dominant life forms of Pakistani halophytes, followed by Herbaceous perennials [79], perennial grass [59], shrubs [46], trees [33], Chaemaephytes [20] and others like aquatic, geophytes, hemi-cryptophytes and small trees are represented by less than 15 individuals (Table 3). The data presented in Figure 2 showed distribution of various life forms in the various geographical regions. In the northern mountains halophytic vegetation is dominated by annuals [47], herbaceous perennials [33] and perennial grasses [29]. Halophytes in Balochistan plateau are also dominated by annuals [41], herbaceous perennials [21], shrubs [20] and perennial grasses [16]. Coastal areas are dominated by herbaceous perennials and shrubs. Most of the halophytes present in the northern mountains are hydrohalophytes [46], followed by xerohalophytes [33], Psammophytes [29], while halophytic vegetation of Balochistan is equally dominated by three types like hydrohalophytes, xerophytes, and xerohalophytes (Figure 3). Hydrohalophytes are the most abundant group among the coastal plants.

Table 2. Type of halophytes distributed in various families

No.	Family	Total #	Hyphal	Xero	Psamm	Xerohal	Chasm	Weedy	Phrea
1	Chenopodiaceae	86	11	20	7	37	–	9	–
2	Poaceae	72	24	1	39	3	–	5	–
3	Cyperaceae	27	27	–	–	–	–	–	–
4	Papilionaceae	23	9	4	3	–	6	–	1
5	Tamaricaceae	16	4	10	–	–	–	–	2
6	Asteraceae	11	–	3	8	–	–	–	–
7	Verbenaceae	10	4	4	1	1	–	–	–
8	Mimosaceae	8	–	7	–	–	–	1	–
9	Brassicaceae	7	4	–	3	–	–	–	–
10	Polygonaceae	7	–	7	–	–	–	–	–
11	Asclepidaceae	6	–	6	–	–	–	–	–
12	Juncaceae	6	6	–	–	–	–	–	–
13	Aizoaceae	5	–	4	1	–	–	–	–
14	Euphorbiaceae	5	–	2	1	–	–	2	–
15	Palmae	5	5	–	–	–	–	–	–
16	Rhizophoraceae	5	5	–	–	–	–	–	–
17	Caryophyllaceae	4	–	–	–	–	–	4	–
18	Convolvulaceae	4	2	–	2	–	–	–	–
19	Plantaginaceae	4	1	2	1	–	–	–	–
20	Plumbaginaceae	4	4	–	–	–	–	–	–
21	Boraginaceae	3	–	1	–	–	–	2	–
22	Malvaceae	3	2	–	1	–	–	–	–
23	Najadaceae	3	3	–	–	–	–	–	–
24	Primulaceae	3	1	2	–	–	–	–	–
25	Rosaceae	3	3	–	–	–	–	–	–
26	Solanaceae	3	–	3	–	–	–	–	–
27	Umbelliferae	3	2	1	–	–	–	–	–
28	Zygophyllaceae	3	–	1	–	2	–	–	–
29	Eleagnaceae	2	2	–	–	–	–	–	–
30	Goodeniaceae	2	–	–	2	–	–	–	–
31	Liliaceae	2	–	–	2	–	–	–	–
32	Myrsinaceae	2	2	–	–	–	–	–	–
33	Nyctaginaceae	2	–	1	1	–	–	–	–
34	Portulacaceae	2	1	1	–	–	–	–	–
35	Residaceae	2	–	2	–	–	–	–	–
36	Salvadoraceae	2	–	2	–	–	–	–	–
37	Scrophulariaceae	2	1	1	–	–	–	–	–
38	Typhaceae	2	2	–	–	–	–	–	–
39	Amaranthaceae	1	–	1	–	–	–	–	–
40	Avicenniaceae	1	1	–	–	–	–	–	–
41	Caesalpinaceae	1	1	–	–	–	–	–	–
42	Cistaceae	1	–	1	–	–	–	–	–
43	Frankeniaceae	1	–	–	1	–	–	–	–
44	Gentianaceae	1	–	–	–	1	–	–	–
45	Hydrocharitaceae	1	1	–	–	–	–	–	–
46	Iridaceae	1	1	–	–	–	–	–	–
47	Lauraceae	1	1	–	–	–	–	–	–
48	Molluginaceae	1	–	–	1	–	–	–	–
49	Moraceae	1	1	–	–	–	–	–	–
50	Orobanchaceae	1	1	–	–	–	–	–	–
51	Punicaceae	1	–	1	–	–	–	–	–
52	Rhamnaceae	1	–	1	–	–	–	–	–
53	Rubiaceae	1	1	–	–	–	–	–	–
54	Ruppiaceae	1	1	–	–	–	–	–	–
55	Salicaceae	1	–	1	–	–	–	–	–
56	Simaroubaceae	1	1	–	–	–	–	–	–
57	Sonneratiaceae	1	1	–	–	–	–	–	–
58	Tiliaceae	1	–	1	–	–	–	–	–

Table 3. Distribution of life form of the halophytes from Pakistan among plant families

No.	Family	Total #	A	AQ	CH	CHN	HP	PG	SH	T2	T4	T(x)	V
1	Chenopodiaceae	86	49	–	6	7	3	–	21	–	–	–	–
2	Poaceae	72	10	–	–	3	–	59	–	–	–	–	–
3	Cyperaceae	27	–	–	–	–	27	–	–	–	–	–	–
4	Papilionaceae	23	6	–	4	1	4	–	4	–	–	4	–
5	Tamaricaceae	16	–	–	3	–	–	–	2	8	1	2	–
6	Asteraceae	11	–	–	4	–	7	–	–	–	–	–	–
7	Verbenaceae	10	–	–	–	2	4	–	4	–	–	–	–
8	Mimosaceae	8	–	–	1	–	–	–	–	–	–	7	–
9	Brassicaceae	7	6	–	1	–	–	–	–	–	–	–	–
10	Polygonaceae	7	7	–	–	–	–	–	–	–	–	–	–
11	Asclepidaceae	6	–	–	–	–	1	–	1	2	–	–	2
12	Juncaceae	6	1	–	–	–	5	–	–	–	–	–	–
13	Aizoaceae	5	4	–	–	–	1	–	–	–	–	–	–
14	Euphorbiaceae	5	3	–	1	–	1	–	1	–	–	–	–
15	Palmae	5	–	–	–	–	–	–	–	–	–	5	–
16	Rhizophoraceae	5	–	–	–	–	–	–	–	–	–	5	–
17	Caryophyllaceae	4	1	–	–	–	3	–	–	–	–	–	–
18	Convolvulaceae	4	–	–	–	–	4	–	–	–	–	–	–
19	Plantaginaceae	4	2	–	–	–	2	–	–	–	–	–	–
20	Plumbaginaceae	4	1	–	–	–	3	–	–	–	–	–	–
21	Boraginaceae	3	–	–	–	–	3	–	–	–	–	–	–
22	Malvaceae	3	–	–	–	–	–	–	–	–	–	3	–
23	Najadaceae	3	–	3	–	–	–	–	–	–	–	–	–
24	Primulaceae	3	1	–	–	–	2	–	–	–	–	–	–
25	Rosaceae	3	–	–	–	–	–	–	–	–	–	–	3
26	Solanaceae	3	–	–	–	–	–	–	3	–	–	–	–
27	Umbelliferae	3	–	1	–	–	1	–	–	1	–	–	–
28	Eleagnaceae	2	–	–	–	–	–	–	1	–	–	1	–
29	Goodeniaceae	2	–	–	–	–	–	–	2	–	–	–	–
30	Liliaceae	2	–	–	–	–	2	–	–	–	–	–	–
31	Myrsinaceae	2	–	–	–	–	–	–	1	–	–	1	–
32	Nyctaginaceae	2	–	–	–	–	–	–	–	–	1	–	1
33	Portulacaceae	2	2	–	–	–	–	–	–	–	–	–	–
34	Residaceae	2	1	–	–	–	1	–	–	–	–	–	–
35	Salvadoraceae	2	–	–	–	–	–	–	–	–	–	2	–
36	Scrophulariaceae	2	–	–	–	–	2	–	–	–	–	–	–
37	Typhaceae	2	–	–	–	–	2	–	–	–	–	–	–
38	Amaranthaceae	1	–	–	–	–	–	–	1	–	–	–	–
39	Avicenniaceae	1	–	–	–	–	–	–	–	–	–	1	–
40	Caesalpinaceae	1	–	–	–	–	–	–	1	–	–	–	–
41	Cistaceae	1	–	–	–	–	–	–	1	–	–	–	–
42	Frankeniaceae	1	1	–	–	–	–	–	–	–	–	–	–
43	Gentianaceae	1	–	–	–	–	1	–	–	–	–	–	–
44	Hydrocharitaceae	1	–	1	–	–	–	–	–	–	–	–	–
45	Iridaceae	1	–	1	–	–	–	–	–	–	–	–	–
46	Lauraceae	1	–	–	–	–	–	–	–	–	–	–	1
47	Molluginaceae	1	1	–	–	–	–	–	–	–	–	–	–
48	Moraceae	1	–	–	–	–	–	–	–	–	–	1	–
49	Orobanchaceae	1	1	–	–	–	–	–	–	–	–	–	–
50	Punicaceae	1	–	–	–	–	–	–	–	1	–	–	–
51	Rhamnaceae	1	–	–	–	–	–	–	–	–	1	–	–
52	Rubiaceae	1	1	–	–	–	–	–	–	–	–	–	–
53	Ruppiaceae	1	–	1	–	–	–	–	–	–	–	–	–
54	Salicaceae	1	–	–	–	–	–	–	–	–	–	1	–
55	Simaroubaceae	1	–	–	–	–	–	–	–	–	–	–	–
56	Sonneratiaceae	1	–	–	–	–	–	–	–	–	–	1	–
57	Tiliaceae	1	–	–	–	–	–	–	–	1	–	–	–
58	Zygophyllaceae	1	–	–	–	–	–	–	2	–	–	–	–

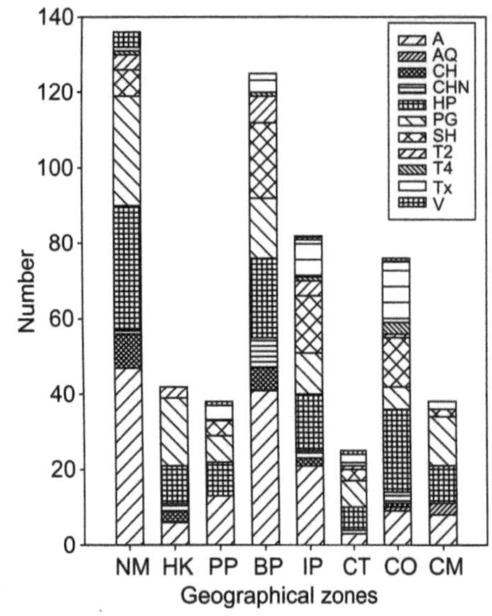

Fig. 2. Life form distribution of halophytes in major geographical regions of Pakistan.

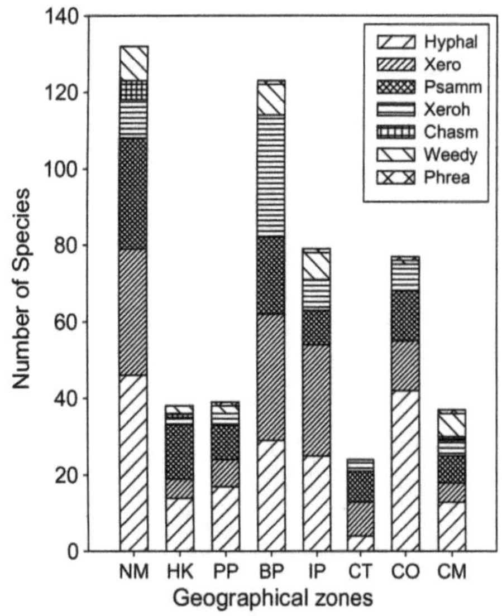

Fig. 3. Distribution of various halophyte types in major geographical regions of Pakistan.

ADAPTATIONS

A number of different mechanisms are used by halophytes to achieve osmotic adjustment, including inorganic ion accumulation, synthesis or accumulation of organic compounds and water loss [138]. Classification schemes have been constructed that attempts to match morphological and physiological characters to specific halophyte habitats and growth strategies [28]. However, all these classifications have little predictive value. Under physical or physiological stress conditions, the leaves of saline plants play an important role and develop certain xeromorphic adaptive characteristics like succulence, reduction in surface area, thick cuticle or a cover of waxy layers on epidermis, hairs on stem and leaves, sunken stomata and salt glands, etc. The succulents often lack the ability to secrete salts but they thwart the rise of salt concentration by an increase of their water content, and they become more and more succulent during their development. *Arthrocnemum, Halogeton, Haloxylon, Heliotropium, Salicornia, Suaeda, Salsola*, and *Zygophyllum* are prominent succulent halophytic genera. There is abrupt reduction in surface area of leaves of some species (e.g. *Salsola imbricata, Trianthema triquetra, Suaeda fruticosa*) during extreme salt stress conditions. *Suaeda fruticosa, Salsola imbricata, Haloxylon stocksii, H. salicornicum, Cressa cretica, Sporobolus arabicus, Urochondra setulosa* and *Aeluropus lagopoides* are characterized by thick cuticle and a cover of waxy layers, while stem and leaves of last five species remain covered with hairs. Only a small number of halophytes are able to excrete salts through glandular cells. Liphschitz and Waisel (98) listed active secreting glands in *Avicennia, Aeluropus, Aegiceras, Limonium, Rhizophora, Ceriops, Bruguiera, Tamarix* and *Reaumuria*. Salt concentration of the growth medium, light, temperature, oxygen, pressure and the presence of metabolic inhibitors are the governing factors of salt excretion. A similar function of salt recretion is ascribed to bladder trichomes of some Chenopodiaceae, for example, *Atriplex* species. The basic role of bladders is the protection of young developing shoots and leaves from toxic salt levels first in the apoplast and subsequently in the symplast.

Tolerance of salinity by halophyte seeds may be expressed either as the ability of un-germinated seeds to tolerate high salinity without loosing viability or the ability of seeds to germinate at high salinities [79]. Seeds of halophytes do not only germinate at higher salinities but also remain viable for long periods of time when immersed in saline water [80]. Halophytes vary in their upper limit of salt tolerance and increase in salinity usually delays their germination [139]. Seeds of salt marsh species like *Atriplex stocksii, A. triangularis, Hordeum jubatum, Polygonum aviculare* and *Zygophyllum simplex* show little germination above 125 mM NaCl [14, 74, 76, 89, 80], however, species like *Aeluropus lagopoides, Haloxylon stocksii, Sporobolus arabicus, Suaeda fruticosa, Limonium stocksii, Triglochin maritima* and *Urochondra setulosa* could germinate up to 500 mM NaCl [61, 62, 75, 78, 82, 83]. A third group of species like *Arthrocnemum macrostachyum, Cressa cretica, Halogeton glomeratus, Kochia scoparia, Salicornia brachiata, Salicornia bigelovii, Salsola iberica* and *Tamarix pentandra* could germinate at 800 mM or higher NaCl concentrations [60, 71, 76, 85, 86, 87, 88, 89, 139]. Species like *A. stocksiii H. stocksii* and *S. fruticosa* could be classified as moderately salt tolerant and *A. macrostachyum* and *C. cretica* as highly salt tolerant. Sharma and Sen [129] observed an extremely fast germination in the seeds of *Haloxylon stocksii* and *H. salicornicum*, occurring within an hour. An ecophysiological adaptive role is assigned to such a phenomenon of germination, which appears like uncoiling of the young embryo out of the testa immediately after contact with water with an unusually high rate of cell elongation, soon after imbibition. Such fast seed germination indicates the adaptive strategy by the plants as the availability of water with reduced NaCl content in soil during the rainy season is for a short duration. Increase in salinity leads to dormancy of seeds in halophytes and glycophytes. More investigations with halophytes [139] have demonstrated that seeds of several species, including *Arthrocnemum macrostachyum, Salicornia bigelovii, Salicornia brachiata, Cressa cretica, Tamarix pentandra, Salsola iberica, Halogeton glomeratus, Kochia scoparia, Aeluropus lagopoides, Atriplex stocksii, Haloxylon stocksii, Sporobolus arabicus, Suaeda fruticosa, Limonium stocksii, Triglochin maritima* and *Urochondra setulosa* remained dormant at high salinity and these will germinate when returned to distilled water [72, 77–83].

Salt tolerance of species vary with the stage of their development. Some species like *Suaeda fruticosa, Haloxylon stocksii, Atriplex stocksii* and *Zygophyllum simplex* were not very highly salt tolerant at germination but showed a high-salinity tolerance at growth stage [74, 76–90]. While other species like *Arthrocnemum macrostachyum* and *Cressa cretica* showed a higher degree of salt tolerance both at germination and growth stages [71, 72].

It has been assumed that survival of plants in saline environments depends upon the altered biochemical relations and on the quantitative ratio between toxic and protective compounds like betaine. Khan et al. [85] while studying *Halopyrum mucronatum, Atriplex stocksii, Haloxylon stocksii* and *Suaeda fruticosa* found high accumulation of betaine with the corresponding increase in salinity. The betaine is said to function as a source of solute for intracellular osmotic adjustment. Betaine accumulation occurs in the tissues of plants exposed to a saline substrate and there is a positive correlation between betaine content and the amount of Na^+ and Cl^- in the cell sap. It is also estimated that about 200 mM L^{-1} plant water or more betaine concentration is needed to successfully achieve osmotic adjustment under saline conditions and most of the Pakistani species tested have betaine concentration higher then this.

Scholander [121] described the presence of an ultrafilter in roots of mangroves of the family Rhizophoraceae, enabling only selective absorption of ions. They may retain a low-internal salinity by means of salt excluding mechanisms in the roots. In this type, sodium and chloride concentrations are higher in xylem sap and do not reach the metabolic cellular environment. Another mechanism of salt regulation in mangroves is salt excretion. In species of *Avicennia* and *Aegiceras*, NaCl concentration in the excreted solution exceeds the NaCl concentration of seawater and this is normally 10 times that of salt exclusion types and also does not reach the metabolic environment [69]. The same holds true for *Aeluropus littoralis, Limonium latifolium* and *Tamarix aphylla*.

The stem and leaf succulent halophytes lack the ability to excrete salt and these accumulate salt in their tissues. They are highly succulent and thwart the rising of salt concentration by a permanent increase of their water content. They become more and more succulent in their development. These are known as cumulative halophytes. Inland halophytes like *Haloxylon stocksii, H. salicornicum, Salsola imbricata, Sesuvium sesuvioides, Suaeda fruticosa, Trianthema triquetra* and *Zygophyllum simplex* lead to thickening in leaves, elongation of cells, higher elasticity of cell walls and smaller relative surface areas, decrease in extensive growth and high water content per unit of surface area. Leaves in some species like *Suaeda fruticosa, Salsola imbricata* and *Trianthema triquetra* are reduced in surface area, when exposed to high-salt content in the soil. Because these lack regulatory mechanisms, salt concentration therefore rises during growing season and when a certain level is reached, the plant dies. Among mangroves species

Avicennia, Ceriops, Rhizophora and *Sonneratia* absorb and accumulate excessive amounts of salts and the leaves become fleshy.

UTILIZATION AND POTENTIALS OF HALOPHYTES

Halophytes can play an important role in local and regional communities in the western, tropical and sub-tropical areas as a source of food, animal feed, chemicals, fine chemicals and other raw materials. Halophytes have their greatest potential not so much in contributing to the world's food supply but primarily in their utilization of the growing areas of saline land for a range of different goals. The most important opportunities relate to reforestation or replanting and ecological recovery of saline areas that have fallen into disuse, coastal development and protection and the production of cheap biomass for renewable energy, climate improvement and CO_2 sequestration. Mangroves besides playing roles in stabilization of coasts and beaches; food chain and life support system; aquaculture; agriculture; and support to development of wild-life sanctuary and recreation areas; also provide tannin, thatching material, fodder, fish poison, food products, medicine and wood for building purposes, fuel, boat and canoe making for the residents of coastal areas.

Food Yielding Halophytes

Of conventional crops, the only species with halophytic ancestors are beets (*Beta vulgaris*) and the date palm (*Phoenix dactylifera*) which can be irrigated with brackish water. The seed bearing species which are used as food include alkali sacaton (*Sporobolus airoides*). Indian almond (*Terminalia catappa*), *Salvadora oleoides* and *S. persica* yield fruits rich in oil and fat. The young leaves and shoots *Salicornia bigellovi, S. brachiata, Sesuvium portulacastrum, Chenopodium album, Portulaca oleracea, Suaeda maritima* have also been used for vegetables, salads and pickles in various parts of the country. *Suaeda fruticosa* is used to prepare a kind of baking soda, which is used in preparation of food. Radicles of *Rhizophora, Bruguiera* and *Ceriops*; tender leaves of *Thespesia populnea, Hibiscus tiliaceus* and fruits or kernels of littoral species such as *Terminalia catappa*.

Forages

In many arid coastal areas where mangroves occur sporadically, but little additional vegetation is available, the foliage of such species as *Avicennia marina, Sonneratia alba, Ceriops tagal* and *Rhizophora mucronata* has served as camel and cattle feed. Among trees, species of *Acacia, Prosopis, Salvadora* and *Zizyphus* are traditional fodder of arid regions. Many species of *Salicornia, Chenopodium, Atriplex, Salsola, Suaeda* and *Kochia* are common fodder shrubs. Among grasses *Leptochloa fusca, Aeluropus lagopoides, Dactyloctenim sindicum, Cynodon dactylon, Paspalum vaginatum, Sporobolus marginatus, Chloris gayana, C. virgata, Echinochloa turnerana, E. colonum* and *Puccinellia distans* are common species found in saline and alkaline areas and used as forages. Aronson et al. (11) recorded 1.26–2.09 kgm^{-2} dry matter and 15.5–39.5% fiber and 10.2–19.5% crude protein in some species of *Atriplex. Kochia indica* has been field tested for domestic livestock and found the good fodder producing with fresh biomass of 8.5 kg per bush from March through August. Kallar grass (*Leptochloa fusca*) has gained much attention as a fodder on salt affected soils (both saline and alkaline) in Pakistan [102]. We could get 46.5 t ha^{-1} green forage when planted in extreme alkali soil ($pH > 10$) for 5 years.

Oil Seeds

Production of vegetable oil from seed-bearing halophytes appears promising. Seeds of various halophytes like *Suaeda fruticosa, Arthrocnemum macrostachyum, Salicornia bigelovii, S. brachiata, Halogeton glomeratus, Suaeda moquinii Kochia scoparia* and *Haloxylon stocksii* possess sufficient quantity of high quality edible oil with unsaturation ranging from 70% to 80% (*141*). Seeds of *Salvadora oleoides* and *S. persica* contain 40–50% fat and are good source of lauric acid. Purified fat is used for soap and candle making and is a potential substitute for coconut oil.

Fuel Wood

More than a billion people in developing countries rely on wood for cooking and heating. Quite often fuel wood is obtained from salt tolerant trees and shrubs, which may include species of *Prosopis, Tamarix, Salsola, Suaeda, Kochia, Capparis* and *Salvadora* (42). In coastal areas the mangroves are used frequently for fuel and timber which has contributed a lot to deforestation of these habitats. Species of *Rhizophora, Ceriops, Avicennia* and *Sonneratia*, are excellent fuel woods and also contribute to form charcoal.

Products of Economic and Common Use

Suaeda, Salicornia, Salsola and *Haloxylon*, a carbonate of soda is obtained in large quantities and used for the soap and glass industry. The stem and leaves of salt-tolerant rushes (species of *Juncus* and *Spinifex*) have been used since ancient times for the manufacture of mats, baskets and cordage. Most of the mangroves species are rich in tannin contents and extraction of tannin from the bark of mangrove species had been one of the major uses of mangrove species.

Medicinal Uses

Many workers have reported the medicinal uses of halophytes while describing the economic importance of plants [42], some of these have been briefed here: *Acanthus volubilis* – leaves are used for dressing boils and wounds. *Capparis decidua* – bark is acrid, laxative, diaphoretic, anthelmintic, useful for cough, asthma and inflammations; fruits useful in cardiac troubles and biliousness. *Juncus rigidus* – it has diuretic properties and used in diarrhea. *Kochia indica* – the plant is considered cardiac and stimulant. *Salsola imbricata* – the plant is considered vermifugal; ash is applied to itches. *S. tetrandra* – it has antispasmodic and anthelmintic properties. *Salvadora persica* and *S. oleoides* – leaves are useful to relieve cough; seed oil in rheumatism and suppositories; stem and roots in toothache, leaf poultice in piles and tumors. *Thespesia populnea* – leaves are used in stomach trouble. *Zygophyllum simplex* – the plant has cardiac, antimicrobial and anthelmintic properties; leaves and seeds are applied in eye disease.

Aronson [5, 6, 9–11] in a recent survey of over 1,600 salt-tolerant plants with economic potential, has identified 290 tree species as being tolerant of 7–8 dS m^{-1} salinity. *Tamarix stricta* has been recorded as yielding 7.2 t DM ha^{-1} yr^{-1}, with a final density of 600 trees ha^{-1} after 5 years. Various species of *Prosopis, Casuarina, Eucalyptus* and *Acacia* have been evaluated for their salinity tolerance and biomass production *Prosopis juliflora* could yield up to 52.3 t ha^{-1} biomass in 6 years. *P. juliflora, Acacia nilotica* and *Casuarina equisetifolia* have been found to be most alkali tolerant and *Tamarix troupii, T. articulata, Prosopis juliflora, Pithecellobium dulce, Parkinsonia aculeata* and *Acacia farnesiana* as of tolerant salinity up to EC 25–35 dS m^{-1}; and *Acacia nilotica, A. tortilis, Casuarina glauca, C. obesa* and *Eucalyptus calmadulensis* up to salinity of 15–25 dS m^{-1}. *Aegiceras corniculata, Avicennia marina, Ceriops tagal* and *Rhizophora mucronata* trees could be grown in the areas with high salinity and low-water table. In recent years research in evaluation of halophytes for land reclamation and landscape management has taken a new dimension.

SUMMARY

Fresh water resources is becoming increasingly limited and that agricultural irrigation systems will steadily increase in salinity in the near future. It is about time to develop sustainable biological production system which could use low-quality saline water for irrigation of halophytic crops in saline lands. Halophytes are non-conventional crops and it would take a little to make people believe that they are good for them. However, there are potentials to extract high-quality edible oil from them and this would not have any problem of acceptability. Same would be true for feed, fiber and forage crops and the role of halophytes in ecologically improving the quality of saline degraded lands. There is no doubt that saline agriculture is here to stay. It would need fine tuning in different ecological situations and modern tissue culture techniques to improve the desired quality of halophytic cash crops. Pakistan has about 380 halophytes and if detailed field and laboratory studies are carried out, I am sure that many potential crops could be found.

REFERENCES

1. Adams, C.D. Flowering plants of Jamaica. University Press, Glasgow, 1972.
2. Agami, M., Eschel, A. and Waisel, Y. Najas marina in Israel: Is it a halophyte or a glycophyte? Physiologia Plantarum 1984; 61:634–6.
3. Ahmed, N. and Chaudhary, G.R. Irrigated Agriculture of Pakistan. Shahzad Nazir, Lahore, 1988.
4. Ali, S.I., Nasir, E. and Qaiser, M., editors. Flora of Pakistan, University of Karachi, Karachi, 1999.

5. Aronson, J. Haloph. Salt tolerant plants for the world – a computerized global data base of halophytes with emphasis on their economic uses. University of Arizona Press, Tucson, USA, 1989.
6. Aronson, J. 1991. The present and future roles of salt tolerant trees and their associated microsymbionts in arid lands. In Choukr-Allah, R., editor. Plant Salinity Research – New Challenges. Proc. Intnl. Conf. Agril. Mgt. Salt Affected Areas, Agadir, Morocco, 1991; 363–76.
7. Aronson, J.A. Halophytes of Central America and the Caribbean Region. Vol.: BGUN-ARI. Ben-Gurion University of the Negev, Beer-Sheva, Israel, Report, 1982.
8. Aronson, J.A. Preliminary report on halophyte and desert plant collection in Northwest India, 1982.
9. Aronson, J.A. Economic halophytes – a global review. In Wickens, G.E., Goodin, J.R. and Field, D.V., editors. Plants for arid lands. George Allen and Unwin, London, 1985; 177–88.
10. Aronson, J.A. Halophytes – new crops for deserts and coastal areas. FAO, CERES, Rome. Unpublished manuscript.
11. Aronson, J.A., Pasternak, A. and Danon, A. Introduction and first evaluation of 120 halophytes under seawater irrigation. In Whitehead, E.E., Hutchinson, C.F., Timmermann, B.N. and Varady R.G., editors. Arid lands: Today and tomorrow. Westview, Boulder, Colorado; Belhaven, London, 1988; 737–46.
12. Ayoub, A.T. Some features of salt tolerance in Senna (*Cassia acutifolia*), in Sudan. In Squires, V.R. and Ayoub, A.T., editors. Tasks for vegetation science, Vol. 32. Kluwer Academic Publisher, Dordrecht, 1994; 297–301.
13. Aziz, I. and Khan, M.A. Physiological adaptations to seawater concentration in *Avicennia marina* from Indus delta, Pakistan. Pak J Bot 2000; 32:151–70.
14. Badger, K.S. and Ungar, I.A. The effects of salinity and temperature on the germination of the inland halophyte *Hordeum jubatum*. Can J Bot 1989; 67:1420–5.
15. Baro, D. and Medrano, F.G. Halophytic flora and vegetation of Tamaulipas, Mexico. Unpublished manuscript, 1985.
16. Barros, M. Las ciperáceas del Estado de Santa Catalina. Sellowia 1960; 12:182–450.
17. Batanouny, K.H. Ecology and flora of Qatar. University of Qatar, Doha, 1981.
18. Batanouny, K.H. Halophytes and halophytic plant communities in the Arab region: Their potential as a rangeland resource. In Squires, V.R. and Ayoub, A.T., editors. Halophytes as a resource for livestock and for rehabilitation of degraded lands. Tasks for vegetation science, Vol. 32. Kluwer Academic Publisher, Dordrecht, 1994; 139–63.
19. Baum, B. The genus *Tamarix*. Israel Academy of Sciences and Humanities, Jerusalem, 1978.
20. Bennert, H.W. and Schmidt, B. On the osmoregulation in *Atriplex hymenelytra* (Torr.) Wats.(Chenopodiaceae). Oecologia 1984; 62:80–84.
21. Bernstein, L., Francois, L.E. and Clark, R.A. Salt tolerance of ornamental shrubs and ground covers. J Am Horticult Sci 1972; 97(4):550–6.
22. Bhandari, M.M. Flora of the Indian desert. Scientific Publishers, Jodhpur, 1978.
23. Boelcke, O. Plantas vasculares de la Argentina nativas y exóticas. Hemisferio Sur, Buenos Aires, 1981.
24. Bond, P. and Goldblatt, P. Plants of the Cape flora: A descriptive catalogue. J S Afri Bot 1984; 13:455.
25. Bor, N.L. Gramineae. In Townsend, C.C. and Guest, E., editors. Flora of Iraq, Vol. 9. Ministry of agriculture and Agrarian Reform, Baghdad, 1968; 434–5.
26. Boyko, H. Principles and experiments regarding irrigation with highly saline and seawater without desalinization. Trans NY Acad Sci Series 1964; II(26):1087–102.
27. Bramwell, D. and Bramwell, Z. Wildflowers of the Canary Islands. Stanley Thorne Ltd, London, 1974.
28. Breckle, S.-W. Temperate deserts and semi-deserts of Afghanistan and Iran. In Goodall, D.-W. and West, N., editors. Ecosystems of the world, Vol. 5. Elsevier, Amsterdam, 1983; 271–319.
29. Breckle, S.-W. Studies on halophytes from Iran and Afghanistan. II. Ecology of halophytes along salt gradients. In Proceedings of the Royal Society of Edinburgh, 1986.
30. Burgess, T.L. and Klein, R.D. Vegetation of the Northern Salt Basin, Hudspeth County, Texas. Unpublished manuscript.
31. Burkart, A. A monograph of the genus *Prosopis* Leguminosae subfam. Mimosoideae. J. Arnold Arboretum 1976; 57:219–49.
32. Bustanai, M., Weisal, Y., Kuller, Z., Guggenheim, J. and Tillman, A.G. The use of saline water for growing Rhodes grass on coastal sand dunes. Hassadeh 1971; 51:603–5.
33. Cano, E. Pasticales naturales de La Pampa, I: Descripcíon de las especies más importantes. Convenio AACREA-Provinica de La Pampa, 1988.
34. Carlton, J.M. A guide to common Florida salt marsh and mangrove vegetation, Vol. 6. Florida Marine Research. St. Petersburg, Florida, 1975.
35. Choukr-Allah, R. The potential of halophytes in the development and rehabilitation of arid and semi-arid zones. In Choukr-Allah, R., Malcolm, C.V., Hamdy, A., editors. Halophytes and biosaline agriculture. Marcel Dekker, Inc., New York, 1996; 3–16.
36. Cintrón, G., Lugo, A.E., Pool D.J. and Morris, G. Mangroves of arid environments in Puerto Rico and adjacent islands. Biotropica 1978; 10(2):110–21.
37. Cooper, A. The effects of salinity and waterlogging on the growth and cation uptake of salt marsh plants. New Phytologist 1982; 90:263–75.
38. Corre, J.J. Flore et Vegetation des milieux littoraux terrestres du Languedoc. Naturalia Monspeliensia 1982; 21:23–48.
39. Corre, J.J. 1985. Institut de Botanique, Université de Montpellier, France. Personal communication to Pasternak, D., in Pasternak, D, 1989.
40. Correll, D.S. and Correll, H.B. Flora of the Bahamas archipelago: Including the Turks and Caicos islands. J. Cramer, Vaduz, Liechtenstein, 1982.
41. Court, D. Succulent flora of southern Africa. A. A. Balkema, Rotterdam, 1981.
42. Dafni, A. Comparative biology of *Prosopis farcta* in four different habitats in Israel. Hebrew University, Jerusalem. Ph.D. dissertation, 1975.
43. Dagar, J.C. Characteristics of halophytic vegetation in India. In Khan, M.A. and Ungar, I.A., editors. Biology of Salt Tolerant Plants. University of Karachi, Karachi, 1995; 255–76.

44. Daoud, H.S. Flora of Kuwait. In Dicotyledoneae Vol. 1, 1985.
45. Den Hartog, C. The sea-grasses of the world. North-Holland, Amsterdam, 1970.
46. Doliner, L.H. and Jolliffe, P.A. Ecological evidence concerning the adaptive significance of the C4 dicarboxylic acid pathway of photosynthesis. Oecologia 1979; 38:23–34.
47. Downton, W.J.S. The occurrence of C_4 photosynthesis among plants. Photosynthetica 1975; 9(1): 96–105.
48. Dubuis, A. and Simmonneau, P. Contilbution á l'étude de la végétation halophile des bassins fermes du Plateau D'Oran. Services des Etudes Scientifiques, Clairbois, Algeria, 1960.
49. Eshel, A. Response of *Suaeda aegyptiaca* to KCl, NaCl and Na_2SO_4 treatments. Physiologia Plantarum 1985; 64:308–15.
50. Farquhar, G.D., Ball, M.C., Von Caemmerer, S. and Roksandic, Z. Effect of salinity and humidity on $d^{13}C$ value of halophytes – evidence for diffusional isotope fractionation determined by the ratio of intercellular atmospheric partial pressure of CO_2 under different environmental conditions. Oecologia 1982; 52:121–4.
51. Feinbrun, N. Flora Palaestina. Part III. Israel Academy of Science, Jerusalem, 1978.
52. Feinbrun, N. Flora Palaestina. Part IV. Israel Academy of Sciences, Jerusalem, 1986.
53. Felger, R.S. and Moser, M.B. Eelgrass (*Zostera maritima* L.) in the Gulf of California: Discovery of its nutritional value by the Seri Indians. Science 1973; 181:355–6.
54. Felker, P., Clark, P.R., Lagg, A.E. and Pratt, P.F. Salinity tolerance of the tree legumes: Mesquite *(Prosopis glandulosa* var. *torreyana, P. velutina* and *P. articulata*), Algarrobo (*P. chilensis*), Kaiwe (*P. pallida*) and Tamarugo (*P. tamarugo*) grown in sand cultures on nitrogen free media. Plant Soil 1981; 61:311–17.
55. Ferguson, C.R. Salt-tolerant plants for south Florida. Florida State Horticultural Society, 1952;306–13.
56. Fischer, M.J. and Skerman, P.J. Salt tolerant forage plants for summer rainfall areas. Reclamat Reveget Res 1986; 5:263–84.
57. Fortunato, R. Medicago. In Correa, M.N., editor. Flora Patagónica 4b:244–51. Colección Científica INTA, Buenos Aires, 1984.
58. Glenn, E.P. and O'Leary, J. Relationship between salt accumulation and water content of dicotyledonous halophytes. Plant Cell Environ 1984; 7:253–61.
59. Glenn, E.P. Relationship between cation accumulation and water content of salt tolerant grasses and a sedge. Plant Cell Environ 1987; 10:205–12.
60. Glenn, E.P., Tanner, R., Miyamoto, M., Fitzsimmons, K. and Boyer, J. Water use productivity and forage quality of the halophyte *Atriplex nummularia* grown on saline waste water in a desert environment. J Arid Environ 1998; 38:45–62.
61. Glenn, E.P., Brown, J.J. and Blumwald, E. Salt tolerance of crop potential of halophytes. Crit Rev Plant Sci 1999; 18:227–55.
62. Gulzar, S. and Khan. M.A. Seed germination of a halophytic grass *Aeluropus lagopoides*. Ann Bot 2001; 87:319–24.
63. Gulzar, S., Khan, M.A. and Ungar, I.A. Effect of temperature and salinity on the germination of *Urochondra setulosa*. Seed Sci Technol 2001; 29:21–29.
64. Gupta, R.K and Saxena. S.K. Resource survey of *Salvadora oleiodes* Decne and *S. persica* L. for non-edible oil in western Rajasthan. Trop Ecol 1968; 9(2):140–52.
65. Heerkloss, B. and Bartolomaeus, W. Experimental studies on germination behavior of cultivated plants at different salinities of the germination media. Archiv für Acker- und Pflanzenbau, Bodenkunde 1980; 24:241–5.
66. Hoffman, G.J. and Shannon, M.C. Relating plant performance and salinity. Rehabilit Reveget Res 1986; 5: 211–25.
67. Hu, C.-C. and Chin, T.-L. Halophytes in China – Floristic distribution and vegetation types. Unpublished manuscript, 1995.
68. Joshi, A.J. Physiological studies on some halophytes. Saurashtra Univ. India. PhD Thesis, 1979.
69. Joshi, A.J. and Iyengar, E.R.R. Economic uses of coastal halophytes. Salt Res Industry 1978; 14(1):18–21.
70. Joshi, G.V., Jamale, B.B. and Bhosale. L.J. 1975. Ion regulation in mangroves. In Walsh, G.E., Snedaker, S.C. and Teas, H.J., editors. Proc. Int. Symp. Biol. and Management of Mangroves. Vol. 2. Int. Food Agric. Sci. Univ. Florida, Gainesville, 1975; 595–607.
71. Kartawinata, K., Adisumarto, S., Soemodihardjo, S. and Tantra, I.G.M. Status pengetahuan hutan bakau Indonesia (the state of knowledge in the Indonesian Mangrove forest), 1979.
72. Khan, M.A. Studies on germination of *Cressa cretica* L. seeds. Pakistan J Weed Sci Res 1991; 4:89–98.
73. Khan, M.A. Comparative influence of salinity and temperature on the germination of subtropical halophytes. In Lieth, H., Moschenko, M., Lohman, M., Koyro, H.-W. and Hamdy, A., editors. Halophyte uses in different climates I: Ecological and ecophysiological studies. Progress in biometeriology, Vol. 13. Backhuys Publishers, The Netherlands, 1999; 77–88.
74. Khan, M.A. and Aziz, I. Salinity tolerance of some mangroves from Pakistan. Wetland Ecol Manage 2001; 9:1–5.
75. Khan, M.A. and Aziz, S. Some aspects of salinity, plant density, and nutrient effects on *Cressa cretica* L. J Plant Nut 1998; 21:769–84.
76. Khan, M.A. and Rizvi, Y. The effect of salinity, temperature and growth regulators on the germination and early seedling growth of *Atriplex griffithii* Moq. var. *stocksii* Boiss. Can J Bot 1994; 72:475–9.
77. Khan, M.A. and Gul, B. High salt tolerance in the germinating dimorphic seeds of *Arthrocnemum indicum*. Int J Plant Sci 1998; 159:826–32.
78. Khan, M.A. and Ungar, I.A. Effects of salinity and temperature on the germination and growth of *Atriplex triangularis* Willd. Am J Bot 1984; 71:481–9.
79. Khan, M.A. and Ungar, I.A. Biology of salt tolerant Plants. Department of Botany, University of Karachi, Pakistan, 1995; 424.
80. Khan, M.A. and Ungar, I.A. Influence of salinity and temperature on the germination of *Haloxylon recurvum*. Ann Bot 1996; 78:547–51.
81. Khan, M.A. and Ungar, I.A. Effect of light, salinity, and thermoperiod on the seed germination of halophytes. Can J Bot 1997; 75:835–41.

82. Khan, M.A. and Ungar, I.A. Effect of thermoperiod on recovery of seed germination of halophytes from saline conditions. Am J Bot 1997; 84:279–83.
83. Khan, M.A. and Ungar, I.A. Germination responses of the subtropical annual halophyte *Zygophyllum simplex*. Seed Sci Technol 1997; 25:83–91.
84. Khan, M.A. and Ungar, I.A. Germination of salt tolerant shrub *Suaeda fruticosa* from Pakistan: Salinity and temperature responses. Seed Sci Technol 1998; 26:657–67.
85. Khan, M.A. and Ungar, I.A. Seed germination and dormancy of *Polygonum aviculare* L. as influenced by salinity, temperature, and gibberellic acid. Seed Sci Technol 1998; 26:107–17.
86. Khan, M.A., Ungar, I.A., Showalter, A.M. and Dewald, H. NaCl – induced accumulation of glycinebetaine in four subtropical halophytes from Pakistan. Physiologia Plantarum 1998; 102:487–92.
87. Khan, M.A., Ungar, I.A. and Showalter, A.M. 1999. The effect of salinity on growth, ion content, and osmotic relations in *Halopyrum mucronatum* (L.) Stapf. J Plant Nut 1999; 22:191–204.
88. Khan, M.A., Ungar, I.A. and Showalter A.M. Effects of sodium chloride treatments on growth and ion accumulation of the halophyte *Haloxylon recurvum*. Commun Soil Sci Plant Nut 2000; 31:2763–74.
89. Khan, M.A., Ungar, I.A. and Showalter, A.M. Growth, water, and ion relationships of a leaf succulent perennial halophyte, *Suaeda fruticosa* (L.) Forssk. J Arid Environ 2000; 45: 73–84.
90. Khan, M.A., Ungar, I.A. and Showalter, A.M. Salt tolerance in the subtropical perennial halophyte *Atriplex griffithii* Moq. var. *stocksii* Boiss. Ann Bot 2000; 85:225–32.
91. Khan, M.A., Gul, B. and Weber, D.J. Salinity and temperature effects on the germination of dimorphic seeds of *Suaeda moquinii*. Aust J Bot 2001; 49:1–8.
92. Lavranos, J. 1986. Bramley, South Africa. Personal communication to Pasternak, D., in Pasternak, D. 1989.
93. Le Houérou, H.N. Browse in Africa – the current state of knowledge. International Livestock Center of Africa, Addis Ababa, Ethiopia, 1980.
94. Le Houérou, H.N. Salt-tolerant plants of economic value in the Mediterranean Basin. Reclamat Reveget Res 1986; 5:319–41.
95. Le Houérou, H.N. Forage halophytes and salt-tolerant fodder crops in the Mediterranean Basin. In Squires, V.R. and Ayoub, A.T., editors. Halophytes as a resource for livestock and for rehabilitation of degraded lands. Tasks for vegetation science, Vol. 32. Kluwer Academic Publisher, Dordrecht, 1994.
96. Lear, R. and Turner, T. Mangroves of Australia. University of Queensland, St. Lucia, 1976.
97. Lieth, A.F. Use of sea water for growth and productivity of halophytes in the Gulf Region. United Arab Emirates University, Faculty of Science, 1994.
98. Lindacher, R. PHANART Datenbank der Gefäßpflanzen Mitteleuropas, Vol. 125. Geobotanisches Institut der ETH, Stiftung Rübel, Zürich. Veröffentlichung des Geobotansichen Institutes, 1995.
99. Liphschitz, N. and Waisel, Y. Adaptation of plants to saline environments: salt excretion and glandular structure. In Sen, D.N. and RajPurohit, K.S., editors. Contribution to the ecology of halophytes. Dr. W. Junk Publishers, The Hague/Boston/London, 1982; 197–214.
100. MacNae, W. and Kalk, M. The ecology of the mangrove swamps at Inhaca Island, Mocambique. J Ecol 1962; 50:17–34.
101. Mahmoud, A., El-Sheikh, A.M. and Isawi, F. Ecology of the littoral salt marsh vegetation at Rabigh on the Red Sea coast of South Arabia. J Arid Environ 1983; 5:35–42.
102. Malcolm C.V. Use of halophyte forages for rehabilitation of degraded lands. In Squires, V.R. and Ayoub, A.T., editors. Tasks for vegetation science, Vol. 32. Kluwer Academic Publisher, Dordrecht, 1994; 25–41.
103. Malik, K.A., Aslam, Z. and Naqvi, M. Kallar grass: a plant for saline land. Nuclear Institute for Agriculture and Biology, Faisalabad, Pakistan, 1986; 93.
104. Martin, W.K. The concise British flora in colour. Book Club Associates, London, 1969.
105. Mason, H.C. A flora of the marshes of California. University of California, Berkeley, 1957.
106. McMillan, C. Salt tolerante within a *Typha* population. Am J Bot 1959; 46:521–6.
107. Menninger, E.A. Seaside plants of the world. Hearthside, New York, 1964.
108. Menzel, U. and Lieth, H. Halophyte Database Vers. 2.0. In Lieth, H., Moschenko, M., Lohman, M., Koyro, H.-W. and Hamdy, A., editors. Halophyte Uses in different climates I: Ecological and Ecophysiological Studies. Progress in Biometeriology, Vol. 13. pp. 77–88. Backhuys Publishers, The Netherlands, 1999.
109. Mepham R.H. and Mepham, J.S. The flora of tidal forests – a rationalization of the use of the term "mangrove". S Afri J Bot 1985; 51:75–99.
110. Mepham R.H. Mangrove floras of the southern continents. Part 1. S Afri J Bot 1983; 2:1–8.
111. Moghaddam, P.R. and Koocheki, A. A comprehensive survey of halophytes in Khorasan province of Iran. Unpublished manuscript, 1999.
112. Morton, J.F. Craft industries from coastal wetland vegetation. Estuarine Processes 1976; 1(2):54–266.
113. Morton, J.F. Wild plants for survival in south Florida. Fairchild Tropical Garden, Miami, Florida, 1982.
114. Moss, H. and Taylor, F. Commercial utilization of veld products of Botswana. Veld Products, Inc., Gaborone, Botswana, 1984.
115. Munz, P. and Keck, D.D. A flora of California and supplement. University of California, Berkeley, 1968.
116. New Mexico Water Resources Research Institute. Evaluation of the salinity tolerance of Russian thistle to determine its potential for forage production using saline irrigation water, Vol. 198. New Mexico State University, Las Cruces. WRRI Report, 1985.
117. O'Leary, J.W., Glenn E.P. and Watson, M.C. Agricultural production of halophytes irrigated with seawater. Plant Soil 1986; 89(1–3):311–22.
118. Osmond, C.B., Bjorkman, O. and Anderson, D. Physiological processes in plant ecology: Towards a synthesis with *Atriplex*. In Ecological studies, Vol. 36. Springer, New York, 1980.

119. Pasternak, D. Fodder production with saline water. The institute for applied research Ben-Gurion university of the Negev. Projekt Report, 1990.
120. Pasternak, D. and Nerd, A. Research and utilization of halophytes in Israel. In Choukr-Allah, R., Malcolm, C.V. and Hamdy, A., editors. Halophytes and biosaline agriculture. Marcel Dekker, Inc., New York, 1996; 325–48.
121. Peltier, J.P. La végétation du Bassin Versant de L'Oued Sous (Maroc). Université Scientifique et Medicaie de Grenoble. Unpublished Ph.D. dissertation, 1982.
122. Qureshi, R.H. and Barrett-Lennard, E.G. Saline agriculture for irrigated land in Pakistan: A handbook. Australian Centre for International Agricultural Research Canberra, Australia, 1998.
123. Quézel, P. La végétation du Sahara. Fischer, Stuttgart, 1965.
124. Raghavendra, A.S. and Das, V.S.R. The occurrence Of C4 photosynthesis: A supplementary list of C4 plants reported during late 1974–mid 1977. Photosynthetica 1978; 12:200–8.
125. Ragonese, A.E. and Covas, G. La flora halofile del sur de la provincia de Santa Fe. Darwiniana 1947;7(3):401–95.
126. Rozema, J. The influence of salinity, inundation and temperature on the germination of some halophytes and non-halophytes. Oecologia Plantarum 1975; 10(4):341–53.
127. Saxena, S.K. and Gupta, R.K. Vegetation of Pachpadra salt basin in Western Rajasthan. J Bombay Natural Hist 1972; 70(1):104–27.
128. Saxena, S.K., Squires, V.R. and Ayoub, A.T. Halophytes of the Indian desert-communities on the various "ranns". In Squires, V.R. and Ayoub, A.T., editors. Tasks for vegetation science, Vol. 32. Kluwer Academic Publisher, Dordrecht, 1994; 211–15.
129. Schulz, A.G. Plantas y frutos comestibles de la región chaquena. Rev Agronóm Noroeste Argeatino 1963; 4:57–83.
130. Sharma, S.K. and Gupta, R.K. Effects of salts on seed germination of some desert grasses. Ann Arid Zone 1971; 10:33–6.
131. Sharma, T.P. and Sen, D.N. A new report on abnormally fast germinating seeds of *Haloxylon* spp. – an ecological adaptation to saline habitat. Curr Sci 1989; 58:382–5.
132. Shishkin, B.K., editor. Flora of the U.S.S.R. In Centrospermae, Vol. 6. Izdatel'stvo Akademii Nauk SSR, Moscow, 1936.
133. Shomer-Ilan, A. Photosynthetic pathways in the flora of Israel and its ecological significance. B. The Gramineae and the Chenopodiaceae in Israel, Vol. 6. Rotem, 1983; 5–17.
134. Shomer-Ilan, A. Tel Aviv University, Israel. Unpublished data, 1985.
135. St. John, H. Classification and distribution of the *Ipomoea pescaprae* group (Convolvulaceae). Botanische Jahrbücher für Systematik, Pflanzengeschichte und Pflanzengeograpbie 1970; 89(4):563–83.
136. Szabolcs, I. Salt affected soils as the ecosystem for halophytes. In Squires, V.R. and Ayoub, A.T., editors. Tasks for vegetation science, Vol. 32. Kluwer Academic Publisher, Dordrecht, 19–24, 1994.
137. The Royal Botanic Garden. Index Kewensis. Computerized data base, 1993.
138. Toutain, G. Éléments d'agronomie Saharienne. Institut National de Recherche Agronomique, Marrakech, Maroc, 1977.
139. Tutin, T.G., Heywood, V.H., Burges, N.A., Valentine, D.H., Walters, S.M. and Webb, D.A. Flora Europaea, Vol. I. Cambridge University, Cambridge, 1964.
140. Ungar, I.A. Ecophysiology of vascular halophytes. CRC Press, Boca Raton, 1991; 209.
141. Ungar, I.A. Seed germination and seed-bank ecology of halophytes. In Kigel, J. and Galili, G., editors. Seed development and germination. Marcel and Dekker, Inc., New York, 1995; 599–627.
142. Vareschi, V. Plantas entre el may e la tierra. Armitano, Caracas, 1979.
143. Weber, D.J., Gul, B., Khan, M.A., Williams, T., Wayman, P. and Warner. S. Composition of vegetable oil from seeds of native halophytic shrubs. In McArthur, E.D. and Fairbanks, D.J., editors. In Proceedings: Shrubland Ecosystem Genetics and Biodiversity, 2000; June 13–15; Provo, UT. Proceedings RMRS-P-000. U.S. Department of Agriculture, Forest Service Rocky Mountain Research Station, Ogden, UT, 2001.
144. Zohary, M. Flora Palaestina. Part I. Israel Academy of Sciences, Jerusalem, 1996.
145. Zohary, M. Geobotanical foundations of the Middle East. Fischer, Stuttgart, 1972.

A comprehensive survey of halophytes in Khorasan province of Iran

P. REZVANI MOGHADDAM AND A. KOOCHEKI

INTRODUCTION

Khorasan province in the North East of Iran is the largest state in the country and occupies one fifth of the land surface of the country which is more than 300,000 km^2. This province has a diversity of climatic conditions and large borders with two main deserts of the country, Kavir Loot and Kavir Namak. The altitude, the location of plateaus, and the deep valleys are some of the most important causes of this diverse climate. Khorasan has four main climatic zones such as steppic, substeppic, subdesertic and mountainous. Mean annual precipitation of this province is less than 220 mm/year. Because of the saline ground water, the presence of a salt layer in the soil, intense irrigation, high evaporation and low precipitation, there is a large saline area in this part of the country. Despite a rich halophytic flora in the province, halophytic communities are poorly known vegetation units.

MATERIALS AND METHODS

In order to study halophyte species in Khorasan which are important as animal feed, fuel, sand dune fixation etc., an attempt was made to survey and evaluate halophyte species in terms of species and family names, local names, distribution and utilization. Botanical information was taken from scientific references, such as Rechinger (1975, 1963, 1964, 1965, 1970, 1972a, 1972b, 1972c, 1972d, 1974, 1977, 1978, 1997), Ghahraman (1976–1998: 18 Volumes), Mobayein (1975, 1980), Rashed Mohasel (1993), Akhani and Ghorbanli (1993), Koocheki (1996) and Koocheki and Mohalati (1994). In some cases, field investigation was also done. A complete list of these species is shown in Table 10.

The Halophyte Families

A list of the halophytes families is shown in Table 1. From this table it is apparent that: (1) A total of 229 halophytic species within 26 families were identified. (2) The ranking order of halophytes families with respect to the numbers of species was Chenopodiaceae 110, Tamaricaceae 26, Poaceae 26, Plumbaginaceae 10,

Table 1. The list of halophyte families

Percentage of total number of halophyte species	Number of species	Family name
0.4	1	Asparagaceae
1.3	3	Asteraceae
0.4	1	Avicenniaceae
0.4	1	Balanophoraceae
1.3	3	Boraginaceae
0.9	2	Brassicaceae
1.7	4	Caryophyllaceae
48.0	110	Chenopodiaceae
0.4	1	Convolvulaceae
0.8	1	Cymodoceaceae
2.2	5	Cyperaceae
2.6	6	Fabaceae
1.3	3	Frankeniaceae
0.4	1	Iridaceae
0.4	1	Juncaceae
0.4	1	Orbachaceae
0.9	2	Plantaginaceae
4.4	10	Plumbaginaceae
11.4	26	Poaceae
3.1	7	Polygonaceae
0.4	1	Primulaceae
0.4	1	Ruppiaceae
0.4	1	Salvadoraceae
1.3	3	Solanaceae
11.4	26	Tamaricaceae
0.4	1	Verbenaceae
3.1	7	Zygophyllaceae

H. Lieth (ed.), Cash Crop Halophytes: Recent Studies, 189–195.

Polygonaceae 7, Zygophyllaceae 7, Fabaceae 6, Cyperaceae 5 other families.

Utilization of Halophytic Species

Halophytes, either native or exotic, are liable to many different economic uses such as: grazing as fodder reserves, fuel, sand dune fixation, erosion control, industrial purposes, vegetable and detergent.

Animal feed

The main economic value of halophytic plant communities is grazing. More than 70% of the halophyte species in Khorasan are used as animal feed. Camels, sheep and goats are the main domestic animals which are grazing halophytic plant communities. The ranking order of halophytes families with respect to the numbers of species used as an animal feed was Chenopodiaceae > Poaceae > Tamaricaceae > other species (Table 2).

Fuel

Nearly 24% of the halophyte species in Khorasan are used as fuel. Among the fuel producing species special mention should be made of two families: Chenopodiaceae and Tamaricaceae, of which the former is grown widely (Table 3).

Sand dune fixation

Halophyte species has an important role in sand dune fixation, partly because of their physiological and

Table 2. List of halophyte families used as animal feed

Percentage of total number of halophyte species	Family name
39	Chenopodiaceae
11	Poaceae
11	Tamaricaceae
9	Other species

Table 3. List of halophyte families used as fuel

Percentage of total number of halophyte species	Family name
8	Chenopodiaceae
7	Tamaricaceae
4	Polygonaceae
4	Solanaceae
1	Other species

Table 4. List of halophyte families used as sand dune fixation

Percentage of total number of halophyte species	Family name
10	Tamaricaceae
7	Chenopodiaceae
4	Poaceae
2	Polygonaceae

Table 5. List of halophyte families used for erosion control

Percentage of total number of halophyte species	Family name
7	Tamaricaceae
6	Chenopodiaceae
3	Poaceae
1	Polygonaceae

Table 6. List of halophyte families used for industrial agent

Percentage of total number of halophyte species	Family name
4	Tamaricaceae
2	Chenopodiaceae
1	Poaceae

Table 7. List of halophyte families used as vegetable

Percentage of total number of halophyte species	Family name
4	Chenopodiaceae
1	Poaceae

Table 8. List of halophyte families used as detergent

Percentage of total number of halophyte species	Family name
2	Chenopodiaceae
1	Tamaricaceae
1	Poaceae

Table 9. List of halophyte families with C_4 photosynthetic pathway

Percentage of total number of halophyte species	Family name
32	Chenopodiaceae 110
7	Poaceae 26
2	Tamaricaceae 26

Table 10. List of halophytes in Khorasan province of Iran. 2: used as animal feed, 3: used as fuel, 4: used as sand dune fixation, 5: used for erosion control, 6: used for industrial agent, 7: used as vegetable, 8: used as detergent, 9: with C_4 photosynthetic pathway

Species	Family name	2	3	4	5	6	7	8	9
Asparagus sp.	Asparagaceae	*	.	*	.	.	.	.	.
Amberboa turanica	Asteraceae	.	.	.	.	.	.	.	.
Artemisia fragrans	Asteraceae	.	*	.	.	.	.	.	.
Helichrysum leucocephalum	Asteraceae	.	.	.	.	.	.	.	.
Avicennia marina	Avicenniaceae	.	.	.	.	.	.	.	.
Cynomorium coccineum	Balanophoraceae	.	.	.	.	.	.	.	.
Heliotropium aucheri	Boraginaceae	.	*	.	.	.	.	.	.
Heliotropium bacciferum	Boraginaceae	*	*	.	.	.	.	.	.
Heliotropium curassavicum	Boraginaceae	.	*	.	.	.	.	.	.
Arabidopsis parvula	Brassicaceae	.	.	.	.	.	.	.	.
Lepidium cartilagineum	Brassicaceae	.	.	.	.	.	.	.	.
Gypsophila perfoliata	Caryophyllaceae	.	.	.	.	.	.	*	.
Spergularia marina	Caryophyllaceae	.	.	.	.	.	.	.	.
Spergularia media	Caryophyllaceae	.	.	.	.	.	.	.	.
Sphaerocoma aucheri	Caryophyllaceae	.	.	.	.	.	.	.	.
Aellenia auricula	Chenopodiaceae	*	*	*	*	.	.	*	*
Agathophora alopecuroides	Chenopodiaceae	*	.	.	.	.	.	.	.
Allenrolfea occidentalis	Chenopodiaceae	*	*	*	*	.	.	.	.
Anabasis annua	Chenopodiaceae	*	.	.	.	.	.	.	.
Anabasis aphylla	Chenopodiaceae	.	.	.	.	.	.	.	*
Anabasis calcarea	Chenopodiaceae	.	.	.	.	.	.	.	*
Anabasis eriopoda	Chenopodiaceae	.	.	.	.	.	.	.	*
Anabasis haussknechti	Chenopodiaceae	.	.	.	.	.	.	.	*
Anabasis salsa	Chenopodiaceae	*	.	.	.	.	.	.	*
Anabasis setifera	Chenopodiaceae	*	.	.	.	.	.	.	*
Arthrocnemum fruticosum	Chenopodiaceae	*	.	*	*	.	.	.	.
Arthrocnemum halocnemoides	Chenopodiaceae	*	.	*	*	.	.	.	.
Atriplex brevifolia	Chenopodiaceae	*	.	.	.	.	.	.	.
Atriplex canescens	Chenopodiaceae	*	*	*	*	*	*	.	*
Atriplex confertifolia	Chenopodiaceae	*	.	.	.	.	.	.	.
Atriplex dimorphostegia	Chenopodiaceae	*	.	*	*	.	*	.	*
Atriplex griffithii	Chenopodiaceae	*	.	*	*	.	*	.	*
Atriplex halimus	Chenopodiaceae	*	*	.	.	.	*	.	*
Atriplex hastata	Chenopodiaceae	*	*	.	.	.	.	.	*
Atriplex hortensis	Chenopodiaceae	*	.	.	.	.	*	.	*
Atriplex inflata	Chenopodiaceae	*	.	*	.	.	.	.	*
Atriplex leucoclada	Chenopodiaceae	*	.	.	.	.	*	.	*
Atriplex moneta	Chenopodiaceae	*	.	.	.	.	.	.	*
Atriplex nitens	Chenopodiaceae	*	.	.	.	.	*	.	*
Atriplex parviflora	Chenopodiaceae	*	.	.	.	.	.	.	.
Atriplex polycarpa	Chenopodiaceae	*	.	.	.	.	.	.	*
Atriplex spongiosa	Chenopodiaceae	*	.	.	.	.	.	.	*
Atriplex tartarica	Chenopodiaceae	*	.	.	.	.	.	.	*
Bassia eriantha	Chenopodiaceae	*	.	.	*	.	.	.	*
Bassia eriophora	Chenopodiaceae	*	.	.	*	.	.	.	*
Bassia hyssopifolia	Chenopodiaceae	*	.	.	.	.	.	.	*
Bienertia cycloptera	Chenopodiaceae	.	.	.	.	.	.	.	*
Brachylepis eriopoda	Chenopodiaceae	*	.	.	.	.	.	.	.
Camphorosma monspeliacum	Chenopodiaceae	.	.	.	.	.	.	.	*
Climacoptera brachiata	Chenopodiaceae	.	.	.	.	.	.	.	*
Climacoptera lanata	Chenopodiaceae	.	.	.	.	.	.	.	*
Climacoptera turcomanica	Chenopodiaceae	.	.	.	.	.	.	.	*
Cornulaca leucacantha	Chenopodiaceae	*	.	.	.	.	.	.	*
Eurotia lanata	Chenopodiaceae	*	.	*	*	.	*	.	.
Gamanthus gamocarpus	Chenopodiaceae	.	.	.	.	.	.	.	*
Halanthium rarifolium	Chenopodiaceae	.	.	.	.	.	.	.	*
Halimione flabellum	Chenopodiaceae	*	.	.	.	.	.	.	*
Halimione verrucifera	Chenopodiaceae	*	.	.	.	.	.	.	*

Table 10. (Cont.)

Species	Family name	2	3	4	5	6	7	8	9
Halimocnemis mollissima	Chenopodiaceae	.	.	.	.	.	.	.	*
Halimocnemis pilosa	Chenopodiaceae	*	.	.	.	.	.	.	*
Halocharis hispida	Chenopodiaceae	*	.	.	.	.	.	.	*
Halocharis sulphura	Chenopodiaceae	*	.	.	.	.	.	.	*
Halocharis violacea	Chenopodiaceae	*	.	.	.	.	.	.	*
Halocnemum strobilaceum	Chenopodiaceae	*	.	*	*	.	.	.	.
Halopeplis perfoliata	Chenopodiaceae	*	.	.	.	.	.	.	.
Halopeplis pygmaea	Chenopodiaceae	*	.	.	.	.	.	.	.
Halostachys caspica	Chenopodiaceae	*	.	.	.	.	.	.	.
Halothamnus auriculus	Chenopodiaceae	.	.	.	.	.	.	.	*
Halothamnus glaucus	Chenopodiaceae	.	.	.	.	.	.	.	*
Halothamnus hierochunticus	Chenopodiaceae	.	.	.	.	.	.	.	*
Halothamnus subaphyllus	Chenopodiaceae	*	.	.	.	.	.	.	*
Halotis occulta	Chenopodiaceae	.	.	.	.	.	.	.	*
Halotis pilifera	Chenopodiaceae	.	.	.	.	.	.	.	*
Haloxylon aphyllum	Chenopodiaceae	*	*	*	*	*	.	.	*
Haloxylon recurvm	Chenopodiaceae	*	*	*	*	*	.	.	*
Haloxylon salicornicum	Chenopodiaceae	*	*	*	.	.	.	.	*
Hypocylix kerneri	Chenopodiaceae	.	.	.	.	.	.	.	.
Kalidium caspicum	Chenopodiaceae	*	.	.	.	.	.	.	.
Kochia americana	Chenopodiaceae	*	.	.	.	.	.	.	.
Kochia brevifolia	Chenopodiaceae	*	.	.	.	.	.	.	.
Kochia georgei	Chenopodiaceae	*	.	.	.	.	.	.	*
Kochia iranica	Chenopodiaceae	*	.	.	.	.	.	.	*
Kochia odontoptera	Chenopodiaceae	*	.	.	.	.	.	.	*
Kochia prostrata	Chenopodiaceae	*	.	*	.	.	.	.	*
Kochia scoparia	Chenopodiaceae	*	.	.	.	.	.	.	.
Kochia sedifolia	Chenopodiaceae	*	.	.	.	.	.	.	*
Panderia pilosa	Chenopodiaceae	.	.	.	.	.	.	.	*
Petrosimonia brachiata	Chenopodiaceae	.	.	.	.	.	.	.	*
Petrosimonia glauca	Chenopodiaceae	.	.	.	.	.	.	.	*
Piptoptera turkestana	Chenopodiaceae	.	.	.	.	.	.	.	*
Salicornia bigelowii	Chenopodiaceae	*	.	.	.	.	.	.	.
Salicornia europaea	Chenopodiaceae	*	.	*	*	*	.	.	.
Salicornia rubra	Chenopodiaceae	*	.	.	.	.	.	.	.
Salsola abarghuensis	Chenopodiaceae	*	.	.	.	.	.	.	*
Salsola arbuscula	Chenopodiaceae	*	.	.	.	*	.	*	.
Salsola baryosma	Chenopodiaceae	*	.	.	.	.	.	.	*
Salsola chorassanica	Chenopodiaceae	*	.	.	.	.	.	.	.
Salsola crassa	Chenopodiaceae	*	*	.	.	.	.	.	*
Salsola dendroides	Chenopodiaceae	*	.	.	.	.	.	.	*
Salsola drummondii	Chenopodiaceae	*	.	.	.	.	.	.	*
Salsola glauca	Chenopodiaceae	*	.	.	.	.	.	.	.
Salsola gossypina	Chenopodiaceae	*	.	.	.	.	.	.	.
Salsola hispidula	Chenopodiaceae	*	.	.	.	.	.	.	.
Salsola incanescens	Chenopodiaceae	*	*	.	.	.	.	.	*
Salsola kali	Chenopodiaceae	*	.	.	*	.	.	.	*
Salsola kerneri	Chenopodiaceae	*	.	.	.	.	.	.	.
Salsola lanata	Chenopodiaceae	*	.	.	.	.	.	.	*
Salsola leptoclada	Chenopodiaceae	*	.	.	.	.	.	.	*
Salsola nitraria	Chenopodiaceae	*	.	.	.	.	.	.	*
Salsola orientalis	Chenopodiaceae	*	.	.	.	.	.	.	*
Salsola rigida	Chenopodiaceae	*	*	*	.	.	.	.	*
Salsola sclerantha	Chenopodiaceae	*	.	.	.	.	.	.	.
Salsola soda	Chenopodiaceae	*	*	.	.	.	.	*	*
Salsola tomentosa	Chenopodiaceae	*	.	.	.	.	.	.	.
Seidlitzia flordia	Chenopodiaceae	*	*	.	.	.	.	*	*
Seidlitzia rosmarinus	Chenopodiaceae	*	*	*	.	*	.	*	*
Suaeda acuminata	Chenopodiaceae	*	*	.	.	.	.	.	.

Table 10. (Cont.)

Species	Family name	2	3	4	5	6	7	8	9
Suaeda aegyptiaca	Chenopodiaceae	*	*	.	.	.	.	.	*
Suaeda altissima	Chenopodiaceae	*	.	.	.	.	.	.	*
Suaeda arcuata	Chenopodiaceae	*	.	.	.	.	.	.	*
Suaeda fruticosa	Chenopodiaceae	*	.	.	.	.	.	.	*
Suaeda heterocarpa	Chenopodiaceae	*	.	.	.	.	.	.	.
Suaeda maritima	Chenopodiaceae	*	*	.	.	.	.	.	.
Suaeda microphylla	Chenopodiaceae	*	*	.	.	.	.	.	*
Suaeda microsperma	Chenopodiaceae	*	.	.	.	.	.	.	*
Cressa cretica	Convolvulaceae	*	.	.	.	.	.	.	.
Thalassodendron ciliatum	Cymodoceaceae	.	.	.	.	.	.	.	.
Bolboschoenus maritimus	Cyperaceae	.	.	.	.	.	.	.	.
Cyperus laevigatus	Cyperaceae	*	.	.	.	.	.	.	*
Cyperus conglomeratus	Cyperaceae	*	.	.	.	.	.	.	*
Scirpus littoralis	Cyperaceae	.	.	.	.	.	.	.	.
Scirpus maritimus	Cyperaceae	.	.	.	.	.	.	.	.
Alhagi maurorum	Fabaceae	*	*	*	.	*	.	.	.
Astragalus kahiricus	Fabaceae	.	*	.	.	.	.	.	.
Melilotus alba	Fabaceae	*	.	.	.	*	.	.	.
Melilotus indica	Fabaceae	*	.	.	.	*	.	.	.
Melilotus officinalis	Fabaceae	*	.	.	.	*	.	.	.
Prosopis farcta	Fabaceae	*	*	.	.	.	.	.	.
Frankenia hirsuta	Frankeniaceae	.	.	.	.	.	.	.	.
Frankenia pulverulenta	Frankeniaceae	*	.	.	.	.	.	.	.
Hypericopsis persica	Frankeniaceae	.	.	.	.	.	.	.	.
Iris spuria	Iridaceae	*	.	.	.	.	.	.	.
Juncus rigidus	Juncaceae	.	.	.	.	.	.	.	.
Cistanche tubulosa	Orobanchaceae	.	.	.	.	.	.	.	.
Plantago crassifolia	Plantaginaceae	*	.	.	.	.	.	.	.
Plantago maritima	Plantaginaceae	.	.	.	.	.	.	.	.
Limonium gmelinii	Plumbaginaceae	.	.	.	.	.	.	.	.
Limonium iranicum	Plumbaginaceae	.	.	.	.	.	.	.	.
Limonium meyeri	Plumbaginaceae	.	.	.	.	.	.	.	.
Limonium reniforme	Plumbaginaceae	.	.	.	.	.	.	.	.
Limonium sogdianum	Plumbaginaceae	.	.	.	.	.	.	.	.
Limonium stocksii	Plumbaginaceae	.	.	.	.	.	.	.	.
Limonium suffruticosum	Plumbaginaceae	.	.	.	.	.	.	.	.
Psylliostachys beludshistanicus	Plumbaginaceae	.	.	.	.	.	.	.	.
Psylliostachys leptostachyus	Plumbaginaceae	.	.	.	.	.	.	.	.
Psylliostachys spicatus	Plumbaginaceae	.	.	.	.	.	.	.	.
Aeluropus lagopoides	Poaceae	*	.	*	*	.	.	.	*
Aeluropus littoralis	Poaceae	*	.	*	.	.	.	.	*
Aeluropus repens	Poaceae	*	.	*	*	.	.	.	*
Agropyron elongatum	Poaceae	*	.	.	.	.	.	.	.
Aristida pennata	Poaceae	*	.	*	.	.	.	.	*
Aristida plumosa	Poaceae	*	.	*	*	*	.	.	*
Astrebla lappacea	Poaceae	*	.	.	.	.	.	.	.
Buchloe dactyloides	Poaceae	*	.	.	.	.	.	.	.
Chloris gayana	Poaceae	*	.	.	.	.	.	.	*
Chloris virgata	Poaceae	*	.	.	.	.	.	.	*
Coelachyrum brevifolium	Poaceae	.	.	.	.	.	.	.	.
Cynodon dactylon	Poaceae	*	.	*	*	.	*	.	*
Dactyloctenium aegyptium	Poaceae	*	.	.	*	.	.	.	.
Dactyloctenium scindicum	Poaceae	*	.	.	*	.	.	.	.
Digitaria adscendens	Poaceae	*	.	.	.	.	.	.	*
Eleusine indica	Poaceae	*	.	.	.	.	.	.	*
Eleusine tristachya	Poaceae	*	.	.	.	.	.	.	*
Enteropogon macrostachya	Poaceae	*	.	.	.	.	.	.	.
Hordeum marinum	Poaceae	*	.	.	*	.	.	.	.
Phragmites australis	Poaceae	*	.	*	*	*	*	.	.

Table 10. (Cont.)

Species	Family name	2	3	4	5	6	7	8	9
Polypogon maritimus	Poaceae	*	.	.	.	.	.	.	.
Porteresia coarctata	Poaceae	*	.	.	.	.	.	.	*
Puccinellia koeieana	Poaceae	*	.	.	.	.	.	.	*
Sporobolus arabicus	Poaceae	*	.	.	.	.	.	.	*
Tetrachne dregei	Poaceae		.	.	.	.	.	.	*
Zoysia macrantha	Poaceae	*	.	.	.	.	.	.	.
Calligonum comosum	Polygonaceae	.	*	.	.	.	.	.	.
Calligonum crinitum	Polygonaceae	.	*	.	.	.	.	.	.
Calligonum leucocladum	Polygonaceae	*	*	.	.	.	.	.	.
Calligonum polygonoides	Polygonaceae	*	*	*	*	.	.	.	.
Calligonum stenopterum	Polygonaceae	.	*	.	.	.	.	.	.
Polygonum aviculare	Polygonaceae	*	*	*	*	.	.	.	.
Pteropyrum olivierii	Polygonaceae	.	*	.	.	.	.	.	.
Glaux maritima	Primulaceae	.	.	.	.	.	.	.	.
Ruppia maritima	Ruppiaceae	.	.	.	.	.	.	.	.
Salvadora persica	Salvadoraceae	*	.	.	.	.	.	.	.
Lycium barbarum	Solanaceae	.	*	.	.	.	.	.	.
Lycium depressum	Solanaceae	.	*	.	.	.	.	.	.
Lycium ruthenicum	Solanaceae	.	*	.	.	.	.	.	.
Reaumuria alternifolia	Tamaricaceae	*	.	.	*	.	.	.	.
Reaumuria cistoides	Tamaricaceae	*	.	*	*	.	.	.	.
Reaumuria fruticosa	Tamaricaceae	*	.	.	*	.	.	.	.
Reaumuria palaestina	Tamaricaceae	*	.	.	*	.	.	.	.
Reaumuria stocksii	Tamaricaceae	*	.	.	*	.	.	.	.
Tamarix kotschyi	Tamaricaceae	*	*	*	.	.	.	.	.
Tamarix mascatensis	Tamaricaceae	*	.	*	.	.	.	.	.
Tamarix androssowii	Tamaricaceae	*	.	*	.	.	.	.	.
Tamarix aphylla	Tamaricaceae	*	*	*	*	*	.	*	.
Tamarix aralensis	Tamaricaceae	*	*	*	.	.	.	.	.
Tamarix aravensis	Tamaricaceae	*	.	*	*	.	.	.	.
Tamarix arceuthoides	Tamaricaceae	*	*	*	.	.	.	.	.
Tamarix gallica	Tamaricaceae	*	*	*	*	*	.	.	.
Tamarix hispida	Tamaricaceae	.	*	*	*	*	.	*	.
Tamarix karakalensis	Tamaricaceae	*	*	*	.	.	.	.	.
Tamarix laxa	Tamaricaceae	*	*	*	*	*	.	.	.
Tamarix leptostachys	Tamaricaceae	*	.	*	.	.	.	.	.
Tamarix macrocarpa	Tamaricaceae	*	*	*	*	.	.	.	.
Tamarix octandra	Tamaricaceae	*	*	*	.	.	.	.	.
Tamarix passerinoides	Tamaricaceae	*	*	*	*	*	.	.	.
Tamarix pentandra	Tamaricaceae	*	*	*	*	*	.	.	.
Tamarix ramosissima	Tamaricaceae	*	*	*	*	*	.	.	.
Tamarix rosea	Tamaricaceae	*	.	*	.	*	.	.	.
Tamarix serotina	Tamaricaceae	*	.	*	.	.	.	.	.
Tamarix szovitsiana	Tamaricaceae	*	*	*	.	.	.	.	.
Tamarix tetragyna	Tamaricaceae	.	*	*	*	*	.	.	.
Acantholippia seriphioides	Verbenaceae	*	.	.	.	.	.	.	.
Malacocarpus crithmifolius	Zygophyllaceae	.	.	.	.	.	.	.	.
Nitraria schoberi	Zygophyllaceae	*	.	.	.	.	.	.	.
Tetradiclis tenella	Zygophyllaceae	.	.	.	.	.	.	.	.
Zygophyllum eichwaldi	Zygophyllaceae	*	*	*	.	.	.	.	.
Zygophyllum fabago subsp. *dolichocarpum*	Zygophyllaceae	*	*	*	.	.	.	.	.
Zygophyllum oxianum	Zygophyllaceae	*	*	*	.	.	.	.	.
Zygophyllum propinquum	Zygophyllaceae	*	*	*	.	*	.	.	*

morphological features, suitable for survival under drought, saline and arid environments. More than 23% of the halophyte species are used as a sand dune fixation in this part of the country (Table 4).

Erosion control

Erosion control is of great importance in arid and desert environment for reducing dust wind and run-off water. Tamaricaceae and Chenopodiaceae families are mostly used as wind break in Khorasan province (Table 5).

Industrial purposes

The halophyte species have been used as industrial and medicinal agents in dry and desert conditions for a long time. More than 9% of the halophytes are used for industrial and medicinal purposes (Table 6).

Vegetable

Vegetative part of some halophyte species are consumed freshly or cooked in some part of the province. Among the vegetable producing species special mention should be made of Chenopodiaceae family (Table 7).

Detergent

Utilization of halophytes as detergent material has a long history in this part of the country. Leaves and roots of some of the halophytes are used as detergent for cleaning and washing purposes (Table 8).

PHOTOSYNTHETIC PATHWAY

Table 9 shows list of C_4 halophyte families which are growing in different saline area of Khorasan province. More than 41% of halophyte families in Khorasan province belong to the C_4 photosynthetic pathway.

A complete list of halophyte species in Khorasan is given in Table 10.

ACKNOWLEDGMENT

We acknowledge with thanks partial financially support of Toos-Ab Consulting Engineers for this project.

REFERENCES

Akhani, H. and Ghorbanli, M. A contribution to the halophytic vegetation and flora of Iran. In Lieth, H. and Al-Masoon, A., editors. Towards the rational use of high salinity tolerant plants, Vol. 1. Kluwer Academic Publishers, Dordrecht, 1993; 35–44.

Ghahraman, A. Iran colour flora, Vol. 1–18. Institute of Forest and Rangelands Research Center, Iran, 1976–1998.

Koocheki, A. The use of halophytes for forage production and combating desertification in Iran. In Choukr-Allah, R., Malcolm, C.V. and Hamdy, A., editors. Halophytes and biosaline agriculture. Marcel Dekker, Inc., New York, 1996; 263–74.

Koocheki, A. and Mohalati, M.N. Feed value of some halophytic range plants of arid regions of Iran. In Squires, V.R. and Ayoub, A.T. editors. Halophytes as a resource for livestock and for rehabilitation of degraded lands. Kluwer Academic Publishers, Dordrecht, 1994; 249–53.

Mobayein, S. Vegetation of Iran, vascular plant flora, Vol. 1. Tehran University Press, 1975.

Mobayein, S. Vegetation of Iran, vascular plant flora, Vol. 2. Tehran University Press, 1980.

Rashed Mohasel, M.H. Vegetation of Khorasan Province. Ferdowsi University of Mashhad Press, 1993.

Rechinger, K.H. Flora Iranica: Convolvulaceae. Akademische Druck- u, Verkagsanstalt Graz, Austria, 1963.

Rechinger, K.H. Flora Iranica: Tamaricaceae. Akademische Druck- u, Verkagsanstalt Graz, Austria, 1964.

Rechinger, K.H. Flora Iranica: Plantaginaceae. Akademische Druck- u, Verkagsanstalt Graz, Austria, 1965.

Rechinger, K.H. (1970). Flora Iranica: Gramineae. Akademische Druck- u, Verkagsanstalt Graz, Austria, 1970.

Rechinger, K.H. Flora Iranica: Frankeniaceae. Akademische Druck- u, Verkagsanstalt Graz, Austria, 1972a.

Rechinger, K.H. Flora Iranica: Juncaceae. Akademische Druck- u, Verkagsanstalt Graz, Austria, 1972b.

Rechinger, K.H. Flora Iranica: Solanaceae. Akademische Druck- u, Verkagsanstalt Graz, Austria, 1972c.

Rechinger, K.H. Flora Iranica: Zygophyllaceae. Akademische Druck- u, Verkagsanstalt Graz, Austria, 1972d.

Rechinger, K.H. Flora Iranica: Plumbaginaceae. Akademische Druck- u, Verkagsanstalt Graz, Austria, 1974.

Rechinger, K.H. Flora Iranica: Iridaceae. Akademische Druck-u, Verkagsanstalt Graz, Austria, 1975.

Rechinger, K.H. Flora Iranica: Polygonacae. Akademische Druck- u, Verkagsanstalt Graz, Austria, 1977.

Rechinger, K.H. Flora Iranica: Cynomoraceae. Akademische Druck- u, Verkagsanstalt Graz, Austria, 1978.

Rechinger, K.H. Flora Iranica: Chenopodiaceae. Akademische Druck- u, Verkagsanstalt Graz-Austria, 1997.

Part V

Future Outlook for Research and Development on Cash Crop Halophytes and Foundation of ISHU, a New Scientific Society

Part V

[illegible]

Introduction to Part V – Gold Medal award

The Concerted Action "Sustainable Utilisation of Halophytes in the Tropical and Dry Regions" lasted from May 1996 until May 2000. It ended at the time when the EXPO 2000 started. This event allowed the members to present their achievements, concepts and future goals conveniently to a large audience.

Until May 2000 the following activities were prepared:

1. A presentation of halophytes and their utilisation in the Botanical Garden of the University of Osnabrueck as part of the OBE 2000, a regional activity mainly covering soil problems of various kinds.
2. A presentation of halophytes in the Hall "Nutrition" at the EXPO 2000 in Hannover, together with commonly used vegetables.
3. Opening to the public the pilot projects of the Concerted Action of the European Community, "Sustainable Utilisation of Halophytes" in Morocco, Tunesia, Egypt, Italy, Spain and Germany.
4. Selling halophyte meals for one week to the students of the University of Osnabrueck.
5. The development of a home page within the frame of the OBE 2000, the University of Osnabrueck and the Institute of Environmental Systems Research: www.usf.uni-osnabrueck.de/projects/expo 2000.

This home page was visited during the lead time of the EXPO 2000 almost 100,000 times with 62,000 visiting from outside of the University and covering all continents. For this achievements received the Concerted Action a Gold Medal from the EXPO 2000 for world wide projects (Figure 1).

A part of the closing activities of the Concerted Action "Sustainable Utilisation of Halophytes in the Tropical and Dry Regions" organised the group a workshop at the Botanical Garden of the University of Osnabrueck, the meeting from which many chapters of this book originated.

During this meeting the new International Society of Halophyte Utilisation (ISHU) was founded. Many activities and individual groups branched out after that meeting. The following chapter summarizes the first two years' activity of the new society. It will continue in many ways and we expect that several more volumes will develop, presenting research and development of halophyte production systems with irrigation using saline and other unconventional waters. We have compiled the list of publications in this volume in order to serve as a resource book for future developments. At the end of this volume we provide a list with all species names reported to be salt tolerant to a certain degree.

Fig. 1. Gold medal for international projects awarded to the Concerted Action project "Sustainable halophyte utilisation in the Mediterranean and subtropical dry regions" by the EXPO 2000 in Hannover, Germany.

H. Lieth (ed.), Cash Crop Halophytes: Recent Studies, 199.

The foundation of an International Society of Halophyte Utilisation (ISHU) 25.06.2002 in Osnabrueck/Germany

Deliberations at the founding session.

The undersigning persons agree to found an international society for the development of halophyte utilisation.

The society will be named the "International Society for Halophyte Utilisation". Out of the founding members we elect as:

President – Helmut Lieth

Vice-President – Hassan El-Shaer

Secretary – Marina Mochtchenko

Treasurer – Vito Sardo.

As members of the governing board we select Atef Hamdy, Xiaojing Liu and Ramon Noriega.

The President is asked to undertake necessary action for a legal registration of the new society.

The official registration can be undertaken with the assistance of the Secretary and the Treasurer.

The President is empowered to undertake all steps to enact the foundation of the society in the intention of the founding members.

Signatures: Helmut Lieth, Vito Sardo, Noomene Sleimi, Chedly Abdelly, Mokhtar Zarrouk, A.J. Joshi, Jose Ramon Noriega, Xiaojing Liu, Hassan El-Shaer

After the founding session the following persons arrived and joint the new society: Cherif Harrouni, Salma Daoud, Ajmal Khan, Benno Boeer, Nebiha Mohammed, Amnon Bustan, Michael DePew.

From this meeting the following resolution was formulated (Table 1). This resolution served successfully as bond for the activities of the members of the society.

Table 1. Resolution to initiate the new international R and D society "International Society of Halophyte Utilisation" (ISHU)

The participants at the final meeting of the EU CA group "Sustainable Utilisation of Halophytes" on occasion of the EXPO 2000 agreed to present a resolution to the public through the internet. After some discussion the following text was adopted:

Resolution

1. The network of experts for halophyte utilisation and the use of unconventional irrigation water developed by the EU CA group be enlarged and structured as a new Association (NGO) or private corporation.
2. The scope of the new organisation will be the promotion of the sustainable utilisation of halophytes through research, application and education.
3. The importance and feasibility of halophyte production systems be demonstrated through pilot projects in relevant countries.
4. Private enterprise be encouraged to implement trial plantations and/or halophyte utilisation projects.
5. A governing board for the new organisation be developed expediently.
6. The new board be asked to have international organisations and/or funding agencies as well as private investors approached to explain the importance of future halophyte utilisation and request funding and support for special projects identified by members of the new association.
7. The new association will issue a biannual or quarterly newsletter.
8. Assist in the creation of environmental awareness groups world wide, related to the future importance of utilising halophytes.
9. Assist in the development of an international gene bank of halophytes.
10. Preliminary Center for the governing board be the institute of Environmental Systems Research at the University of Osnabrueck/Germany.[1]

Contact address: Marina.Moschenko@usf.uni-osnabrueck.de or Mochtchenko@t-online.de; Helmut.Lieth@usf.uni-osnabrueck.de or Helmut.Lieth@t-online.de

Signed by: N. Sleimi/Tunisia, C. Abdelly/Tunisia, M. Zarrouk/Tunisia, V. Sardo/Italy, A. J. Joshi/India, J. R. Noriega/Mexico, X. Liu/China, H. M. El Shaer/Egypt, Ch. Harrouni/Morocco, S. Daoud/Morocco, M. A Khan/Pakistan, B. Boeer/UAE, Nebiha Mohamed/Ethiopia-UAE, A. Bustan/Israel, M. DePew/USA, H. Lieth/Germany.

Persons interested in the newly developing activity are kindly requested to contact the preliminary headquarter via e-mail to marina.moschenko@usf.uni-osnabrueck.de

[1]Since 2001 the preliminary center is moved to the following address: Marina Mochtchenko, Johann Heinrich-Lieth. Str. 2, D-52525 Kuerten-Duerscheid, Tel/Fax: 49 2207 706920. The old address may still be used.

H. Lieth (ed.), Cash Crop Halophytes: Recent Studies, 201.

Activities of ISHU

Summary of the activities of ISHU and the Concerted Action project "Sustainable Utilisation of Halophytes in the Tropical and Dry Regions".

Actions undertaken by the Steering Committee since the foundation in May 2000.

Due to the intensive workload during the years 2000 and 2001 it was impossible to get the people together for the registration procedure. The organisational and scientific work, however, began immediately.

In 15–22 September 2001 was the first Congress organised in Huanghua, China by Xiaojing Liu. The results of the meeting were published in 2002 in the book "Proceedings of International Symposium on Halophyte Utilisation and Regional Sustainable Development of Agriculture" compiled by Xiaojing Liu and Helmut Lieth. The table of content is shown in Table 1.

In 2002 two meetings were organised. January 2002 in Doha/Qatar the meeting entitled "QUEST 2002 for MENA" was organised by Benno Boeer. The abstract volume of that meeting, prepared by H. Lieth,U. Herpin, B. Boer and A. Kotb was published prior to the meeting. From this Volume we present in Table 2 the statements by Boer et al.

The final wording of these recommendations in form of an activity programme is presently in preparation by the Supreme Counsel of the Environment of the State of Qatar.

April 2002 a meeting in Cairo/Egypt was organised by Hassan El-Shaer as a starting point for halophyte utilisation in inland semideserts as fodder bases for sheep and goats. The proceedings of this conference are still in preparation.

Due to the intense work load the governing board decided to delay the official registration and have the administrative work executed within the frame of the Institute for Environmental Systems Research of the University of Osnabrueck where the President and the Secretary had their offices at that time. Their activities were documented in the home page listed below.

RECENT ACTIVITIES FOR RESEARCH AND DEVELOPMENT WORK BY ISHU MEMBERS

The wars in Afghanistan and Iraq have greatly affected research possibilities in the Near East region. This included the cooperative projects on halophytes and saline irrigation systems. The editorial work for this volume and especially for the Quest conference in Doha/Qatar was also delayed.

The International Center for Biosaline Agriculture (ICBA) in Dubai maintained its work at home and opened new cooperative projects in countries outside of the war-affected area. See ICBA Strategic plan 2000–2004, Annual report 2000 and *Biosalinity News* vol. 3. 2002 and volume 4. 2003.

The working group in Osnabrück was able to continue a project in the Amudarjev Delta region of Usbekistan which they had started in 2001 as an INTAS–Aralsee project: see homepage <www.usf.uni-osnabrueck.de/projects/aral>.

Our member, Dr. Boeer continued his work also as member of the UNESCO Doha office, mostly outside the war-affected region.

Our member Ramon Noriega cooperated with the Ocean Desert Enterprise (ODE), Amsterdam in developing a saline systems project in the Colorado Delta area of Mexican Baja California.

The evaluation meeting for the project was organised by Jeanette Höck in April 2003, during which the working group presented the plan for the development of an entire saline systems-based community whose income stems initially mainly from agriculture and forestry and some services to a nearby nature reserve. The ecological sustainability and economical feasibility was presented in two report volumes: Colorado Delta Proposal (March 2003) and Reader peer meeting (April 2003). The reports demonstrate how the group has planned the land use and community development. They plan to get their

H. Lieth (ed.), Cash Crop Halophytes: Recent Studies, 203–207.

income initially mainly through

- *Salicornia bigelovii* production and other salinity tolerant vegetables later on;
- Wood production with local *Tamarix* species, *Prosopis* and a foreign *Eucalyptus* species;
- Melioration work for the nearby nature reserve;
- Improvement of fisheries and aquaculture in general.

While this volume appears on the market we see that all members of ISHU are continuing their work for the promotion of the Cashcrop Halophyte Concept. For the Former Osnabrück group, Dr. Huchzermeyer at the Botanical Institute of the University of Hannover is presently collecting plans and proposals for future grant applications. For the Agadir/Morokko working group Dr. Harrouni plans an "Island system" for halophyte production, where the main energy is generated by wind.

Electronic addresses

1. E-mails:
Al Attar, Dr. Mohammad, ICBA Dubai / UAE <icba@biosaline.org.ae>
Taha, Dr. Faisal, ICBA Dubai / UAE <icba@biosaline.org.ae>
Lieth, Dr. Helmut, USF, University of Osnabrueck Germany <Helmut.Lieth@usf.uni-osnabrueck.de>
Mochtchenko, Mrs. Marina, ISHU at Lambda Digitec Kuerten/Germany <Mochtchenko@t-online.de>
Van Dieren, Dr. Wouter, IMSA Amsterdam/ Netherlands <ODE@imsa.nl>
Hoek, Mrs. Jeanette, ODE Amsterdam/Netherlands <ODE@imsa.nl>
Noriega, ing..Jose Ramon, Ensenada, Baja California, Mexico <ODE@imsa.nl>
Hamdy, Prof. Atef, IAMB. Valenzano/Bari, Italy <hamdy@iamb.it>
Chedly, Dr. Abdelly, INRST, Hamam Lif/Tunesia <chedly.abdelly@inrst.rnrt.tn>
Casador Dra.Isabel Univ. Lisboa Lisbon/Portugal <icacador@fe.ul.pt>
Harrouni,Dr. Cherif, IAV Hassan II Agadir/Morocco <c.harrouni@iavcha.ac.ma>
El Shaer, Dr. Hassan, DRC El Mataria, Cairo/Egypt <h-elshaer@yahoo.com>
Sardo, Dr. Vito, Univ. Catania/Italy <Sardov@unict.it>
Boeer, Dr. Benno, Unesco regional office Doha/Qatar <b.boer@memo.unesco.org>
Cluesener-Godt, Dr. Miguel, Unesco Paris/France <mcluesener-godt@unesco.org>
Huchzermeyer, Dr. Bernhard, Univ. Hannover/ Germany <huchzermeyer@botanik.uni-hannover.de>
Depew, Mr. Michael, ETS Pineland,FL/ USA <saltturf@etsturf.com>
Tadano, Dr. Toshiki, Tokio/Japan <Tadano@nodai.ac.jp>
Liu, Dr. Xiaojing, IAMCAS Shijizhuang, Hebei/ China <xjliu@ms.sjziam.ac.cn>
Razakov, Dr. Rustam, Technc. University Tashkent/ Usbekistan <Razak@econet.org.net>
Karimov, Dr. Bakthiyor, Inst. Waterproblems Tashkent/Uzbekistan <Karimov@dostlink.uz>
Joldasova, Dra Ilja, Karakalpak Acad. Science, Nukus/Uzbekistan <zholdas@online.ru>
Bustan, Dr. Amnon, Ben Gurion Univ. Beer Sheva/ Israel <abustan@bgumail.bgu.ac.il>
Koyro, Dr. H.-W., Uni-Giessen/ Germany <hans-werner.koyro@bot2.bio.uni-giessen.de>
Khan, Dr. Ajmal, Uni-Karachi/Pakistan <ajmalk@super.net.pk>

2. Internet pages
<www.biosaline.org>
<www.usf.uni-osnabrueck.de/projects>
<www.unesco.org/Doha/>

Literature

Al-Attar, Mohammad 2002 Strategic plan 200–2004 70 p. Dubai International Center for Biosaline Agriculture.
Biosaline News 2002 vol. 3 Dubai ICBA
Biosaline News 2003 Vol. 4 Dubai ICBA
International Center for Biosaline Agriculture 2001 Annual Report 1421/1422H 58 p. Dubai ICBA
Lieth, H and Schlueter, M. 2002 Periodic Report INTAS-Aralsea 00-1039 16p. website: <www.usf.uni-osnabrueck.de/reports>
Ocean Desert Enterprises 2003 Colorado Delta Project. A biosaline solution for saline desert with a shallow water table. 47 p. ODCO 134 Amsterdam ODE
Ocean Desert Enterprises 2003 Reader Peer meeting ODE/SSM April 2003 152 p. Amsterdam RSSR025 ODE

Table 1. Table of Contents from the proceedings of the International Symposium on Halophyte Utilisation and Regional Sustainable Development of Agriculture

目　录

Table 1. (Cont.)

Table 2. Statement for Quest 2002 for Mena by the conference steering committee B. Boer, A. Almuftha, N. Al-Muraikhi, F. Alqrimli, E.A. Elhag and A. Kotb

The Arabian Region and the Middle East are among many regions of the world currently facing a process of rapid development together with a lack of environmental management capacity.
Adverse environmental impacts include:

- Loss of the natural and cultural heritage
- Habitat loss and fragmentation
- Pollution of water, soil, air and biota
- Depletion of resources
- Declining primary and secondary productivity
- Declining biological diversity

These impacts will lead to declining economically essential resources, which will result in undesired costs and losses in the resource based economies, which in turn will cause the loss of jobs and income.

Good environmental research is the prerequisite towards all disciplines of environmental management, such as environmental education, environmental legislation, environmental protection and restoration, environmental development and environmental law enforcement.

Numerous organisations have been established and activities have been initiated in each of the regions countries, towards the improvement of environmental management. These are essential to minimise adverse environmental impacts, as well as to allow for sustainable human development.

The QUEST Conference (Qatar United for the Environment, Science & Technology) aims to contribute to the improvement of all disciplines, which are necessary towards environmental management. The conference brings together environmental experts from Arabia, Africa, Asia, Australia, Europe, North America and South America with expertise relevant for the Middle East and North Africa.

The QUEST Conference will culminate, via scientific and technical presentations and workshops, in the development of seven recommendations :

- Women's contribution towards environmental management
- Marine Resources
- Terrestrial Resources
- Freshwater supply and saline Water utilisation
- Industrial pollution and Ecotoxicology
- Protected areas and Sustainable development
- Environmental legislation, standards and regulations.

Publications of the Concerted Action members relevant to the program

Abdelly, C., Zid, E., Hajji, M. and Grignon, C. Biomass production and nutrition of *Medicago* species associated with halophytes on the edge of sebkha in Tunisia. In Choukr-Allah, R., Malcolm, C.V. and Hamdy, A., editors. Halophytes and biosaline agriculture. Marcel Dekker, Inc. New York, USA, 1996; 313–24.

Abdelly, C. Mécanismes d'une association de luzernes spontanées et de halophytes pérennes en bordure de sebkha. Thèse de Doctorat d'Etat, Faculté des Sciences de Tunis, Université de Tunis II. 1997, 290 pp.

Abdelly, C., Sleimi, N., Soltani, A., Hajji, M. and Grignon, C. Improvment perspectives of pastures in strongly salted zones. In Lieth, H., Hamdy, A. and Koyro, H.-W., editors. Actes de la Conférence Internationale Water Management, Salinity and Pollution Control Towards Sustainable Irrigation in the Mediterranean Region. Session spéciale: Salinity problems and halophytes use, 1997; 99–118.

Abdelly, C., Lachaâl, M. and Grignon, C. Importance of microheterogeneity of salinity and fertility for maintenance of plant diversity. In Lieth, H. et al., editors. Progress in biometeorology, Vol. 13, Halophyte uses in different climates I. Backhuys Publishers, Leiden, The Netherlands, 1999; 65–76.

Al-Jaloud, A.A. Use of saline water for irrigation in Saudi Arabia. In: Halophyte uses in different climates II. In Lieth, H., editor. Progress in Biometeorology, Vol. 14. Halophyte crop development for different climates. Backhuys Publishers, Leiden, The Netherlands, 1999; 105–19.

Aziz, I. and Khan, M.A. Physiological adaptations to seawater concentration in *Avicennia marina* from Indus delta, Pakistan. Pak J Bot (in press).

Aziz, I. and Khan, M.A. Effect of salinity on the growth, ionic composition and water relations of *Ceriops tagal* from Miani Hor, Pakistan. Aquatic Bot 2000 (in press).

Belligno, A., La Loggia, F., Sambuco, F. and Sardo, V. Experiences on Vetivergrass Salinity Tolerance of Plants. In Proceedings of the Special Session of Non-conventional Water Resources Practices and Management, IAV Hassan II, Rabat, Morocco, 1999.

Böer, B. New wetland plants in the UAE. Tribulus 7.1, 1997; 22–23.

Böer, B. Saltwater plants for agriculture along the Arabian Gulf – an ecologically valuable concept (In German: Salzwasserpflanzen für die Landwirtschaft am Arabischen Golf – Ein ökologisch wertvolles Konzept). Kurier, 1997.

Böer, B. An introduction to habitat management techniques in Abu Dhabi. Internal Research Report No. 15. Environmental Research and Wildlife Development Agency, Abu Dhabi, 1997; 28 pp.

Böer, B. Anthropogenic factors and their potential impacts on the terrestrial habitats of Abu Dhabi. Internal Research Report No. 16. Environmental Research and Wildlife Development Agency, Abu Dhabi, 1997; 17 pp.

Böer, B. and Gliddon, D. Abu Dhabi's coastal zone – conservation requirements and land use potentials. In International Conference on Water Management, Salinity and Pollution Control towards Sustainable Irrigation in the Mediterranean Region. Instituto di Bari, Italy, 22–26 September 1997.

Böer, B. and Gliddon, D. The vegetation of Abu Dhabi. Internal Research Report No. 7. Environmental Research and Wildlife Development Agency, Abu Dhabi, 1997; 46 pp.

Böer, B. Anthropogenic Factors and their potential impacts on the sustainable development of Abu Dhabi's terrestrial biological resources. Int J Sustain Develop World Ecol 1998 (in press).

Böer, B. New wetland plants for the UAE: *Potamogeton pectinatus* L. and *Zannichellia palustris* L. Tribulus, 1998 (in press).

Böer, B. and Gliddon, D. Mapping of coastal ecosystems and halophytes in Abu Dhabi. Aust J Freshwater Marine Biol (in press).

Caçador, I., Vale, C. and Catarino, F. Accumulation of Zn, Pb, Cu and Ni in sediments between roots of the Tagus estuary salt marshes. Port Estuarine Coastal Shelf Sci 1996; 42:393–403.

Caçador, I., Mascarenhas, I. and Mascarenhas, P. Seasonal variations of biomass of *Spartina maritima*, *Halimione portulacoides* and *Arthrocnemum fruticosum* in Tagus estuary salt marshes. Port Estuarine Coastal Shelf Sci 1997; 43:234–40.

Caçador, I., Vale, C. and Catarino, F. Relationships between metal concentration and organic matter content in the Tagus estuary salt marsh sediments. Marine Environ Res 1997; 32:80–87.

Caçador, I., Mascarenhas, I. and Mascarenhas, P. Biomass of *Spartina maritima*, *Halimione portulacoides* and *Arthrocnemum fruticosum* in Tagus estuary salt marshes. Marine Environ Res 1999; 40:33–43.

Caçador, I., Vale, C. and Catarino, F. Relationships between metal concentrations and organic matter content in the Tagus Estuary salt marsh sediments. In Lieth, H., editor. Halophyte uses in different climates I. Backhuys Publishers, Leiden, The Netherlands, 1999a; 105–12.

Caçador, I., Vale, C. and Catarino, F. Seasonal variation of Zn, Pb, Cu and Cd concentrations in the root-sediment system of *Spartina maritima* and *Halimione portulacoides* from Tagus estuary sallt marshes. Marine Environ Res 2000; 49:279–90.

Cutore, L., La Loggia, F. and Sardo, V. A quick test to assess salinity tolerance of plants. In Proceedings of the Special Session of Non-conventional Water Resources Practices and Management, IAV Hassan II, Rabat, Morocco, 1999.

H. Lieth (ed.), Cash Crop Halophytes: Recent Studies, 209–212.

El-Shaer, H.M. Rangelands as feed resources in the Egyptian desert: management and improvement. In Proceedings of the International Conference on Desert Development in the Arab Gulf Countries, State of Kuwait, 23–26 March 1996.

El-Shaer, H.M. Practical approaches for improving utilization of feed resources under extensive production system in sinai. In Proc Inter Symp. on Systems of sheep and goat production, Bella, Italy, 25–27 October 1997.

El-Shaer, H.M. Sustainable utilisation of halophytic plant species as livestock fodder in Egypt. In Water management, salinity and pollution control towards sustainable irrigation in the Mediterranean region: Salinity problems and halophyte use. Tecnomack Publication, Bari, 1997; 171–85.

El-Shaer, H.M. and Kandil, H.M. Potential of *Atriplex* species as fodder shrubs under the arid conditions of Egypt. In Lieth, H., editor. Halophyte uses in different climates II. Halophyte crop development for different climates, Vol. 14, Progress in biometeorology. Backhuys Publishers, Leiden, The Netherlands, 1999; 87–95.

El-Shaer, H.M., Fahmy, A.A., Abdul-Aziz, G.M., Shalaby, A.S. and Abd El Gawad, A.M. Nutritional Evaluation of some halophytic silages fed to goats in Sinai. In Proceedings of the International Conference on Goats, 15–18 May 2000, Tourse, France, 2000.

Germanà, C., Cutore, L. and Sardo, V. Assessing tolerance to irrigation water salinity in five woody plants. In Proceedings of the Special Session on Non-conventional Water Resources, Practices and Management, Rabat, Morocco, 1999.

Germanà, C., Cutore, L. and Sardo, V. Assessing tolerance to irrigation water salinity in five woody plants. In Proceedings of the Special Session of Non-conventional Water Resources, Practices and Management, IAV Hassan II, Rabat, Morocco, 1999.

Gulzar, S., Khan, M.A. and Ungar, I.A. Effect of temperature and salinity on the germination of *Urochondra setulosa*. Seed Sci Technol 2000 (in press).

Hamdy, A. and Lieth, H. Advance short course on saline irrigation: halophyte production and utilisation. 6–15 April, 1999. Agadir, Morocco Bari-Valenzano, IAM Bari, 1999; 363 pp.

Hamdy, A., Lieth, H., Todorović, M. and Moschenko, M. Halophyte uses in different climates II. In Lieth, H., editor. Progress in biometeorology, Vol. 14. Halophyte crop development for different climates. Backhuys Publishers, Leiden, The Netherlands, 1999.

Harrouni, M.C., Daoud, S., El Alami, A., Debbagh, B. et al. Responses of some halophytes to seawater irrigation in Morocco. In Lieth, H., editor. Halophyte uses in different climates II. Progress in biometeorology, Vol. 14. Halophyte crop development for different climates. Backhuys Publishers, Leiden, The Netherlands, 1999; 57–77.

Khan, M.A. and Ungar, I.A. Germination responses of the subtropical annual halophyte *Zygophyllum simplex*. Seed Sci Technol 1997; 25:83–91.

Khan, M.A. and Ungar, I.A. Alleviation of seed dormancy in the desert forb *Zygophyllum simplex* L. from Pakistan. Ann Bot 1997; 80:395–400.

Khan, M.A. Seed germination characteristics of *Halopyrum mucronatum*: a coastal dune grass from Pakistan. Can J Bot 1998; 67:455–61.

Khan, M.A. and Aziz, S. Some aspects of salinity, plant density, and nutrient effects on *Cressa cretica* L.J. Pl Nut 1998; 21:769–84.

Khan, M.A., Ungar, I.A. and Gul, B. Action of compatible osmotica and growth regulators in alleviating the effect of salinity on the germination of dimorphic seeds of *Arthrocnemum indicum* L. Int J Pl Sci 1998; 159:313–17.

Khan, M.A., Ungar, I.A. and Showalter, A.M. Growth, water, and ion relationships of a leaf succulent perennial halophyte, *Suaeda fruticosa* (L.) Forssk. Ann Bot 1998.

Khan, M.A., Ungar, I.A. and Showalter, A.M. The effect of salinity on growth, ion content, and osmotic relations in *Halopyrum mucronatum* (L.) Stapf. J Pl Nut 1998.

Khan, M.A., Ungar, I.A., Showalter, A.M. and Dewald, H. NaCl – induced accumulation of glycinebetaine in four subtropical halophytes from Pakistan. Physiol Plant 1998; 102:487–92.

Khan, M.A. and Ungar, I.A. Germination of salt tolerant shrub *Suaeda fruticosa* from Pakistan: Salinity and temperature responses. Seed Sci and Technol 1998; 26:657–67 (UK).

Khan, M.A., Ungar, I.A. and Showalter, A.M. The effect of salinity on growth, ion content, and osmotic relations in *Halopyrum mucronatum* (L.) Stapf. J Pl Nut 1999; 22:191–204.

Khan, M.A. Comparative influence of salinity and temperature on the germination of subtropical halophytes. In Lieth, H., Moschenko, M., Lohman, M., Koyro, H.-W. and Hamdy, A., editors. Halophyte uses in different climates I: ecological and ecophysiological studies. Progress in biometeriology, Vol. 13. Backhuys Publishers, Leiden, 1999, 77–88.

Khan, M.A., Ungar, I.A. and Showalter, A.M. Salt tolerance in the subtropical perennial halophyte *Atriplex griffithii* Moq. var. *stocksii* Boiss. Ann Bot 2000; 85:225–32.

Khan, M.A., Ungar, I.A. and Showalter, A.M. Growth, water, and ion relationships of a leaf succulent perennial halophyte, *Suaeda fruticosa* (L.) Forssk. J Arid Environ 2000 (in press).

Koyro, H.-W., Wegmann, L., Lehmann, H. and Lieth, H. Adaptation of the mangrove *Laguncularia racemosa* to high NaCl salinity. In Moschenko, M., Menzel, U. and Lieth, H., editors. Proceedings of the Al Ain meeting, report volume No.1, CEC concerted action Sustainable halophyte utilisation in the Mediterranean and subtropical dry regions, 1996; 15–16.

Koyro, H.-W. Ultrastructural and physiological changes in root cells of *Sorghum* plants (*Sorghum bicolor X S. sudanesis* cv. Sweet Sioux) induced by NaCl. J Exp Bot 1997; 48:693–706.

Koyro, H.-W. and Huchzermeyer, B. Starting a reference system for the biochemical evaluation of salt stress effects on C4-type plants. In Lieth, H., editor. Annual report about Sustainable halophyte utilisation in the Mediterranean and subtropical dry regions. CEC concerted action Sustainable halophyte utilisation in the Mediterranean and subtropical dry regions, 1997; 23–5.

Koyro, H.-W., Wegmann, L., Bezerra-Neto, E. and Lieth, H. Influence of sodium chloride on the physiology of the mangrove *Laguncularia racemosa* L. In Congresso Brasiliero de Ciencia do solo, 26, 1997. Rio de Janeiro, Anais. Brazil: Sociedade Brasileira de Ciencia do Solo (ed.), arq297.doc, CD-ROM, 1997; 5 pp.

Koyro, H.-W. and Huchzermeyer, B. Progress of the physiological studies relevant for the CA. In Lieth, H., editor. Annual report CEC concerted action Sustainable halophyte utilisation in the Mediterranean and tropical dry regions, 1997; 26–9.

Koyro, H.-W., Wegmann, L., Lehmann, H. and Lieth, H. Adaptation of the mangrove *Laguncularia racemosa* to high NaCl salinity. In Lieth, H., editor. Annual report CEC concerted action Sustainable halophyte utilisation in the Mediterranean and tropical dry regions, 1997; 43–44.

Koyro, H.-W. Sustainable use of halophytes under artificial conditions in a sea-water irrigation system. In Lieth, H., Moschenko, M., Menzel, U., Braun, G. and Koyro, H.-W., editors. Brochure about the sustainable use of halophytes in Mediterranean and subtropical dry regions, 1997; 6–8.

Koyro, H.-W. and Huchzermeyer, B. The physiological response of *Beta vulgaris* ssp. *maritima* to sea water irrigation. In Lieth, H., Hamdy, A. and Koyro, H.-W., editors. Water management, salinity and pollution control towards sustainable irrigation in the Mediterranean region: salinity problems and halophyte use. Tecnomack Publications, Bari, 1997; 29–50.

Koyro, H.-W., Wegmann, L., Lehmann, H. and Lieth, H. Physiological mechanisms and morphological adaptation of *Laguncularia racemosa* to high salinity. In Lieth, H., Hamdy, A. and Koyro, H.-W., editors. Water management, salinity and pollution control towards sustainable irrigation in the mediterranean region: salinity problems and halophyte use. Tecnomack Publishers, Bari, 1997; 51–78.

Koyro, H.-W. and Huchzermeyer, B. Physiological studies of halophytes under artificial conditions in a sea-water irrigation system. In Lieth, H., editor. Short activity report about Sustainable halophyte utilisation in the Mediterranean and subtropical dry regions. CEC concerted action Sustainable halophyte utilisation in the Mediterranean and subtropical dry regions, 1997, 52 pp.

Koyro, H.-W., Lieth, H. and Moschenko, M. Comparison tables for salinty, measurements within the Concerted Action IC18CT96-0055 Sustainable use of halophytes. In Lieth, H., editor. Short activity report about Sustainable halophyte utilisation in the Mediterranean and subtropical dry regions, Annex 4. CEC concerted action Sustainable halophyte utilisation in the Mediterranean and subtropical dry regions, 1997; 55–7.

Koyro, H.-W. Preliminary comparison tables of the mineral content of the water used by the partners of the CA project. In Lieth, H., Hamdy, A. and Koyro, H.-W., editors. Short activity report (15.7.1997–15.1.1998) of the EU concerted action Sustainable halophyte utilisation in the Mediterranean and subtropical dry regions. University of Osnabrueck, 1998; 32–34.

Koyro, H.-W. and Lieth, H. Salinity conversion table. 2nd enlarged edn. © H. Lieth, Osnabrueck, 1998 (ISSN 09336-3114).

Koyro, H.-W. and Huchzermeyer, B. Salt and drought stress effects on metabolic regulation in maize. In Pessarakli, M., editor. Handbook of plant and crop stress. Marcel Dekker, Inc., New York, Basel, 1999; 843–78.

Koyro, H.-W. and Huchzermeyer, B. Influence of high NaCl-salinity on growth, water and osmotic relations of the halophyte *Beta vulgaris* ssp. *maritima*. Development of a quick check. In Lieth, H., Moschenko, M., Lohmann, M., Koyro, H.-W. and Hamdy, A., editors. Progress in biometeorology, Vol. 13. Halophyte uses in different climates I. Ecological and ecophysiological research. Backhuys Publishers, Leiden, 1999; 89–104.

Koyro, H.-W., Wegmann, L., Lehmann, H. and Lieth, H. Adaptation of the mangrove *Langunculari*a *racemosa* to high NaCl salinity. In Lieth, H., Moschenko, M., Lohmann, M., Koyro, H.-W. and Hamdy, A., editors. Progress in biometeorology, Vol. 13. Halophyte uses in different climates I. Ecological and ecophysiological research. Backhuys Publishers, Leiden, 1999; 43–64.

Koyro, H.-W. and Huchzermeyer, B. Salt and drought stress effects on metabolic regulation in maize. In Pessarakli, M., editor. Handbook of plant and crop stress, 2nd edn. Marcel Dekker Inc., New York, USA, 1999; 843–78.

Koyro, H.-W. and Huchzermeyer, B. Plant and crop stress: salt and drought stress effects on metabolic regulation in maize. In Hamdy, A. and Lieth, H., editors. Saline irrigation: halophyte production and utilization. Tecnomack Publications, Bari, 1999; 251–322.

Koyro, H.-W. Effect of high NaCl-salinity on plant growth, leaf morphology, and ion composition in leaf tissues of *Beta vulgaris* ssp. *maritima*. J Appl Bot 74:67–73.

Lieth, H., Moschenko, M., Menzel, U., Lohmann, M., Braun, G. and Koyro, H.-W. Sustainable use of halophytes under artificial conditions in a sea-water irrigation system. Lieth, H., editor. Brochure about the sustainable use of halophytes in Mediterranean and subtropical dry regions. 1997; 15 pp.

Lieth, H. und Koyro, H.-W. Dauerhafte Nutzung von Halophyten im Mittelmeerraum und in subtropischen Trockenregionen. In: Forschungsförderung der Europäischen Union, Beispiele aus Niedersachsen, Vol. 2. EG-Hochschulbuero Hannover/Hildesheim (ed.), Auflage, 1997; 48–49.

Lieth, H., Hamdy, A. and Koyro, H.-W., editors. Water management, salinity and pollution control towards sustainable irrigation in the Mediterranean region Bari (Valenzano), Italy, 1997; 209 pp.

Lieth, H. Sustainable halophyte utilisation in the Mediterranean and subtropical dry regions. In: Markert B., editor. INTECOL 2000 Research activities. Zittau, IHI, 1998; 59–66.

Lieth, H. Hamdy, A and Koyro, H.-W. Agroecology and sustainability of cashcrop halophytes. In Farina, editor. Perspectives in ecology. Backhuys Publishers, Leiden, The Netherlands, 1998; 309–14.

Lieth, H., Moschenko, M., Lohmann, M., Hamdy, A. and Koyro, H.-W. Halophyte uses in different climates I. Ecological and ecophysiological studies. In Lieth, H., editor. Progress in biometeorology, Vol. 13. Backhuys Publishers, Leiden, The Netherlands, 1999; 258 pp.

Lieth, H. Nachhaltige Nutzung von Salzpflanzen. In Biologie in der Schule 49. 2000. Heft 2. Pädagogischer Zeitschriftenverlag GmBH & Co. 2000; 120–6 (ISSN 0406-3317).

Lieth, H. and Lohmann, M. Cashcrop Halophytes – for future halophyte growers. 2000 (ISSN 09336-3114 No. 20).

Lieth, H., Zeilinger, Z. and Papenbrock, J. Developing cash crops from halophytes. In Projects for Saline Agriculture, International seminar on April 10–12, 2000, Islamabad, Pakistan. Kluwer publisher, The Netherland (in press).

Menzel, U. and Lieth, H. Halophyte Database vers.2.0 1999. In Lieth, H., editor. Halophyte uses in different climates II. Progress in biometeorology, Vol. 14. Halophyte crop development for different climates. Backhuys Publishers, Leiden, The Netherlands, 1999; 127–33.

Merlo, L., Morgana, B., Pennisi, M. and Sardo, V. A. Probabilistic approach in modelling biomass production. In Proceedings 9th European Bioenergy Conference – Copenhagen, 1996.

Merlo, L., Morgana, B., Sardo, V. and e La Loggia, F. Experiences with giant reeds and perennial C_4 grasses in Sicily. In Proceedings of the X European Biomass Conference, Wuerzburg, 1998.

Sardo, V. Experiences with halophytes and non-conventional crops in Sicily. In Water management, salinity and pollution control towards sustainable irrigation in the Mediterranean region. Bari, 22–26 settembre 1997.

Sardo, V. Irrigation with saline and marginal waters: past, present and planned experiences. In Water management, salinity and pollution control towards sustainable irrigation in the Mediterranean region, Bari, 22–26 settembre 1997.

Sardo, V. Selecting local spontaneous vegetation to combat desertification in the Mediterranean environment. In Proceedings of the International Seminar on Indicators for assessing desertification in the Mediterranean, Porto Torres, 1998.

Sardo, V. Irrigating with saline waters: the result of the first year's experience. In Proceedings of the Advanced Short Course on Saline Irrigation: Halophyte Production and Utilisation, Agadir, Morocco, 1999.

Sardo, V. Establishment of an experimental field for saline irrigation in Sicily. In Hamdy et al., editors. Halophyte uses in different climates II, Vol. 14. Backhuys publishers, Leiden, The Netherlands, 1999.

Sleimi N., Abdelly, C., Soltani, A. and Hajji, M. Biomass production and mineral nutrition of *Suaeda fruticosa* grown on high salinity medium in connection with nutrient availability. In Lieth, H. et al., editors. Progress in biometeorology, Halophyte uses in different climates I, Vol. 13. Backhuys Publishers, Leiden, The Netherlands, 1999; 7–29.

Additional suggested reading

Achmad, A. and San Pietro, A., editors. Prospects of biosaline research. Department of Botany, University of Karatchi, Karatchi, 1982.

Boyko, H., editor. Saline irrigation for agriculture and forestry. Dr. W. Junk, The Hauge, 1968, 325 pp.

Choukr-Allah, R., Malcolm, C.V. and Hamdy, A., editors. Halophytes and biosaline agriculture. Marcel Dekker, New York, 1996, 400 pp.

Environmental Research Laboratory. Ornamental Halophytes for landscape use. University of Arizona ERL Tucson, 1987; unpaginated.

Glenn, E., Miyamoto, S., Moore, D., Brown. J., Thompson, T.L. and Brown, P.J. Arid Environ 1997; 36:711–30.

Hamdy, A., editor. Proceedings. of Advanced Short Course on Sustainable use of nonconventional water resources in the Mediterranean region. CIHEAM/MAI-Bari and ICARDA (Syria). 18–30 April 1998, Aleppo, Syria, 1998.

Hamdy, A., Lieth, H., Todorović, M. and Moschenko, M., editors. Halophyte uses in different climates 2. Halophyte crop development: Pilot studies. Progress in Biometeorology, Vol. 14. Backhuys Publishers, Leiden, 1999, 144 pp.

Huiskes, A.H.L., Blom, C.W.P.M. and Rozema, J., editor. Vegetation between land and sea. Dr. W. Junk, The Hauge, 340 pp.

Huiskes, A.H.L. comp. A contribution to the diversification of the production of vegetable crops by research on the cultivation methods and selection of halophytes. EU AIR Report, Center for Marine Research, Yerseke, Netherlands, 1994; 273 pp.

Koyro, H.-W. and Huchzermeyer, B. Influence of high NaCl-salinity on growth, water and osmotic relations of the halophyte *Beta vulgaris* ssp. *maritima*. Development of a quick check. In Lieth, H., Moschenko, M., Lohmann, M., Koyro H.-W. and Hamdy A., editors. Progress in Biometeorology Vol 13. Halophyte uses in different climates I. Ecological and ecophysiological research. Backhuys Publishers, Leiden, 1999; 89–104.

Koyro, H.-W. and Lieth, H. Annex 3: Conversion table for commonly used salinity measures in water, soil and plants or animals. In: Lieth, H., Moschenko, M., Lohmann, M., Koyro, H.-W. and Hamdy, A., editors. Halophyte uses in different climates 1. Ecological and ecophysiological studies. Progress in Biometeorology Vol. 13. Backhuys Publishers, Leiden,1999; 258 pp.

Lieth, H. and Al Massoum A.A., editors. Towards the rational use of high salinity tolerant plants, Vol.1 of the Proc. of the Al Ain Conf.1990,T:VS Vol. 27. Kluwer Academic. Publishers, Dordrecht, 1993a; 521 pp.

Lieth, H., Al Massoum A.A., editors. Towards the rational use of high salinity tolerant plants, Vol. 2 of the Proc. of the Al Ain Conf. 1990,T: VS Vol. 28. Kluwer Academic Publishers, Dordrecht, 1993; 448 pp.

Lieth, H., Moschenko, M., Lohmann, M., Koyro, H.-W. and Hamdy, A., editors. Halophyte uses in different climates 1. Ecological and ecophysiological studies. Progress in biometeorology, Vol. 13. Backhuys Publishers, Leiden, 258 pp.

Lieth, H., Hamdy, A. and Koyro, H.-W., editors. Water management, salinity and pollution control towards sustainable irrigation in the mediterranean region. Instituto Agronomico Mediterraneo, Valenzano-Bari, 1997; 209 pp.

Lieth, H., Berlekamp, J., Fuest, S. and Riediger, S., editors. Climate diagram world atlas on CD. CD series: climate and biosphere. Backhuys Publishers, Leiden, 1999.

Menzel, U. and Lieth, H. Annex 4: Halophyte Database Vers. 2. In: Lieth, H., Moschenko, M., Lohmann, M., Koyro, H.-W. and Hamdy, A., editors. Halophyte uses in different climates 1. Ecological and ecophysiological studies. Progress in biometeorology, Vol. 13. Leiden, Backhuys Publishers, 1999; 258 pp.

Rozema, J. Biology of halophytes. In Choukr-Allah et al., editors. 1996; 17–30.

Pasternak, D. comp. Fodder production with Saline water. Beer-Sheva/Israel Institutes for applied research, Ben Gurion University of the Negev, 1990; 173 pp.

Pasternak, D. and Nerd, A. Research and utilisation of halophytes in Israel. In Choukr-Allah et al., editors. 1996; 325–48.

Sen, D. N. and Rajpurohit, K.S. editors. Contributions to the ecology of halophytes. T:VS Vol. 2. Junk publishers, The Hague, 1982, 272 pp.

Squires, V.R. and Ayoub, A.T., editors. Halophytes as a resource for livestock and for rehabilitation of degraded lands. Kluwer Academic Publisher, Dordrecht, 1994, 315 pp.

Teas, H.J., Grossenbacher, C. and Karafel, R. Mangroves as a crop in the United States Choukr-Allah et al., editors. 1996; 297–312.

Whitehead, E.E., Hutchinson, C.F., Timmermann, B.N., Varady, R.G., editors. Arid lands today and tomorrow. In Proc. Intl. Res. & Dev. Conf., Tucson AZ, Westview Press, Colorado, 1425, 1425 pp.

Yensen, N.C., NyPa® (unpublished). Plants for salty soils. NyPa, Inc., Tucson, Arizona.

H. Lieth (ed.), Cash Crop Halophytes: Recent Studies, 213.

Table of contents for the four volumes of the "Halophytes in Different Climates"

PROGRESS IN BIOMETEOROLOGIE
Series Editor H. Lieth
Volume 13
HALOPHYTE USES IN DIFFERENT CLIMATES I
Ecological and Ecophysiological Studies
Edited by H. Lieth, M. Moschenko, M. Lohman,
H.-W. Koyro and A. Hamdy
Backhuys Publishers, Leiden, 1999

H. Lieth (ed.), Cash Crop Halophytes: Recent Studies, 215–220.

PROGRESS IN BIOMETEOROLOGIE

Series Editor H. Lieth

Volume 14

HALOPHYTE USES IN DIFFERENT CLIMATES II

Halophyte Crop Development: Pilot Studies

Edited by A. Hamdy, H. Lieth, M. Todorović and M. Moschenko

Backhuys Publishers, Leiden, 1999

HALOPHYTES IN DIFFERENT CLIMATES VOLUME IV
CASH CROP HALOPHYTES
for future halophyte growers
2nd edition prepared by Helmut Lieth and Marina Mochtchenko
on behalf of the International Society of Halophyte Utilization
for the QUEST Conference for the Middle East and North Africa
in Doha/Qatar – January 27–30, 2002

Table of content **page**

PROGRESS IN BIOMETEOROLOGIE
Series Editor H. Lieth
Volume 15
HALOPHYTE USES IN DIFFERENT CLIMATES III
Computer-Aided Analysis of Socio-Economic Aspects of the Sustainable Utilisation of Halophytes
Backhuys Publishers, Leiden, 2001

Halophyte Database Vers. 2.0 in alphabetical order including some updates

UWE MENZEL AND HELMUT LIETH

Interest in halophytes has grown during recent years. Their potential as crops and other useful plants was proposed by the CA members (see Lieth and Moschenko 1998). This has triggered research, especially in regions with a shortage of freshwater and a need for an increase in fodder production or landscape protection.

While the members of the EU Concerted Action "Sustainable Utilisation of Halophytes" gradually increased their research potential for halophytic plants and saline irrigation, the task arose to compile the most complete list of halophytes in existence. This task was accepted by Uwe Menzel and coworkers in the coordinator's office who compiled as much literature about halophytes as possible. After the termination of the EU CA, U. Menzel continued the collection of halophyte data. He compiled initially two different lists. First, he compiled all halophytes listed in the literature as crop plants or plants with other uses by humans. This list is published in Vol. 2 of the 1998 INTECOL report (Menzel and Lieth, 1999). Second, he evaluated all the literature so far available to him on halophytic species. While doing this he discovered large differences in the number of species different authors were quoting as halophytes. We are continuously working on this species lists. We present at this time the list in alphabetical order, as required by many colleagues not familiar with plant taxonomy.

A closer checking revealed that different authors had differing opinions about the salinity tolerance of even the same species and the salinity level which would qualify a species to be a halophyte. This fact requires continuous revision, and we ask our colleagues to send us their comments if they find errors or have different opinions.

Several authors had started the compilation of halophytes and provided with it some qualitative physiological information. While all these attempts were very useful, we found that many papers had shortcomings as far as distinguishing synonyms, geographical origin and physiological properties were concerned. We therefore considered it to be necessary to continue the compilation of an authentic list of halophytes which could be used as a basic data bank in order to complement it with various properties of each individual halophytic species.

We present here the third published version of this data bank with the desire to have this list double-checked for possible errors and/or for suggestions as to what property categories we should collect and add in the future. Comments of users to the e-mail adresses at the end are welcome. The list has been provided in our Internet page <www.usf.uni-osnabrueck.de/expo2000>. Users were requested to identify themselves before drawing data from this list. This has not worked satisfactorily so far. Corrections on the Internet page required too much work and negotiations. We present, therefore, this time the data in alphabetical order with the capability to add more comments. If anyone finds it necessary to correct or amend data he or she should do it and send us a copy of the amendments via e-mail or diskette to the addresses listed at the end.

The listings on the CD-ROM provided with this book contain over 2600 names of halophytic species which U. Menzel checked for synonyms and spelling errors. As nomenclature standard we used the *Index Kewensis*, as we did previously (Menzel and Lieth 1999). The CD-ROM contains several lists. The core list follows the sequence used in taxonomic papers. In addition to that we added an alphabetical listing. Only the alphabetical list is attached to this printed text.

The literature searched so far has been added following the tabulation in the core list. The citations are

H. Lieth (ed.), Cash Crop Halophytes: Recent Studies, 221–250.

numbered and the number is shown at the relevant species. The numbers in the alphabetical list refer to the page number in the respective list on the CD-ROM. Please draw our attention to omissions and newly published papers. A numbering system has been used to cross-reference the authors of the papers with the numbers used in the species list. Further explanations are shown in the following chapter.

We accepted the definitions from the authors in the published literature to consider a species as salt-tolerant. This has caused the acceptance of a low level of salinity tolerance for species to qualify for inclusion in our list. See the values in the relevant columns. We hope that in time we can add for each species the salinity tolerance range. Our own level of salinity tolerance for a halophytic species is somewhat higher. We have set it at the level where common crops cannot grow any more (see Lieth et al. 1999 and several other contributions in this volume).

All readers and users of this list are kindly requested to report corrections required, additional species or qualitative properties to be added, in an attempt to provide an up-to-date internet listing.

REFERENCES

Aronson, A. 1990. Haloph "A data base of salt tolerant plants of the world". In: Pasternak, D. (ed.) Fodder production with saline water, Project report Jan. 1982 – Dec. 1989, BGUN-ARI-35–90.

Lieth, H.; Moschenko, M. (eds.) 1998. Short activity report for the period 15.7.1997–15.1.1998 of the EU concerted action IC18CT96–0055 "Sustainable halophyte utilisation in the Mediterranean and subtropical dry regions", USF, University of Osnabrück, 66+6pp.

Menzel, U.; Lieth, H. 1999. Tabulation of Halophytes reported as utilized in different publications and handbooks. In: Hamdy, A., Lieth, H., Todoroviá, M., Moschenko, M. (eds.): Halophyte uses in different climates. Vol. 2. Progress in Biometeorology 14, Backhuys Publishers, Leiden, The Netherlands, 127–133.

Menzel, U., Lieth, H. 1999. Halophyte Database, 9–168. In: Lieth, H., Hamdy, A. (eds.): Saline irrigation: Halophyte production and utilization. Agadir, Morocco.

CONTENTS OF THE LISTINGS ON THE ATTACHED CD-ROM

The **core** data base lists the species in taxonomic order and is structured in three columns:

Column 1 contains taxonomic information according to entries taken from *Index Kewensis*, see reference number 473. Plants with * are not in the Index.

Column 2 provides the highest reported salinity tolerance level in â salinity for species where this value was found in the literature. Value with * = Chloride in root area.

Column 3 gives the reference number under which the relevant author and/or paper can be found in the reference list.

The **literature** number 473 is the *Index Kewensis*. When a species has this one number only, we found only the synonym.

The **alphabetical** list is added for easier locating of species names prior to searching in the taxonomic listing. We have added in this list the page number where each species can be found in the taxonomic list. We supply two kinds of lists, one as a .doc (Microsoft Word) file and one as a PDF file.

Please note that the page numbers given in the .doc file may differ from the page numbers in the taxonomic list depending on the age of the PC. The PDF file gives the page numbers correctly.

USING THE CORE LIST IN TAXONOMIC ORDER

The following chapter explains the difficulty in obtaining the valid name among synonyms in the *Index Kewensis* for a name found in the literature.

The name *Sesuvium sesuvioides* leads according to the rules of the *Index Kewensis* to two other valid names which in turn lead again to other names. Since we cannot decide ourselves which is the valid name we enter this species under 3 different names as shown below.

Sesuvium sesuvioides {(Fenzl) Verd.} (Fenzl) Verdc. = **FICODIACEAE** *Diplochonium sesuvioides* {Fenzl / = *Trianthema hydaspica* Edgew.}; { Wawra & Peyr. / = *Sesuvium digynum* Welw. ex Oliver }

SEARCHING IN THE *INDEX KEWENSIS*

AIZOACEAE *Sesuvium sesuvioides* {(Fenzl) Verd. literature author is different from *Index Kewensis* author} (Fenzl) Verdc.

Under *Sesuvium sesuvioides* one finds in the *Index* the plant with the publ. author: (Fenzl) Verdc. with the remark *Diplochonium sesuvioides* in field notes. = **FICODIACEAE *Diplochonium sesuvioides***; under D. s. you can find two plants, D. s. Fenzl and also D. s. Wawra & Peyr.; this means {Fenzl … } and also {Wawra & Peyr. …}

{Fenzl / = *Trianthema hydaspica* Edgew.}; { Wawra & Peyr. / = *Sesuvium digynum* Welw. ex Oliver} {... / = means under the entry from D. s. under publ. author Fenzl one finds in notes: = *Trianthema hydaspica* and under D. s. publ. author Wawra & Peyr. = *Sesuvium digynum*. Under the entry from T. h. you find now the publ. author Edgew. under S. d. the publ. author Welw. ex Oliver.

These data were taken from the *Index Kewensis on Compact Disc*. The data in their original form remain under copyright of the Trustees of the Royal Botanic Gardens, Kew.

```
Record Type    Spx
Vol/Supp       S13
Family         AIZOACEAE
Genus          SESUVIUM
Species        sesuvioides
Publ. Author   (Fenzl) Verdc.
Publication    in Kew Bull 1957, 349 (1957)
Notes          =Diplochonium sesuvioides.
```

```
Record Type    Spx
Vol/Supp       OV1F2
Family         FICOIDACEAE
Genus          DIPLOCHONIUM
Species        sesuvioides
Publ. Author   Fenzl,
Publication    Nov. Stirp. Dec. 58 (1839)
Notes          =Trianthema hydaspica.
```

```
Record Type    Spx
Vol/Supp       OV1F2
Family         FICOIDACEAE
Genus          DIPLOCHONIUM
Species        sesuvioides
Publ. Author   Wawra & Peyr.
Publication    Sert. Beng. 24
Notes          =Sesuvium digynum.
```

```
Record Type    Spx
Vol/Supp       OV2F4
Family         FICOIDACEAE
Genus          TRIANTHEMA
Species        hydaspica
Publ. Author   Edgew.
Publication    in Journ. Linn. Soc.
        vi. (1862) 203.
Notes          Ind. or.; Afr. trop.
```

```
Record Type    Spx
Vol/Supp       OV2F4
Family         FICOIDACEAE
Genus          SESUVIUM
Species        digynum
Publ. Author   Welw. ex Oliver,
Publication    Fl. Trop. Afr. ii.
        586
Notes          Afr. trop.
```

INDEX IN ALPHABETICAL ORDER

The page numbers added refer to the pages where the species are found in the taxonomic list on the CD-ROM included in the inside back cover of this volume.

F

I

O

P

R

Tasks for vegetation science

1. E.O. Box: *Macroclimate and plant forms.* An introduction to predictive modelling in phytogeography. 1981 ISBN 90-6193-941-0
2. D.N. Sen and K.S. Rajpurohit (eds.): *Contributions to the ecology of halophytes.* 1982 ISBN 90-6193-942-9
3. J. Ross: *The radiation regime and architecture of plant stands.* 1981 ISBN 90-6193-607-1
4. N.S. Margaris and H.A. Mooney (eds.): *Components of productivity of Mediterranean-climate regions.* Basic and applied aspects. 1981 ISBN 90-6193-944-5
5. M.J. Müller: *Selected climatic data for a global set of standard stations for vegetation science.* 1982 ISBN 90-6193-945-3
6. I. Roth: *Stratification of tropical forests as seen in leaf structure* [Part 1]. 1984 ISBN 90-6193-946-1
   For Part 2, see Volume 21
7. L. Steubing and H.-J. Jäger (eds.): *Monitoring of air pollutants by plants.* Methods and problems. 1982 ISBN 90-6193-947-X
8. H.J. Teas (ed.): *Biology and ecology of mangroves.* 1983 ISBN 90-6193-948-8
9. H.J. Teas (ed.): *Physiology and management of mangroves.* 1984 ISBN 90-6193-949-6
10. E. Feoli, M. Lagonegro and L. Orlci: *Information analysis of vegetation data.* 1984 ISBN 90-6193-950-X
11. Z. Sv̂esták (ed.): *Photosynthesis during leaf development.* 1985 ISBN 90-6193-951-8
12. E. Medina, H.A. Mooney and C. Vzquez-Ynes (eds.): *Physiological ecology of plants of the wet tropics.* 1984 ISBN 90-6193-952-6
13. N.S. Margaris, M. Arianoustou-Faraggitaki and W.C. Oechel (eds.): *Being alive on land.* 1984 ISBN 90-6193-953-4
14. D.O. Hall, N. Myers and N.S. Margaris (eds.): *Economics of ecosystem management.* 1985 ISBN 90-6193-505-9
15. A. Estrada and Th.H. Fleming (eds.): *Frugivores and seed dispersal.* 1986 ISBN 90-6193-543-1
16. B. Dell, A.J.M. Hopkins and B.B. Lamont (eds.): *Resilience in Mediterranean-type ecosystems.* 1986 ISBN 90-6193-579-2
17. I. Roth: *Stratification of a tropical forest as seen in dispersal types.* 1987 ISBN 90-6193-613-6
18. H.-G. Dässler and S. Börtitz (eds.): *Air pollution and its influence on vegetation.* Causes, Effects, Prophy´laxis and Therapy. 1988 ISBN 90-6193-619-5

Tasks for vegetation science

19. R.L. Specht (ed.): *Mediterranean-type ecosystems.* A data source book. 1988 ISBN 90-6193-652-7
20. L.F. Huenneke and H.A. Mooney (eds.): *Grassland structure and function.* California annual grassland. 1989 ISBN 90-6193-659-4
21. B. Rollet, Ch. Hägermann and I. Roth: *Stratification of tropical forests as seen in leaf structure*, Part 2. 1990 ISBN 0-7923-0397-0
22. J. Rozema and J.A.C. Verkleij (eds.): *Ecological responses to environmental stresses.* 1991 ISBN 0-7923-0762-3
23. S.C. Pandeya and H. Lieth: *Ecology of Cenchrus grass complex.* Environmental conditions and population differences in Western India. 1993 ISBN 0-7923-0768-2
24. P.L. Nimis and T.J. Crovello (eds.): *Quantitative approaches to phytogeography.* 1991 ISBN 0-7923-0795-X
25. D.F. Whigham, R.E. Good and K. Kvet (eds.): *Wetland ecology and management.* Case studies. 1990 ISBN 0-7923-0893-X
26. K. Falinska: *Plant demography in vegetation succession.* 1991 ISBN 0-7923-1060-8
27. H. Lieth and A.A. Al Masoom (eds.): *Towards the rational use of high salinity tolerant plants*, Vol. 1: Deliberations about high salinity tolerant plants and ecosystems. 1993 ISBN 0-7923-1865-X
28. H. Lieth and A.A. Al Masoom (eds.): *Towards the rational use of high salinity tolerant plants*, Vol. 2: Agriculture and forestry under marginal soil water conditions. 1993 ISBN 0-7923-1866-8
29. J.G. Boonman: *East Africa's grasses and fodders.* Their ecology and husbandry. 1993 ISBN 0-7923-1867-6
30. H. Lieth and M. Lohmann (eds.): *Restoration of tropical forest ecosystems.* 1993 ISBN 0-7923-1945-1
31. M. Arianoutsou and R.H. Groves (eds.): *Plant-animal interactions in Mediterranean-type ecosystems.* 1994 ISBN 0-7923-2470-6
32. V.R. Squires and A.T. Ayoub (eds.): *Halophytes as a resource for livestock and for rehabilitation of degraded lands.* 1994 ISBN 0-7923-2664-4
33. T. Hirose and B.H. Walker (eds.): *Global change and terrestrial ecosystems in monsoon Asia.* 1995 ISBN 0-7923-0000-0
34. A. Kratochwil (ed.): *Biodiversity in Ecosystems: Principles and case studies of different complexity levels.* 1999 ISBN 0-7923-5717-5
35. C.A. Burga and A. Kratochwil (ed.): *Biomonitoring: General and applied aspects on regional and global scales.* 2001 ISBN 0-7923-6734-0

Tasks for vegetation science

36. H.-J. Barth and B. Böer (eds.): *Sabkha Ecosystems.* Volume I: The Arabian Peninsula and Adjacent Countries. 2001 ISBN 1-4020-0504-0
37. R. Ahmad and K.A. Malik (eds.): *Prospects for Saline Agrigulture.* 2002 ISBN 1-4020-0620-9
38. H. Lieth and M. Mochtchenko (eds.): *Cashcrop Halophytes Recent Studies.* 10 Years after the Al Ain Meeting. 2003 ISBN 1-4020-1202-0

KLUWER ACADEMIC PUBLISHERS – DORDRECHT / BOSTON / LONDON

FSC
www.fsc.org
MIX
Papier aus verantwortungsvollen Quellen
Paper from responsible sources
FSC® C105338

If you have any concerns about our products,
you can contact us on
ProductSafety@springernature.com

In case Publisher is established outside the EU,
the EU authorized representative is:
Springer Nature Customer Service Center GmbH
Europaplatz 3, 69115 Heidelberg, Germany

Printed by Libri Plureos GmbH
in Hamburg, Germany